BASIC
ENGINEERING
ELECTROMAGNETICS

BASIC ENGINEERING ELECTROMAGNETICS

An Applied Approach

Richard L. Coren

Drexel University

PRENTICE HALL, ENGLEWOOD CLIFFS, N.J. 07632

Library of Congress Cataloging-in-Publication Data

COREN, RICHARD L.
    Basic engineering electromagnetics.

    Includes index.
    1. Electromagnetism.   I. Title.
QC760.C67 1989   537   88-17906
ISBN 0-13-060369-4

Editorial/production supervision and
    interior design: Joseph Scordato
Cover design: 20/20 Services, Inc.
Manufacturing buyer: Mary Noonan

© 1989 by Prentice-Hall, Inc.
A Division of Simon & Schuster
Englewood Cliffs, New Jersey 07632

Printed in the United States of America
10   9   8   7   6   5   4   3   2

ISBN 0-13-060369-4

Prentice-Hall International (UK) Limited, *London*
Prentice-Hall of Australia Pty. Limited, *Sydney*
Prentice-Hall Canada Inc., *Toronto*
Prentice-Hall Hispanoamericana, S.A., *Mexico*
Prentice-Hall of India Private Limited, *New Delhi*
Prentice-Hall of Japan, Inc., *Tokyo*
Simon & Schuster Asia Pte. Ltd., *Singapore*
Editora Prentice-Hall do Brasil, Ltda., *Rio de Janeiro*

To Elaine
For her great encouragement
and her unbelievable patience

CONTENTS

PREFACE

This text takes a two-tier approach to the study of statics. Chapters 3, 4, and 5 present the fundamental concepts and symmetries of electrostatics and magnetostatics, including material properties, but with limited emphasis on problem solving. More detailed analytic techniques are then presented in Chapters 6 (electrostatics) and 7 (magnetostatics) and in Chapter 8, dealing with computer methods for both cases. This division has the advantage that the first learning is reinforced by the second. In addition, by skipping over Chapters 6, 7, 8, or using them selectively, the book can be used for a one term course in field theory without conceptual omission, yet leaving time to cover current important topics in field dynamics. This has worked successfully at Drexel University for those students who take the single, 40 hour course.

The pedagogical approach is from Maxwell's Equations, facilitating rapid entry into the subject, as needed for the short course. These are immediately specialized to the static cases so that fundamental logical development can take place without having the student overwhelmed by complexity. Throughout the book the presentation strongly emphasizes physical reasoning and symmetry arguments, and the illustrative examples are selected to allow the drawing of conclusions without extended discussions, in cases where the students' developing insight allows them to understand and accept the generalizations.

For the majority of students, taking a two term course, the material on statics can be covered in either the sequential order presented, or in the traditional format of electrostatics (Chapters 3 and 6), magnetostatics (Chapters 4 and 7), quasistatics (Chapter 5). In the latter case it will only be necessary to assign the last sections of Chapter 5 (boundary conditions) between Chapters 3 and 6. Both approaches have been successful at Drexel University where the course is taken in the fourth year of a five year undergraduate program.

In either the short or long course an effort should be made to introduce some of the material from Chapter 8 (Numerical Methods). This represents one of the exciting developing areas of electromagnetics and is presented in more depth than in most texts at this level. To facilitate this, the spreadsheet method of solving problems by the finite difference method, which requires no programming *per se*, is given in some detail.

As in any course some trade-offs must also be made in covering dynamics. After treating plane waves (Chapter 9) there are chapters on surface interactions (Chapter 10), transmission lines (Chapter 11), and waveguides (Chapter 12). The important properties of Striplines and Microstrip are found in Chapter 13. Some knowledge of antennas is important to a modern engineer and Chapter 14 presents dipole and aperture antennas and array theory.

Acknowledgement is certainly due the Drexel University students, who worked with the early versions of my notes. A special note of appreciation goes to Tim Brophy for his painstakingly close reading and discussions of style, grammar, clarity, and consistency. Ted Kubasak kindly supplied the computer program in the final appendix. Chapter 8 on Numerical Methods was mostly written by Dr. Jeevan Hoole of Harvey Mudd College in Claremont, CA, as were sections in Chapter 12, though I made such extensive changes and additions that, like the rest of the book, and in spite of all the help I had from students, editors, and reviewers, I am fully responsible for any errors.

Richard L. Coren

Philadelphia, PA

BASIC
ENGINEERING
ELECTROMAGNETICS

1

VECTOR ANALYSIS

1.1 POINTS IN SPACE

In this text we will be discussing the effects of charges and currents in physical systems and as they exist and vary in space and time. To do so we need a means of describing positions in space. Since our physical world requires the specification of three coordinates to locate a point in space, we establish the three normal axes shown in Fig. 1.1. The independent variables along these axes are denoted by x, y, and z, and each point in this three-dimensional space represents a point with coordinates (x, y, z). A function of the three coordinates is a surface drawn on these three-dimensional axes. Figure 1.1 shows a family of surfaces in space, on each of which the function $f(x, y, z)$ has a different value. An example of such a function might be the temperature in a room, where the surfaces shown are the constant-temperature, or isothermal, suraces.

The coordinate axes shown in Fig. 1.1 are mutually orthogonal lines, so that a constant-coordinate surface (a surface on which x or y or z is constant) is a plane parallel to the other two axes and perpendicular to the axis of the constant coordinate. This rectangular coordinate system (x, y, z) is also referred to as the cartesian coordinate system, after René Descartes, a renowned philosopher and mathematician of the eighteenth century, who invented analytic geometry. Cartesian coordinates are not the only way to indicate a point in space and it is often preferable to use other coordinate systems. To illustrate these, consider the two-dimensional case.

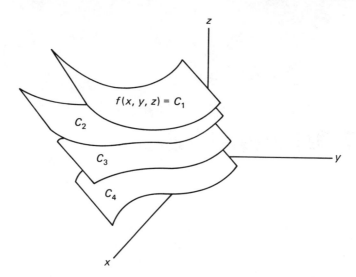

Figure 1.1   Representation in three-dimensional space.

1.2 TWO-DIMENSIONAL COORDINATES

Figure 1.2 shows a point P in the plane that is perpendicular to the z axis. The point is located by its rectangular coordinates (x, y) and by its *polar* coordinates (ρ, ϕ). If a line is drawn from the origin of the coordinate system to P, then ρ is the length of the line, and ϕ is the angle it subtends to a reference axis, which is taken to be the x axis. Either polar or rectangular descriptions are sufficient to specify P. From the figure, it is seen that the coordinates are related by

$$\rho^2 = x^2 + y^2 \qquad \tan \phi = y/x \qquad z = z \tag{1.1a, b}$$

$$y = \rho \sin \phi \qquad x = \rho \cos \phi \qquad z = z \tag{1.1c, d}$$

The surfaces of constant ρ are circular cylinders about the z axis, which is perpendicular to the $\rho\phi$ plane, and the surfaces of constant ϕ are planes containing the z

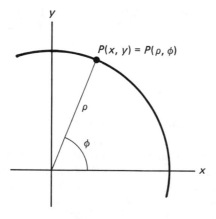

Figure 1.2   Polar and rectangular coordinates.

axis. Note that just as the x = constant and y = constant surfaces (planes) are perpendicular to each other, so also are the ρ = constant and ϕ = constant surfaces (circular cylinders and radial planes, respectively).

The rectangular and polar coordinate systems will be used throughout this text, but the reader should know that there are a host of other coordinate descriptions that can be conveniently used when the geometry of a particular problem conforms to their constant-coordinate surfaces. One such system, the elliptic cylindrical system, is shown in Fig. 1.3. Here the curves of constant u are confocal ellipses and the curves of constant v are confocal hyperbolas; these curves are actually sections of cylindrical surfaces parallel to the z axis. As before, the surfaces are orthogonal. This coordinate system can be related to the rectangular system through the transformations

$$x = c \cosh u \cos v \qquad y = c \sinh u \sin v \qquad z = z \qquad (1.2a, b)$$

where c is a magnitude that defines the scale of the ellipses and hyperbolas.

The circular and elliptic cylindrical systems are examples of *curvilinear* coordinates, in which some of the coordinates do not vary along straight lines, and in which some of the coordinate variables are not lengths (e.g., ϕ and v are angles). In rectangular coordinates, a change of the variable y corresponds to an equal change of length: $dl_y = dy$. This is not always true in curvilinear systems and, as a result, care must be taken when writing displacements in these systems. In the polar system, there is no problem with respect to ρ, since this is a linear variable and is a length; the length of an infinitesimal displacement along ρ is $dl_\rho = d\rho$. However, the change $d\phi$ is not itself a length.

The length corresponding to $d\phi$ can be obtained from the fundamental definition of an angle. For a circle of radius r with a central angle ϕ, we have $\phi = s/r$, as shown in Fig. 1.4, where s is the circumferential arc length included in ϕ. From

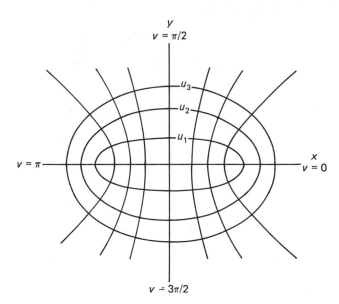

Figure 1.3 Elliptical and rectangular coordinates.

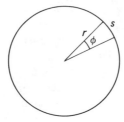

Figure 1.4   Definition of the angle ϕ.

this we see that, in circular cylindrical coordinates, the arc length is given by $dl_\phi = ds = \rho\, d\phi$. The factor that converts the coordinate change to a length (here the factor is ρ) is called the *metric* of that coordinate. If we denote the general metric by h, then in circular cylindrical coordinates,

$$h_\rho = 1 \qquad h_\phi = \rho \qquad h_z = 1 \qquad\qquad (1.3\text{a, b, c})$$

In the elliptic cylindrical system, the metrics are

$$h_u = c\,(\cosh^2 u - \cos^2 v/\sinh^2 u)^{1/2} \qquad h_v = c\,(\cosh^2 u - \cos^2 v/\sin^2 u)^{1/2}$$

$$h_z = 1 \qquad (1.4\text{a, b, c})$$

1.3 THREE-DIMENSIONAL COORDINATE SYSTEMS

Figure 1.5 shows the three-dimensional rectangular coordinate system with variables (x, y, z). Since its coordinate variables are lengths, the metrics are $h_x = 1$, $h_y = 1$, and $h_z = 1$, respectively. For circular and elliptic cylindrical systems, the metrics

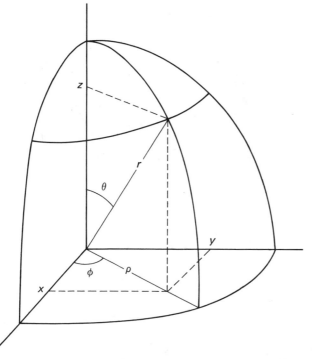

Figure 1.5   Relationships between rectangular, circular–cylindrical, and spherical coordinates.

are given by Eqs. (1.3) and (1.4). The full circular cylindrical (polar) system, with coordinates (ρ, ϕ, z), is also shown in Fig. 1.5.

Figure 1.5 also shows an important alternative to the rectangular and circular cylindrical systems, which is known as the spherical coordinate system. Here the coordinates are r, the distance of a point P from the origin; θ, the angle measured from the polar axis (the z axis); and ϕ, the planar angle measured between the projection of r onto the base plane, which is normal to the polar axis, and a reference direction in that plane (the x axis). It is clear that the projection of r onto the base plane corresponds to ρ of circular cylindrical coordinates, and that the angle ϕ is also the same. The projections on the figure yield relations between spherical and circular cylindrical coordinates:

$$\rho = r \sin \theta \qquad z = r \cos \theta \qquad \phi = \phi \qquad\qquad (1.5\text{a, b})$$

and between the spherical and cartesian coordinates:

$$x = \rho \cos \phi = r \sin \theta \cos \phi \qquad y = \rho \sin \phi = r \sin \theta \sin \phi \qquad z = r \cos \theta$$
$$(1.6\text{a, b, c})$$

Example 1.1

The equation of a sphere with unit radius and whose origin is located at $(x, y, z) = (0, 0, 1)$ is given in rectangular coordinates by

$$x^2 + y^2 + (z - 1)^2 = 1$$

Determine its equation in circular cylindrical and in spherical coordinates.

By direct substitution, the cylindrical equation is

$$\rho^2 + (z - 1)^2 = 1$$

Further substitution into this result gives

$$(r \sin \theta)^2 + (r \cos \theta - 1)^2 = 1$$

Squaring and collecting terms yields

$$r = 2 \cos \theta$$

Note in Example 1.1 that it is not generally necessary to specify which equation corresponds to which coordinate system, as our notation makes the coordinate combinations unique.

From the definition of angle and from the relations illustrated in Fig. 1.5 for the spherical system, it is seen that

$$h_r = 1 \qquad h_\theta = r \qquad h_\phi = r \sin \theta \qquad\qquad (1.7\text{a, b, c})$$

1.4 VECTORS

Figure 1.1 shows the functions $f(x, y, z) = C$ for several values of C, and it was suggested that this might represent the temperature distribution in a region of space. A physical quantity such as temperature or mass density, which varies with position, is relatively simple to describe by a function like f, which gives the magnitude at each point. Magnitude is its only attribute; it is a *scalar* quantity. In our study of electro-magnetics, we will be dealing with *vectors*, that is, quantities that require three de-scriptors. The reader has likely met vector quantities such as force or velocity. To specify a force, one must give its magnitude (in units in which it is measured, e.g., newtons) and its direction (which generally requires specifying two angles with re-spect to some reference directions). The vector direction is as important as the mag-nitude, since the acceleration of a body on which it acts is in the direction of the force. Alternatively, a force can be specified by giving its projections, along three mutually perpendicular axes. We shall show both methods.

To clearly indicate a vector quantity, we adopt the convention of using bold-face type, for example, **V**. An italic symbol is a scalar, so that the magnitude of a vector is $V = |\mathbf{V}|$, that is, neglecting direction. We have pointed out that a vector de-scription requires references of direction. Figure 1.6 shows an arbitrary vector **V** whose length is drawn equal to its magnitude (i.e., if it represents a speed of 2.5 m/sec, then it is 2.5 units long), and whose orientation is that of its direction of action.

For convenience of our analysis, we have translated the coordinate axes to the origin of **V**. Also shown are three reference vectors of constant *unit magnitude* and having fixed directions, each lying along one of the rectangular coordinate axes; they are denoted by $\mathbf{a}_x$, $\mathbf{a}_y$, and $\mathbf{a}_z$. **V** has *components* V_x, V_y, and V_z, each equal to the perpendicular projection of **V** onto the respective axis. The length and direction of **V** can be duplicated by traveling a length V_x parallel to $\mathbf{a}_x$, which we denote by writing the couple $V_x\mathbf{a}_x$, and then adding $V_y\mathbf{a}_y$ and $V_z\mathbf{a}_z$. Thus, we have

$$\mathbf{V} = V_x\mathbf{a}_x + V_y\mathbf{a}_y + V_z\mathbf{a}_z \tag{1.8a}$$

which expresses the vector in terms of its components parallel to the unit basis vec-tors $\mathbf{a}_x$, $\mathbf{a}_y$, and $\mathbf{a}_z$. Each component may be a function of position, so that the vector **V** varies in magnitude and direction from point to point as, for example, when **V** represents the fluid velocity in a turbulent channel. A scalar magnitude that varies with position is called a *scalar field,* and a vector quantity that varies with position is called a *vector field.* In this text we will be studying electric and magnetic vector fields.

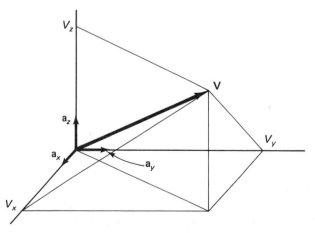

Figure 1.6   Cartesian components of a vector.

Another representation of **V** is shown in Fig. 1.7, where the vector is imbedded in a spherical coordinate system. As in the rectangular system, the unit vectors are directed along the coordinate directions. $\mathbf{a}_\theta$ and $\mathbf{a}_\phi$ are tangent to the spherical surface at the location of the vector and are oriented in the directions of increasing θ and increasing ϕ, respectively. As with the rectangular system, we have

$$\mathbf{V} = \mathbf{a}_r V_r + \mathbf{a}_\theta V_\theta + \mathbf{a}_\phi V_\phi \qquad (1.8b)$$

Note that the unit vectors in the spherical coordinate system, as with all curvilinear systems, are *not* constant vectors; they are constant in magnitude (unity) but not in direction. It is apparent that as θ and ϕ change, so do the directions of the tangents to the sphere, and the directions of $\mathbf{a}_\theta$ and $\mathbf{a}_\phi$ will vary. This can be seen by using Fig. 1.7 to write the spherical unit vectors in terms of $\mathbf{a}_x$, $\mathbf{a}_y$, and $\mathbf{a}_z$, the cartesian unit vectors of Fig. 1.6, which are truly constant vectors since they always lie along the same directions. As an aid in this calculation, Fig. 1.8(a) shows a z-axis section through Fig. 1.7, and Fig. 1.8(b) is a projection onto the xy plane. From these we can find

$$\mathbf{a}_r = \mathbf{a}_\rho \sin \theta + \mathbf{a}_z \cos \theta = \mathbf{a}_x \sin \theta \cos \phi + \mathbf{a}_y \sin \theta \sin \phi + \mathbf{a}_z \cos \theta$$
$$(1.9a)$$

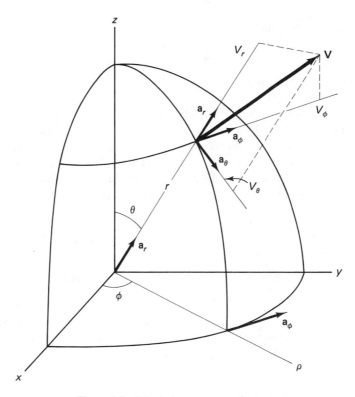

Figure 1.7   Spherical components of a vector.

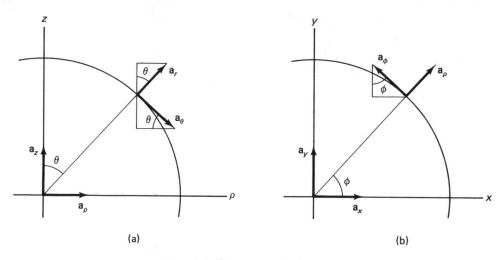

Figure 1.8 Components of unit vectors.

$$\mathbf{a}_\theta = \mathbf{a}_\rho \cos\theta - \mathbf{a}_z \sin\theta = \mathbf{a}_x \cos\theta \cos\phi + \mathbf{a}_y \cos\theta \sin\phi - \mathbf{a}_z \sin\theta$$
$$(1.9\text{b})$$

$$\mathbf{a}_\phi = -\mathbf{a}_x \sin\phi + \mathbf{a}_y \cos\phi \tag{1.9c}$$

1.5 VECTOR ALGEBRA

In dealing with vectors, it will be necessary to perform certain algebraic operations. To do this most simply, and to point out similarities of form among the various coordinate systems, consider the unit reference vectors to be $\mathbf{a}_1$, $\mathbf{a}_2$, and $\mathbf{a}_3$, corresponding to coordinates x_1, x_2, and x_3, respectively, where $(1, 2, 3)$ can be (x, y, z) or (ρ, ϕ, z) or (ρ, θ, ϕ), etc. Two vectors that have components in this generalized coordinate system are $\mathbf{A} = \mathbf{a}_1 A_1 + \mathbf{a}_2 A_2 + \mathbf{a}_3 A_3$ and $\mathbf{B} = \mathbf{a}_1 B_1 + \mathbf{a}_2 B_2 + \mathbf{a}_3 B_3$.

Addition of Vectors.   The sum of vectors $\mathbf{A}$ and $\mathbf{B}$ is the directional summation, illustrated in Fig. 1.9(a). As the vectors are added linearly, their components also add, so that

$$\mathbf{A} + \mathbf{B} = \mathbf{a}_1(A_1 + B_1) + \mathbf{a}_2(A_2 + B_2) + \mathbf{a}_3(A_3 + B_3) \tag{1.10a}$$

Subtraction of Vectors.   This is an extension of addition with the understanding that the negative of a vector means a positive one whose direction is reversed. Then, since $-\mathbf{B} = -\mathbf{a}_1 B_1 - \mathbf{a}_2 B_2 - \mathbf{a}_3 B_3$, we have

$$\mathbf{A} - \mathbf{B} = \mathbf{A} + (-\mathbf{B}) = \mathbf{a}_1(A_1 - B_1) + \mathbf{a}_2(A_2 - B_2) + \mathbf{a}_3(A_3 - B_3)$$
$$(1.10\text{b})$$

This is illustrated in Fig. 1.9(b).

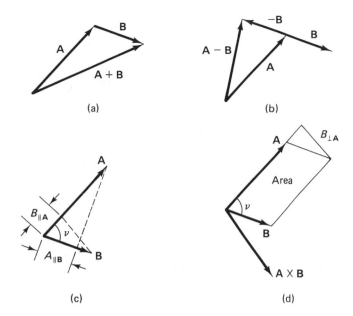

Figure 1.9 Algebra of vectors.

Multiplication of Vectors.   Two types of multiplication are possible, depending on how one accounts for the vector directions. If the angle between the directed vectors, from $\mathbf{A}$ to $\mathbf{B}$, is ν, then the product can have either magnitude $|\mathbf{A}|\,|\mathbf{B}|\cos\nu$ or magnitude $|\mathbf{A}|\,|\mathbf{B}|\sin\nu$.

From Fig. 1.9(c), $|\mathbf{A}|\,|\mathbf{B}|\cos\nu = A(B\|\mathbf{A}) = (A\|\mathbf{B})B$. Since there is no unique orientation left in this product (it might be parallel to either $\mathbf{A}$ or $\mathbf{B}$), its directional property is surrendered, resulting in a scalar. It is therefore called a *scalar product,* written $\mathbf{A}\cdot\mathbf{B} = |\mathbf{A}|\,|\mathbf{B}|\cos\nu$. Since the result is a scalar magnitude, it follows that $\mathbf{A}\cdot\mathbf{B} = \mathbf{B}\cdot\mathbf{A}$. Note that if two nonzero vectors have $\mathbf{A}\cdot\mathbf{B} = 0$, then they must be orthogonal.

The second type of product is $|\mathbf{A}|\,|\mathbf{B}|\sin\nu = A(B\perp\mathbf{A}) = (A\perp\mathbf{B})B$. In either case, the product is numerically equal to the area (base times height) of the parallelogram defined by the vectors $\mathbf{A}$ and $\mathbf{B}$. Since it represents an area, we can define a unique axis for the product, to be perpendicular to that area. The direction of the area normal can be along either orientation of the axis (into or out of the plane of Fig. 1.9d) and the choice is a matter of convention. The *vector product* is written $\mathbf{A}\times\mathbf{B} = \mathbf{n}|\mathbf{A}|\,|\mathbf{B}|\sin\nu$, where $\mathbf{n}$ is defined according to the right-hand rule, that is, when the right-hand fingers curl from $\mathbf{A}$ to $\mathbf{B}$, the thumb is directed along $\mathbf{n}$. (It would seem that mathematicians have simply deferred to that 85 percent of the population who are right handed. Note also that an ordinary screw advances along $\mathbf{n}$ when turned according to the right-hand rule.) Since the right-hand rule specifies direction from the first vector to the second, it follows that $\mathbf{B}\times\mathbf{A} = -\mathbf{A}\times\mathbf{B}$.

While we will not deal with the matter here, the necessity to define the direction of $\mathbf{A}\times\mathbf{B}$ is one manifestation of the fact that some properties of this cross-product vector are different from those of vectors that are not the result of such a multiplication. Vectors **A** and **B** are said to be *polar* vectors, whereas **A** × **B** is an *axial* vector.

If two nonzero vectors have **A** × **B** = 0, then they are parallel.

Example 1.2

Prove that **B** · [(**A** + **B**) × **A**] = 0.

$$\mathbf{B} \cdot [(\mathbf{A} + \mathbf{B}) \times \mathbf{A}] = \mathbf{B} \cdot [\mathbf{A} \times \mathbf{A} + \mathbf{B} \times \mathbf{A}] = \mathbf{B} \cdot \mathbf{B} \times \mathbf{A} = \mathbf{B} \cdot \mathbf{n} = 0$$

The first two equalities assume the distributive law of vector and scalar multiplication, that (**A** + **B**) × **C** = **A** × **C** + **B** × **C** and **C** · (**A** + **B**) = **C** · **A** + **C** · **B**. We have also used **A** × **A** = 0, since sin 0° = 0 and the fact that **B** × **A** = **n** is perpendicular to both **B** and **A**, so that **B** · **n** = 0.

It will be useful to apply the multiplication definitions to the unit vectors of the coordinate systems. As before, the unit reference vectors are $\mathbf{a}_1$, $\mathbf{a}_2$, and $\mathbf{a}_3$. Since these vectors lie along the coordinate axes, they are mutually orthogonal. It therefore follows that

$$\mathbf{a}_i \cdot \mathbf{a}_j = 0 \quad \text{for } i \neq j \qquad \mathbf{a}_i \cdot \mathbf{a}_i = 1 \quad \text{for } i = 1, 2, 3 \qquad (1.11a, b)$$

$$\mathbf{a}_i \times \mathbf{a}_j = \mathbf{a}_k \qquad\qquad\qquad\qquad\qquad\qquad\qquad (1.11c)$$

where $i \rightarrow j \rightarrow k \rightarrow i$ are cyclically permuted in order $1 \rightarrow 2 \rightarrow 3 \rightarrow 1$.

Equation (1.11c) is a statement of what is meant by a right-hand coordinate system. In almost all engineering work, and particularly in this text, we must be careful to always observe the convention of using right-hand coordinate systems. For the systems studied so far, the right hand sequences are

For the rectangular system: $x \rightarrow y \rightarrow z$
For the polar system: $\rho \rightarrow \phi \rightarrow z$
For the elliptical system: $u \rightarrow v \rightarrow z$
For the spherical system: $r \rightarrow \theta \rightarrow \phi$

In view of these relations between the basis vectors, the *dot* product (scalar product) and *cross* product (vector product) can be found for the arbitrary vectors **A** and **B**:

$$\mathbf{A} = A_x \mathbf{a}_x + A_y \mathbf{a}_y + A_z \mathbf{a}_z \qquad \mathbf{B} = B_x \mathbf{a}_x + B_y \mathbf{a}_y + B_z \mathbf{a}_z$$

For the scalar product we have, sequentially taking each component of **A**,

$$\begin{aligned}
\mathbf{A} \cdot \mathbf{B} &= (A_x \mathbf{a}_x + A_y \mathbf{a}_y + A_z \mathbf{a}_z) \cdot (B_x \mathbf{a}_x + B_y \mathbf{a}_y + B_z \mathbf{a}_z) \\
&= A_x \mathbf{a}_x \cdot (B_x \mathbf{a}_x + B_y \mathbf{a}_y + B_z \mathbf{a}_z) \\
&\quad + A_y \mathbf{a}_y \cdot (B_x \mathbf{a}_x + B_y \mathbf{a}_y + B_z \mathbf{a}_z) \\
&\quad + A_z \mathbf{a}_z \cdot (B_x \mathbf{a}_x + B_y \mathbf{a}_y + B_z \mathbf{a}_z) \\
&= A_x B_x + A_y B_y + A_z B_z \qquad\qquad\qquad\qquad (1.12)
\end{aligned}$$

The scalar product of two vectors is equal to the sum of the products of their parallel components.

For the vector product, we sequentially multiply by each component of **A**:

$$(A_x\mathbf{a}_x + A_y\mathbf{a}_y + A_z\mathbf{a}_z) \times (B_x\mathbf{a}_x + B_y\mathbf{a}_y + B_z\mathbf{a}_z)$$
$$= A_x\mathbf{a}_x \times (B_x\mathbf{a}_x + B_y\mathbf{a}_y + B_z\mathbf{a}_z)$$
$$+ A_y\mathbf{a}_y \times (B_x\mathbf{a}_x + B_y\mathbf{a}_y + B_z\mathbf{a}_z)$$
$$+ A_z\mathbf{a}_z \times (B_x\mathbf{a}_x + B_y\mathbf{a}_y + B_z\mathbf{a}_z)$$
$$= A_xB_y\mathbf{a}_z + A_xB_z(-\mathbf{a}_y)$$
$$+ A_yB_x(-\mathbf{a}_z) + A_yB_z\mathbf{a}_x$$
$$+ A_zB_x\mathbf{a}_y + A_zB_y(-\mathbf{a}_x)$$

Collecting parallel terms gives the vector components

$$\mathbf{A} \times \mathbf{B} = \mathbf{a}_x(A_yB_z - A_zB_y) + \mathbf{a}_y(A_zB_x - A_xB_z) + \mathbf{a}_z(A_xB_y - A_yB_x) \qquad (1.13a)$$

The cyclic symmetry of this form is made clear through a useful mnemonic of expanding the determinant

$$\mathbf{A} \times \mathbf{B} = \begin{vmatrix} \mathbf{a}_x & \mathbf{a}_y & \mathbf{a}_z \\ A_x & A_y & A_z \\ B_x & B_y & B_z \end{vmatrix} \qquad (1.13b)$$

1.6 TRIPLE PRODUCTS

A useful extension of the previous discussion is to the product of three vectors. A *scalar* triple product is $\mathbf{A} \cdot (\mathbf{B} \times \mathbf{C})$; we have already come across special cases of this in Example 1.2. This triple product has a simple geometric interpretation. Recall that the vector product $\mathbf{B} \times \mathbf{C}$ has a magnitude equal to the area of the parallelogram demarked by **B** and **C** and is a vector along **n**, which is perpendicular to their plane. The scalar product $\mathbf{A} \cdot \mathbf{n}$ gives the component of **A** along **n**, that is, it gives the perpendicular height of **A** above that plane. Therefore, as seen in Fig. 1.10, the scalar

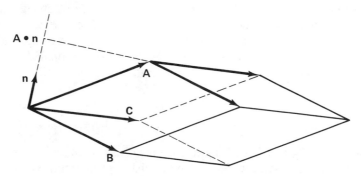

Figure 1.10   Interpretation of scalar triple product.

triple product is base area times height; it equals the scalar volume of the parallelepiped formed by the three vectors. As a consequence of this, we see that the order in which one takes the dot and cross products is immaterial, so that the triple product can be written simply as $(\mathbf{A}, \mathbf{B}, \mathbf{C})$. The sequential order of the vectors must be maintained in order to avoid introduction of a negative sign from the vector product. By extending Eq. (1.13b), we note that the scalar triple product can be expressed as

$$(\mathbf{A}, \mathbf{B}, \mathbf{C}) = \begin{vmatrix} A_x & A_y & A_z \\ B_x & B_y & B_z \\ C_x & C_y & C_z \end{vmatrix} \qquad (1.14)$$

Another important identity is obtained from the *vector* triple product:

$$\mathbf{A} \times (\mathbf{B} \times \mathbf{C}) = \mathbf{B}(\mathbf{A} \cdot \mathbf{C}) - \mathbf{C}(\mathbf{A} \cdot \mathbf{B}) \qquad (1.15)$$

The proof of this relation is rather long and is omitted. Note, however, that $\mathbf{B} \times \mathbf{C}$ is along $\mathbf{n}$, normal to $\mathbf{B}$ and $\mathbf{C}$. Then, $\mathbf{A} \times \mathbf{n}$ is normal to $\mathbf{n}$, that is, it is a vector back in the plane of $\mathbf{A}$ and $\mathbf{B}$. Equation (1.15), therefore, presents the vector product as the sum of vectors parallel to $\mathbf{B}$ and $\mathbf{C}$. The product appearing along $\mathbf{B}$ is $CA_{\|\mathbf{C}}$, and that along $\mathbf{C}$ is $BA_{\|\mathbf{B}}$. From this we realize that the sequential order of taking the cross product must be maintained, that is, the location of the parentheses on the left side of Eq. (1.15).

1.7 GRADIENT

In Fig. 1.1 we showed the functions $f(x, y, z) = C$ for several values of C, and suggested that this might represent the temperature in a region of space. Let us explore this further. For the function (temperature distribution) $T = T(x, y, z)$, we know that heat flows from hot surfaces to cooler surfaces. Furthermore, it is known that the quantity of heat is proportional to the rate at which T changes with position. Therefore, let us examine this rate of change. For a general function $T(x, y, z)$, we have, along each axis, a heat flow proportional to the rate of change along that axis. The rate of change along the x axis is $T'_x = \partial T/\partial x$; along the y axis we have the rate of change $T'_y = \partial T/\partial y$; and along z it is $T'_z = \partial T/\partial z$. Just as the entire heat flow is the sum of these components, since they are in orthogonal directions, so the three derivatives are also parts of the entire rate of change, which must be given by

$$T' = \frac{\partial T}{\partial \ell} = [(T'_x)^2 + (T'_y)^2 + (T'_z)^2]^{1/2} \qquad (1.16a)$$

where $\ell = |\boldsymbol{\ell}|$, and $\boldsymbol{\ell}$ is a length along the direction in which T changes most rapidly, that is, along which the derivative of T has its maximum value; the vector along $\boldsymbol{\ell}$, with magnitude $\partial T/\partial \ell$, is called the *gradient* of T (written *grad T*).

Since the gradient has both magnitude [Eq. (1.16a)] and direction (along ℓ), it must be a vector, and it can be written

$$\text{grad } T = \mathbf{a}_x T_x' + \mathbf{a}_y T_y' + \mathbf{a}_z T_z'$$

$$= \mathbf{a}_x \frac{\partial T}{\partial x} + \mathbf{a}_y \frac{\partial T}{\partial y} + \mathbf{a}_z \frac{\partial T}{\partial z} \qquad (1.16\text{b})$$

as illustrated in Fig. 1.11.

A shorthand notation is commonly used for the expression of Eq. (1.16b) by factoring T to give

$$\text{grad } T = \left(\mathbf{a}_x \frac{\partial}{\partial x} + \mathbf{a}_y \frac{\partial}{\partial y} + \mathbf{a}_z \frac{\partial}{\partial z} \right) T \qquad (1.17\text{a})$$

It is understood that the function T is to be substituted in the blank spaces of the derivatives. In this expression, grad has been written as a vector-differential operator, which is denoted by the symbol of the inverted Greek letter delta (Δ), that is, by ∇, and called the *Nabla* operator, or del.

$$\nabla \equiv \mathbf{a}_x \frac{\partial}{\partial x} + \mathbf{a}_y \frac{\partial}{\partial y} + \mathbf{a}_z \frac{\partial}{\partial z} \qquad (1.17\text{b})$$

While this operator has a particularly concise form in rectangular coordinates, we must remember that it consists of derivatives over displacements along the coordinate axes, that is, changes with respect to distance. In curvilinear systems, these distances involve the metrics, so that, for the general function F,

$$\text{gradient } F = \nabla F = \sum_1^3 \mathbf{a}_i \frac{1}{h_i} \frac{\partial F}{\partial x_i} \qquad (1.17\text{c})$$

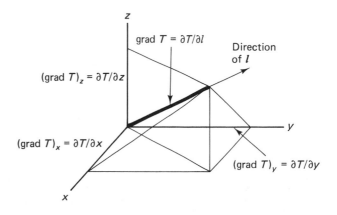

Figure 1.11   Relation of the gradient of $T(x, y, z)$ and its partial derivatives.

Example 1.3

Find grad T in a region of space that has an axis of symmetry and where the temperature is given by

$$T = T_0\rho^2\left(z - \frac{1}{2}\right) \qquad z > \frac{1}{2} \text{ cm}$$

Constant-temperature curves are shown in Fig. 1.12. This is a section through the true temperature distribution, which is a figure of revolution about the z axis. The gradient involves rates of change with respect to position, so that the metrics must be used along each coordinate axis. Therefore, in circular cylindrical coordinates,

$$\frac{\partial T}{\partial \ell_\rho} = \frac{\partial T}{\partial \rho} \qquad \frac{\partial T}{\partial \ell_\phi} = \frac{1}{\rho}\frac{\partial T}{\partial \phi} \qquad \frac{\partial T}{\partial \ell_z} = \frac{\partial T}{\partial z}$$

For the specified function, we see that T has no derivative with respect to ϕ and that

$$\frac{\partial T}{\partial \ell_\rho} = \frac{\partial T}{\partial \rho} = T_0 2\rho\left(z - \frac{1}{2}\right) \qquad \frac{\partial T}{\partial \ell_z} = \frac{\partial T}{\partial z} = T_0\rho^2$$

Then

$$\nabla T = \mathbf{a}_\rho 2T_0\rho\left(z - \frac{1}{2}\right) + \mathbf{a}_z T_0\rho^2$$

At a representative point, taken to be $\rho = 1$ and $z = 1.5$, the value $T = T_0$ and

$$\text{grad}\frac{T}{T_0} = (2\mathbf{a}_\rho + \mathbf{a}_z) \text{ cm}^{-1}$$

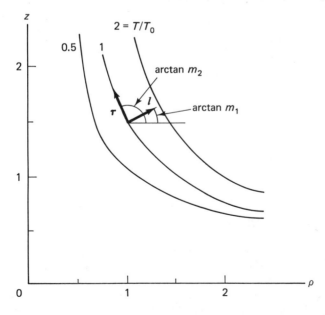

Figure 1.12 Constant-T curves, showing 1, the direction of the gradient of T.

The orientation of the gradient of T at this point is shown in Fig. 1.12. If m_1 is the slope of this vector, then, in terms of its components,

$$m_1 = (\nabla T)_z / (\nabla T)_\rho = \rho/2 \left(z - \frac{1}{2} \right)$$

For comparison, consider the line which is tangent to the $T = $ constant curve, that is, along the tangent τ.

$$m_2 = \frac{\partial z}{\partial \rho} (T = \text{const.}) = -2 \frac{T}{T_0} \rho^3$$

Note that

$$m_1 m_2 = -\frac{(T/T_0)}{\left(z - \frac{1}{2} \right) \rho^2} = -1$$

where the last equality follows from the expression for the constant-temperature curve. The reader will recall from analytic geometry that two lines are orthogonal when the product of their slopes is -1.

In considering the intersection of two lines, we avoid carrying unnecessary constants by moving the coordinate origin to the point of intersection. Figure 1.13 shows $\mathbf{V}_1$, a vector along one of the lines.

$$\mathbf{V}_1 = \Delta x_1 \mathbf{a}_x + \Delta y_1 \mathbf{a}_y = \Delta x_1 [\mathbf{a}_x + (\Delta y/\Delta x)_1 \mathbf{a}_y]$$

$$= \Delta x_1 [\mathbf{a}_x + m_1 \mathbf{a}_y]$$

where m_1 is the slope of the line. There is a similar vector $\mathbf{V}_2$ along the other line through the origin.

$$\mathbf{V}_2 = \Delta x_2 [\mathbf{a}_x + m_2 \mathbf{a}_y]$$

Multiplying these vectors gives

$$\mathbf{V}_1 \cdot \mathbf{V}_2 = (\Delta x_1 \Delta x_2)(1 + m_1 m_2)$$

If the vectors are orthogonal, we have $\mathbf{V}_1 \cdot \mathbf{V}_2 = 0$, so that $m_1 m_2 = -1$.

The last result of Example 1.3 illustrates the fact that the maximum rate of change of a function of position (its gradient) is perpendicular to the surface in which the function is constant, or in which it does not change at all. Our derivation

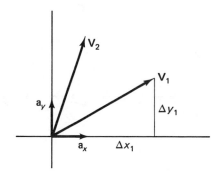

Figure 1.13   Vector slope.

of ∇T has emphasized that, in any direction, the rate of change of the function T is the component of grad T along that direction. Since there is no change within the $T = $ constant surface, there can be no component of ∇T in that surface, so that ∇T is normal to $T = $ constant.

Example 1.4

Consider the equation of the family of planes $F = \dfrac{x}{2} + y + z$. Prove that ∇F is perpendicular to the planes.

The lower plane shown in Fig. 1.14 is $F = 1$; its intersection with the axes are at $x = 2$, $y = 1$, and $z = 1$. The figure also shows the arrow representing

$$\nabla F = \frac{\mathbf{a}_x}{2} + \mathbf{a}_y + \mathbf{a}_z$$

which is a vector of magnitude $[(1/2)^2 + 1^2 + 1^2]^{1/2} = 3/2$. To show that ∇F is normal to the plane $F = 1$, we demonstrate that it is perpendicular to two nonparallel lines that lie in the plane. We take the lines $L_1 = -2\mathbf{a}_x + \mathbf{a}_z$ and $L_2 = \mathbf{a}_y - \mathbf{a}_z$, both of which have $\nabla \mathbf{F} \cdot \mathbf{L}_i = \nabla F_x L_x + \nabla F_y L_y + \nabla F_z L_z = 0$.

1.8 INTEGRALS

In performing integrals of scalar and vector quantities, the integration is carried out either along a specific path, over a surface, or through a volume. It is necessary to examine what is meant by each of these processes.

Contour Integrals.   The line, or contour, integral of a vector $\mathbf{V}$ along a path denoted by ℓ is written as $\int \mathbf{V} \cdot \boldsymbol{\tau}\, d\ell$, where $\boldsymbol{\tau}$ is a unit vector, tangent to path ℓ at each point. The integrand $\mathbf{V} \cdot \boldsymbol{\tau} = V_{\text{component along path}}$ is summed along ℓ. The reader has met this situation when calculating the work done by a force $\mathbf{V}$ as it moves through a distance ℓ.

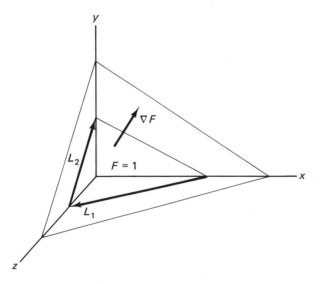

Figure 1.14 Family of planes.

Example 1.5

Take $\mathbf{V} = y\mathbf{a}_x$ and evaluate $\oint \mathbf{V} \cdot \boldsymbol{\tau} d\ell$ around the rectangular path shown in Fig. 1.15.

The circle on the integral indicates that the path of integration is a closed loop. In performing a closed integration, we observe the convention of keeping the enclosed area to the left as we travel about the contour, that is, we evaluate the integral along the path as we move in a counterclockwise direction (or in the direction of increasing angle). The closed-loop integral of $\mathbf{V}$ around the perimeter is called the *circulation* of $\mathbf{V}$ about the path; if $\mathbf{V}$ were a velocity, the circulation would give the total flow about the contour.

For this contour integral, the path must be separated into its constituent parts along the four sides of the rectangle, each of which is described by different parameters. Starting from the origin,

$$\oint \mathbf{V} \cdot \boldsymbol{\tau} d\ell = \int_0^X y\mathbf{a}_x \cdot \mathbf{a}_x \, dx \Big|_{y=0} + \int_0^Y y\mathbf{a}_x \cdot \mathbf{a}_y \, dy \Big|_{x=X} + \int_X^0 y\mathbf{a}_x \cdot \mathbf{a}_x \, dx \Big|_{y=Y}$$

$$+ \int_0^Y y\mathbf{a}_x \cdot (-\mathbf{a}_y) \, dy \Big|_{x=0} = 0 + 0 - YX + 0 = -XY$$

where the parameters denoting the path segments have been written below each integrand as a reminder of the portion being considered. Thus, the first integral is characterized by $y = 0$ and the second integral by $x = X$. Note that in the last integral, $\boldsymbol{\tau}$ is in the direction of traversing the path ($\boldsymbol{\tau} = -\mathbf{a}_y$) and the limits are in the positive direction, that is, from their minimum to their maximum values. In contrast, in the next-to-last integral, $\boldsymbol{\tau}$ is in the positive direction, that is, along $+\mathbf{a}_x$, which is opposite to the direction of travel, and the limits account for the direction of travel. The direction can be included *either* in the unit vector or in the limits, but care must be taken not to do both, that is, not to introduce a double negative.

The first integral is zero because its integrand contains the zero factor y. The second and fourth integrals have $\boldsymbol{\tau} = \pm\mathbf{a}_y$, so that the vectors $\mathbf{V}$ and $\boldsymbol{\tau}$ are orthogonal and the scalar products are zero. The third reduces to the value $-YX$.

Example 1.6

For $\mathbf{V} = \mathbf{a}_\phi$, find the circulation of $\mathbf{V}$ about the circle $\rho = R$ and $z = 0$.

Since $\boldsymbol{\tau} = \mathbf{a}_\phi$ and $d\ell_\phi = \rho \, d\phi$, the integral is

$$\oint \mathbf{V} \cdot \boldsymbol{\tau} d\ell = \int_0^{2\pi} \mathbf{a}_\phi \cdot \mathbf{a}_\phi \rho \, d\phi \Big|_{\rho=R} = 2\pi R$$

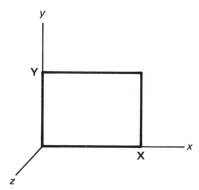

Figure 1.15 Path for the contour integral.

Example 1.7

Given the vector field $\mathbf{V} = \mathbf{a}_y/x$, evaluate the line integrals from $(0, 0)$ to $(2, 4A)$: (a) along the curve $y = Ax^2$, and (b) along the piecewise linear path from points $(0, 0)$ to $(2, 0)$ to $(2, 4A)$.

(a) To integrate along the curve, we need expressions for an infinitesimal arc length, $d\ell$, and for the unit vector $\boldsymbol{\tau}$ tangent to the curve.

As shown in Fig. 1.16, an infinitesimal length along a curve is

$$d\ell = (dx^2 + dy^2)^{1/2} = dx(1 + m^2)^{1/2}$$

where $m = dy/dx$ is the slope of the curve. From this and the figure, we have, for the unit vector,

$$\boldsymbol{\tau} = \frac{d\boldsymbol{\ell}}{|d\boldsymbol{\ell}|} = \frac{\mathbf{a}_x\, dx + \mathbf{a}_y\, dy}{d\ell} = \frac{\mathbf{a}_x + m\mathbf{a}_y}{(1 + m^2)^{1/2}}$$

Multiplying by $d\ell$ gives

$$\boldsymbol{\tau}\, d\ell = dx(\mathbf{a}_x + m\mathbf{a}_y)$$

For the parabola, $m = dy/dx = 2Ax$, so that

$$\int \mathbf{V} \cdot \boldsymbol{\tau}\, d\ell = 2A \int_0^2 dx = 4A$$

(b) For the piecewise linear path,

$$\int \mathbf{V} \cdot \boldsymbol{\tau}\, d\ell = \int \mathbf{V} \cdot \mathbf{a}_x\, dx \Big|_{y=0} + \int \mathbf{V} \cdot \mathbf{a}_y\, dy \Big|_{x=2} = \frac{1}{2} \int_0^{4A} dy = 2A$$

The first integral in the center is zero because $\mathbf{V} \cdot \mathbf{a}_x = 0$. Note that the integrals along the two paths lead to different values.

Surface Integrals.   The integral of a vector $\mathbf{V}$, over a surface denoted by S, is written as $\int \mathbf{V} \cdot \mathbf{n}\, dS$, where $\mathbf{n}$ is a unit vector, normal to the surface at each point. The integrand $\mathbf{V} \cdot \mathbf{n} = V_\perp$ is summed over S. The integral $\int \mathbf{V} \cdot \mathbf{n}\, dS$ is

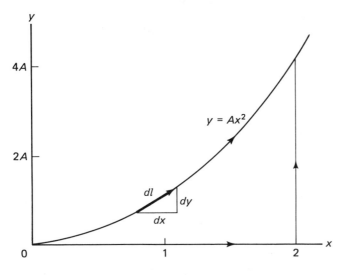

Figure 1.16   Integration along a curved path.

called the *flux* of **V** through the surface S; it measures the total amount of **V** flowing across the surface. The direction of **n** is along the right-hand thumb when the fingers curl about the area in the direction that one would take the enclosing contour.

Example 1.8

Determine the flux of the vector

$$\mathbf{V} = \mathbf{a}_x \frac{y^2}{z^2} + \mathbf{a}_y \frac{z^2}{x^2} + \mathbf{a}_z \frac{x^2}{y^2}$$

through the area of Fig. 1.15.

Here $\mathbf{n} = \mathbf{a}_z$, so that $\mathbf{V} \cdot \mathbf{n} = V_z = x^2/y^2$, and

$$\int \mathbf{V} \cdot \mathbf{n}\, dS = \int_{y=0}^{Y} \int_{x=0}^{X} \left(\frac{x^2}{y^2}\right) dx\, dy = -\frac{X^3}{3Y}$$

Example 1.9

Evaluate the surface integral of $\mathbf{V} = \mathbf{a}_r/r$ over the entire surface of a sphere of radius R centered at the origin.

Since this surface is closed, there is no bounding contour, so we must adopt the universal convention of taking the unit normal to be directed outward from the enclosed volume. Then $\mathbf{n} = \mathbf{a}_r$, and

$$\oint \mathbf{V} \cdot \mathbf{n}\, dS = \oint \left(\frac{\mathbf{a}_r}{r}\right) \cdot \mathbf{a}_r r^2 \sin\theta\, d\theta\, d\phi = 4\pi R$$

Figure 1.17 shows the derivation of the element of surface area in terms of the spherical coordinate angles. dS is the product of the two infinitesimal sides, both of which have the form

$$\text{length} = \text{radius} \times \text{angle}$$

giving

$$dS = (r\, d\theta)(r\, \sin\theta\, d\phi)$$

Note that, on the sphere, $r = R$, $0 \le \theta \le \pi$, and $0 \le \phi \le 2\pi$.

1.9 VECTOR INTEGRATION

Observe that the integrands of all the integrals in the previous section are scalars resulting from the dot product of **V** with $\boldsymbol{\tau}$ or **n**. One also encounters integrals where the integrand is a vector. In this event, not only is the magnitude of the integrand a function of the integration variable, but its direction can change as well. Since integration corresponds to a summation process, we therefore need to perform vector sums of changing vectors. In certain cases, the symmetry properties of a particular problem allow sufficient simplification that the integration can be done with relative ease. In general, however, the procedure that must be followed is to reduce the integrand to its rectangular components. Then, since the cartesian-vector directions are constant, for each component, the remaining integrals involve scalars.

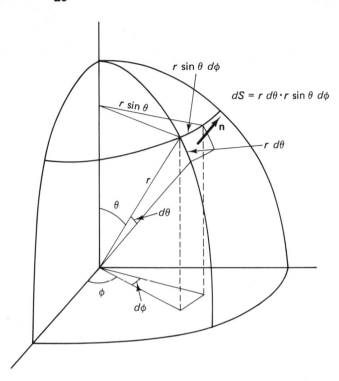

Figure 1.17   Spherical area element.

Example 1.10

Find the volume integral of $\mathbf{V} = \mathbf{a}_r r$ through the hemisphere with $r = 1$ and $z \geq 0$.

From Fig. 1.17, we have $dv = dr\, dS = r^2 \sin\theta\, d\theta\, d\phi\, dr$, so the desired integral is

$$\int \mathbf{V}\, dv = \int_{r=0}^{1} \int_{\theta=0}^{\pi/2} \int_{\phi=0}^{2\pi} r\mathbf{a}_r r^2 \sin\theta\, d\phi\, d\theta\, dr$$

This cannot be carried out directly because the unit vector $\mathbf{a}_r$ is not constant; its direction changes with θ and ϕ. From Eq. (1.9a), $\mathbf{a}_r$ can be written as the sum of three cartesian unit vectors, so that

$$\int \mathbf{V}\, dv = \mathbf{a}_x \int r \sin\theta \cos\phi\, dv + \mathbf{a}_y \int r \sin\theta \sin\phi\, dv + \mathbf{a}_z \int r \cos\theta\, dv \ .$$

Since the cartesian unit vectors are constant, they have been removed from the integrals, leaving three scalar integrals to be evaluated.

Rather than solve these three integrals, we can take advantage of the axial symmetry of the hemisphere and of $\mathbf{V}$, with respect to the z axis, to simplify the problem. For this, convert $\mathbf{a}_r$ to polar coordinates, again using Eq. (1.9a):

$$\int \mathbf{V}\, dv = \mathbf{a}_z \int r \cos\theta\, dv + \int \mathbf{a}_\rho r \sin\theta\, dv$$

Here the variable vector $\mathbf{a}_\rho$ cannot be taken out of the integral. Figure 1.18 represents any section containing the z axis, cut through the hemisphere. It shows two points, P and Q, in the same section and with the same r and θ, but with $\Delta\phi = 180°$. The $\mathbf{a}_\rho$ at P and Q are oppositely directed and in the rightmost integrand above, the $r \sin\theta$, is the same at both

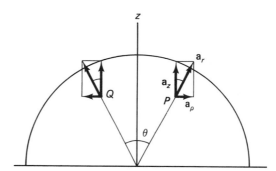

Figure 1.18 Vector components using axial symmetry. The figure shows any plane containing $\mathbf{a}_z$.

points. As a result, in the vector sum represented by the integral, their $\mathbf{a}_\rho$ contributions are equal and opposite, and they cancel. Every point in the hemisphere has an opposite point, corresponding to P and Q, so that the entire $\mathbf{a}_\rho$ integral is zero.

As a result of axial symmetry, we therefore have only the integral

$$\int \mathbf{V}\, dv = \mathbf{a}_z \int_{r=0}^{1} \int_{\theta=0}^{\pi/2} \int_{\phi=0}^{2\pi} r^3 \sin\theta \cos\theta \, d\phi \, d\theta \, dr = \mathbf{a}_z \frac{\pi}{4}$$

If we do not recognize the axial symmetry and, instead, evaluate the three cartesian integrals, the first two are equal zero. Recognition of the existence of symmetry, however, lends great insight into a problem and greatly simplifies its solution.

1.10 CURL OF A VECTOR

For a vector function of position, that is, a vector field, we have already discussed the contour integral about a closed path: $\oint \mathbf{V} \cdot \boldsymbol{\tau} d\ell$. This *circulation* of $\mathbf{V}$ about the path tells us something about its rotational behavior. It is of interest to examine the integral as the path shrinks to zero, so that the curling, or vortex property, of $\mathbf{V}$ can be specified at each point. Care should be taken if we are to do our evaluation so that the result is applicable in any coordinate system, because we saw earlier that the path lengths $d\ell_i$ depend on the metrics of the system and that the unit vectors $\mathbf{a}_i$ vary in direction as the coordinates change. In cartesian coordinates, however, the metrics are all unity, so that $d\ell_i = dx_i$, and the situation is greatly simplified. We will, therefore, consider this case first.

The contour in Fig. 1.19 consists of four infinitesimal segments parallel to the coordinate axes at x, y, $x + dx$, and $y + dy$. Moving in a counterclockwise direction from point 0, we have

$$\oint \mathbf{V} \cdot \boldsymbol{\tau} d\ell = [V_x\, dx]_{\text{path } a} + [V_y\, dy]_{\text{path } b} + [V_x\, dx]_{\text{path } c} + [V_y\, dy]_{\text{path } d} \qquad (1.18)$$

For such small path elements, each component integral has been reduced to the product of its integrand, at the path point, times the displacement. We take the vector at point 0 to be $\mathbf{V} = V_x\mathbf{a}_x + V_y\mathbf{a}_y + V_z\mathbf{a}_z$. Because V changes with position, we note that

$$[V_x\, dx]_{\text{path } a} = V_x\, dx \qquad (1.19a)$$

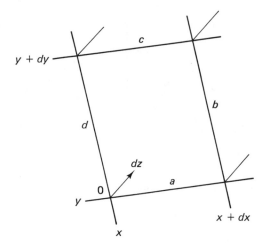

Figure 1.19   Infinitesimal contour path.

$$[V_y\,dy]_{\text{path }b} = [V_y + \left(\frac{\partial V_y}{\partial x}\right) dx]\,dy \tag{1.19b}$$

$$[V_x\,dx]_{\text{path }c} = [V_x + \left(\frac{\partial V_x}{\partial y}\right) dy](-dx) \tag{1.19c}$$

$$[V_y\,dy]_{\text{path }d} = V_y(-dy) \tag{1.19d}$$

Substituting Eqs. (1.19) into Eq. (1.18), and canceling terms, yields

$$\oint \mathbf{V} \cdot \boldsymbol{\tau}\,d\ell = \left(\frac{\partial V_y}{\partial x} - \frac{\partial V_x}{\partial y}\right) dx\,dy \tag{1.20}$$

Note that $dx\,dy$ is the area enclosed by the path and, following our earlier practice, we assign this area to its normal direction, so that $dx\,dy \rightarrow \mathbf{a}_z\,dS$, since the unit vector normal to the xy plane is the $\mathbf{a}_z$ coordinate vector.

In keeping with our stated goal, we wish to take the limit so that the path encloses only a point. From Eq. (1.20), we see that it is dS that must approach zero and that the rotation or curl at the point must be expressed as the circulation per unit area:

$$(\text{curl of } \mathbf{V})_z = \lim_{dS \to 0} \frac{\oint \mathbf{V} \cdot \boldsymbol{\tau}\,d\ell}{dS} = \left(\frac{\partial V_y}{\partial x} - \frac{\partial V_x}{\partial y}\right) \tag{1.21}$$

It should be clear that we could have selected an area $dy\, dx \rightarrow \mathbf{a}_x\, dS$ or $dz\, dx \rightarrow \mathbf{a}_y\, dS$ to obtain different components of the curl of $\mathbf{V}$; they can be obtained by cyclically permuting the subscripts $x \rightarrow y \rightarrow z \rightarrow x$. These are components of the vector

$$
\begin{aligned}
\text{Curl } \mathbf{V} = \mathbf{a}_x &\left(\frac{\partial V_z}{\partial y} - \frac{\partial V_y}{\partial z} \right) \\
+ \mathbf{a}_y &\left(\frac{\partial V_x}{\partial z} - \frac{\partial V_z}{\partial x} \right) \\
+ \mathbf{a}_z &\left(\frac{\partial V_y}{\partial x} - \frac{\partial V_x}{\partial y} \right)
\end{aligned}
\tag{1.22a}
$$

This form can be obtained by expanding the determinant

$$
\text{curl } \mathbf{V} =
\begin{vmatrix}
\mathbf{a}_x & \mathbf{a}_y & \mathbf{a}_z \\
\dfrac{\partial}{\partial x} & \dfrac{\partial}{\partial y} & \dfrac{\partial}{\partial z} \\
V_x & V_y & V_z
\end{vmatrix}
\tag{1.22b}
$$

Equation (1.22b) is reminiscent of Eq. (1.12b) for the vector product

$$
\mathbf{D} \times \mathbf{V} =
\begin{vmatrix}
\mathbf{a}_x & \mathbf{a}_y & \mathbf{a}_z \\
D_x & D_y & D_z \\
V_x & V_y & V_z
\end{vmatrix}
\tag{1.23a}
$$

Replacing $\mathbf{D}$ by corresponding elements of the ∇ operator from Eq. (1.16b),

$$
\mathbf{D} \rightarrow \nabla = \mathbf{a}_x \frac{\partial}{\partial x} + \mathbf{a}_y \frac{\partial}{\partial y} + \mathbf{a}_z \frac{\partial}{\partial z}
\tag{1.23b}
$$

yields

$$
\text{curl } \mathbf{V} \rightarrow \nabla \times \mathbf{V}
\tag{1.23c}
$$

The curl of a vector expressed in curvilinear coordinates is derived in Appendix 1. There the coordinate metrics h_1, h_2, and h_3 must be included and the result can be expressed as

$$
\text{curl } \mathbf{V} = \left(\frac{1}{h_1 h_2 h_3} \right)
\begin{vmatrix}
h_1 \mathbf{a}_1 & h_2 \mathbf{a}_2 & h_3 \mathbf{a}_3 \\
\dfrac{\partial}{\partial x_1} & \dfrac{\partial}{\partial x_2} & \dfrac{\partial}{\partial x_3} \\
h_1 V_1 & h_2 V_2 & h_3 V_3
\end{vmatrix}
\tag{1.24}
$$

It is clear from comparison of Eqs. (1.24) and (1.23) that it is only in the rectangular coordinate system that the curl of a vector has the simple form of an apparent vector cross product of the differential operator with the vector. Nonetheless, the notation curl $\mathbf{V} \equiv \nabla \times \mathbf{V}$ is almost universally used, though the correct form must be taken from Eq. (1.24).

Example 1.11

Find the curl of

$$V = \frac{-\mathbf{a}_x y + \mathbf{a}_y x}{(x^2 + y^2)^{1/2}}$$

$$\nabla \times \mathbf{V} = \begin{vmatrix} \mathbf{a}_x & \mathbf{a}_y & \mathbf{a}_z \\ \dfrac{\partial}{\partial x} & \dfrac{\partial}{\partial y} & \dfrac{\partial}{\partial z} \\ \dfrac{-y}{(x^2 + y^2)^{1/2}} & \dfrac{x}{(x^2 + y^2)^{1/2}} & 0 \end{vmatrix}$$

$$= \mathbf{a}_z \frac{\partial}{\partial x}\left[\frac{x}{(x^2 + y^2)^{1/2}}\right] + \frac{\partial}{\partial y}\left[\frac{y}{(x^2 + y^2)^{1/2}}\right]$$

$$= \mathbf{a}_z \frac{y^2 + x^2}{(x^2 + y^2)^{3/2}}$$

$$- \frac{\mathbf{a}_z}{[x^2 + y^2]^{1/2}}$$

From Eqs. (1.1a) and (1.9c), we note that vector **V** of this exercise can also be written using circular cylindrical coordinates:

$$\mathbf{V} = -\mathbf{a}_x \frac{y}{\rho} + \mathbf{a}_y \frac{x}{\rho} = -\mathbf{a}_x \sin\phi + \mathbf{a}_y \cos\phi = \mathbf{a}_\phi$$

so that Eq. (1.24) becomes

$$\nabla \times \mathbf{V} = \left(\frac{1}{\rho}\right) \begin{vmatrix} \mathbf{a}_\rho & \rho\mathbf{a}_\phi & \mathbf{a}_z \\ \dfrac{\partial}{\partial \rho} & \dfrac{\partial}{\partial \phi} & \dfrac{\partial}{\partial z} \\ 0 & \rho & 0 \end{vmatrix} = \frac{\mathbf{a}_z}{\rho}$$

which is the same as our previous result.

The result of Example 1.11 illustrates that the curl of a vector is the same regardless of the coordinate system in which it is evaluated; curl is a physical property of a vector field and is invariant with regard to the manner in which it is expressed. This is also apparent from the integral form of Eq. (1.21). As a result of its invariance, we have the freedom to evaluate the curl in any convenient coordinate system.

1.11 STOKES' THEOREM

A generalization of Eq. (1.21) can be obtained by relaxing the limit that $dS \to 0$. Multiplying Eq. (1.21) by dS yields

$$\oint_{\text{around } dS} \mathbf{V} \cdot \boldsymbol{\tau} \, d\ell = (\nabla \times \mathbf{V})_{\perp \, dS} \, dS = (\nabla \times \mathbf{V}) \cdot \mathbf{n} \, dS \qquad (1.25)$$

where **n** is the unit vector normal to the element of surface about which the contour integral is taken. Figure 1.20 has a large surface S bounded by the curve C and shows the elemental areas of which it is composed.

Let us carry the contour integral on the left-hand side of Eq. (1.25) about the perimeters of all these small areas and then add the results. The inserts to the figure show that the τ are oppositely directed on the common sides of adjacent regions. As a result, the individual path integrals give opposite contributions on their common sides and cancel when they are added. The exceptions to this cancellation are on the perimeter curve C, where the integrals are found on only one side. Therefore, after adding the integrals around all the infinitesimal areas, we are left with only the contributions on C, the perimeter of S. Also, adding the surface contributions on the right-hand side of Eq. (1.25) yields the integral over the entire surface bounded by C. We then have

$$\oint \mathbf{V} \cdot \boldsymbol{\tau}\, d\ell = \int (\nabla \times \mathbf{V}) \cdot \mathbf{n}\, dS \qquad (1.26)$$

where it is understood that the contour integral is over the bounding curve C and the surface integral is over the entire surface bounded by C. It is important to realize that it has not been necessary to specify the surface S, which has the curve C as its boundary. Equation (1.26) is remarkable in that S can be any surface terminated by C. In the cases to be studied in this text, this general property is associated with the relation of the curl of the vector fields to the field sources. Equation (1.26) is *Stokes' theorem.*

Example 1.12

Verify Stokes' theorem for $\mathbf{V} = \mathbf{a}_\phi$, where C is a circle of radius R, centered at the origin.

The infinitesimal circumferential length of a circle is $d\ell = \rho\, d\phi$, and is directed along $\mathbf{a}_\phi$, so that

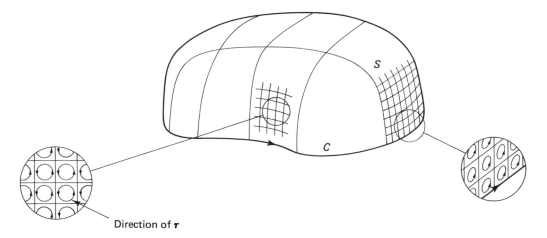

Direction of $\boldsymbol{\tau}$

Figure 1.20   Stokes' theorem.

$$\oint \mathbf{V} \cdot \boldsymbol{\tau}\, d\ell = \int_0^{2\pi} \mathbf{a}_\phi \cdot \mathbf{a}_\phi \rho\, d\phi \Big|_{\rho=R} = 2\pi R$$

From Example 1.11, we already have $\nabla \times \mathbf{V} = \mathbf{a}_z/\rho$, so that

$$\int \mathbf{n} \cdot (\nabla \times \mathbf{V})\, dS = \int_0^R \mathbf{a}_z \cdot \frac{\mathbf{a}_z}{\rho} 2\pi\rho\, d\rho = 2\pi R$$

where the element of area is the annular ring: $dS = 2\pi\rho\, d\rho$, as shown on Fig. 1.21.

Example 1.13

Verify Stokes' theorem for the contour of Fig. 1.22, with $\mathbf{V} = \mathbf{a}_\phi/\rho$.

Equation (1.24) gives $\nabla \times \mathbf{V} = 0$, so that the surface integral of the curl is identically zero.

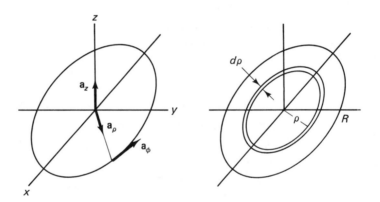

Figure 1.21   Circulation about a circular path.

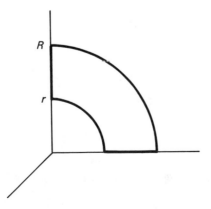

Figure 1.22   Path of integration.

By definition, the contour integral is positive when the enclosed area is kept to the left, which means that we should proceed in a counterclockwise direction about the perimeter. By starting at point $(r, 0)$, the circulation is

$$\oint \mathbf{V} \cdot \boldsymbol{\tau} \, d\ell = \int_r^R \frac{\mathbf{a}_\phi}{\rho} \cdot \mathbf{a}_\rho \, d\rho \bigg|_{\phi=0} + \int_0^{\pi/2} \frac{\mathbf{a}_\phi}{\rho} \cdot \mathbf{a}_\phi \rho \, d\phi \bigg|_{\rho=R}$$

$$+ \int_R^r \left(\frac{\mathbf{a}_\phi}{\rho}\right) \cdot \mathbf{a}_\rho \, d\rho \bigg|_{\phi=\pi/2} + \int_{\pi/2}^0 \left(\frac{\mathbf{a}_\phi}{\rho}\right) \cdot \mathbf{a}_\phi \rho \, d\phi \bigg|_{\rho=r}$$

$$= 0 + \pi/2 + 0 - \pi/2 = 0$$

The first and third integrals have vanished because the unit vectors are orthogonal.

1.12 DIVERGENCE THEOREM

The flux of a vector across an area was discussed earlier and given by $\int \mathbf{V} \cdot \mathbf{n} \, dS$, where $\mathbf{n}$ is the unit normal to the surface. Let us take $\mathbf{V}$ to represent a physical, conserved quantity, that is, one that cannot be created or destroyed, for example, where $\mathbf{V}$ is the mass of water flow. If the surface is closed, that is, completely enclosing a volume, the outward flow of $\mathbf{V}$ has to originate from the volume within S. We can imagine that there is a function giving the sources of $\mathbf{V}$ at each point of the volume, whose strength is denoted by the name "divergence of $\mathbf{V}$," to imply that $\mathbf{V}$ is moving away from, or out of, each point. Then the total divergence of $\mathbf{V}$ from the volume, obtained by integrating over all these sources, must equal the flux across the bounding surface:

$$\int \text{div } \mathbf{V} \, dv = \oint \mathbf{V} \cdot \mathbf{n} \, dS \tag{1.27}$$

where the loop on the surface integral indicates that S is a closed surface. Equation (1.27) is known as the *divergence theorem*. It is a powerful identity for studying vector field behavior. The quantity div $\mathbf{V}$ is only understood in the qualitative sense just discussed; Eq. (1.27) is its definition, and it must be studied in more detail.

To help visualize this relationship, let us again consider that $\mathbf{V}$ is a velocity field, with units of m/sec. If we multiply by μ, the mass density, then $\mu\mathbf{V}$ is $(g/m^3) \cdot (m/sec) = (g/sec)/m^2$, which is the mass per second crossing each unit area. When the right-hand side of Eq. (1.27) is evaluated, this is multiplied by dS (with units of m^2), giving units of g/sec. Thus, we are dealing with the total flux, or mass flow rate across the closed surface denoted by S. (The dot product $\mu\mathbf{V} \cdot \mathbf{n}$ guarantees that the mass measured is that which crosses the area; any flow tangential to S is of no concern.) The quantity [div $\mu\mathbf{V}$] on the left-hand side of the equation must have the units of the right-hand side (g/sec) divided by the volume, or $(g/m^3)/$ sec; it is the rate at which the *density* changes at each point within the volume, or the rate at which the mass diverges from (leaves) each infinitesimal point volume, hence, the name *divergence*.

It will be useful to obtain an explicit expression for the divergence of a vector, as defined by Eq. (1.27). Since it is a point function, that is, it has a value at each point, we take the limit of Eq. (1.27) when the volume and surface reduce to zero so as to contain only a single point:

$$\text{div } \mathbf{V} = \lim_{dv \to 0} \frac{\oint \mathbf{V} \cdot \mathbf{n}\, dS}{dv} \tag{1.28}$$

As we did with Stokes' theorem, we will limit our immediate discussion to cartesian coordinates, leaving the more general derivation to Appendix 1.

Consider the infinitesimal volume element between coordinate surfaces, as shown in Fig. 1.23. To calculate the integral of Eq. (1.28), we note the convention that $\mathbf{n}$ is positive when taken directed out of an enclosed volume, whereas other directions are positive when taken along increasing x_i. Therefore, the flux of V through the left yz face (containing point 0) is

$$\mathbf{V} \cdot \mathbf{n}\, dS_{\text{left face}} = -V_{x(\text{left})}\, dy\, dz = -V_x\, dy\, dz \tag{1.29a}$$

Through the right yz face, the flux is

$$\mathbf{V} \cdot \mathbf{n}\, dS_{\text{right face}} = V_{x(\text{right})}\, dy\, dz = \left[V_x + \frac{\partial V_x}{\partial x} dx \right] dy\, dz \tag{1.29b}$$

Adding Eqs. (1.29a) and (1.29b) gives the total outward flux of V in the x direction. Dividing this by $dv = dx\, dy\, dz$, and letting $dv \to 0$, yields

$$(\text{div } \mathbf{V})_x = \lim_{dv \to 0} \frac{\left(\oint \mathbf{V} \cdot \mathbf{n}\, dS \right)_x}{dv} = \frac{\partial V_x}{\partial x} \tag{1.29c}$$

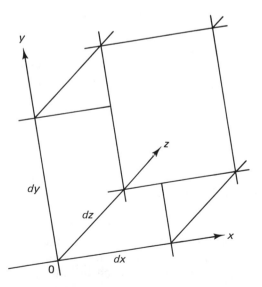

Figure 1.23 Infinitesimal volume element.

Contributions to the divergence of V from the other faces of the volume can be found by permuting the subscripts. The total is

$$\text{div } V = \frac{\partial V_x}{\partial x} + \frac{\partial V_y}{\partial y} + \frac{\partial V_z}{\partial z} = \nabla \cdot V \tag{1.30}$$

where, for the last equality, we have treated the ∇ operator of Eq. (1.16b) as though it were a true vector, and assumed that $\nabla \cdot V$ corresponds to a scalar product of vectors.

In curvilinear coordinates, the divergence is derived in Appendix 1.

$$\text{div } V = \frac{1}{h_1 h_2 h_3} \left[\frac{\partial}{\partial x_1} (V_1 h_2 h_3) + \frac{\partial}{\partial x_2} (V_2 h_3 h_1) + \frac{\partial}{\partial x_3} (V_3 h_1 h_2) \right] \tag{1.31}$$

We see that the notation div $V \to \nabla \cdot V$ does not apply in curvilinear coordinates. However, because of the simple form of Eq. (1.30), it is common to use $\nabla \cdot V$ to denote the divergence of V, even in those systems where such simplicity does not exist and where Eq. (1.31) must be employed.

Example 1.14

Find $\nabla \cdot V$ for

$$V = \frac{x\mathbf{a}_x + y\mathbf{a}_y + z\mathbf{a}_z}{x^2 + y^2 + z^2}$$

Use of Eq. (1.30) yields

$$\nabla \cdot V = \frac{(x^2 + y^2 + z^2 - 2x^2)}{(x^2 + y^2 + z^2)^2} + 2 \text{ cycles of permuted terms in } x, y, z$$

$$= \frac{1}{x^2 + y^2 + z^2}$$

This vector can also be written in spherical coordinates as $V = \mathbf{r}/r^2 = \mathbf{a}_r/r$, so that, using Eq. (1.31),

$$\nabla \cdot V = \frac{1}{r^2 \sin \theta} \frac{\partial}{\partial r} (V_r r^2 \sin \theta) = \frac{1}{r^2} \frac{\partial r}{\partial r} = \frac{1}{r^2}$$

which agrees with the rectangular form.

Example 1.15

Verify the divergence theorem for $V = \mathbf{a}_r$, where S is the surface of a sphere of radius R, centered at the origin of coordinates.

From Eq. (1.31), we have $\nabla \cdot V = 2/r$. The spherical volume element is $dv = r^2 \sin \theta \, d\theta \, d\phi \, dr$, and since $\nabla \cdot V$ is a function only of r, the integration over ϕ and θ can be done immediately, giving $dv = 4\pi r^2 \, dr$, which is the spherical shell shown in Fig. 1.24. Then

$$\int \nabla \cdot V \, dv = \int_0^R \frac{2}{r} 4\pi r^2 \, dr = 4\pi R^2$$

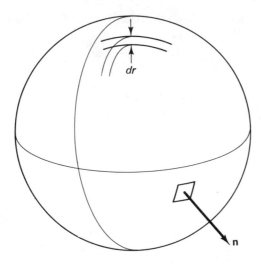

Figure 1.24 Volume of a spherical shell.

Although the divergence decreases as r increases (div $\mathbf{V} \sim 1/r$), the volume increases more rapidly ($v \sim r^3$), so that the integral is proportional to R^2. For the surface integral, we have

$$\oint \mathbf{V} \cdot \mathbf{n}\, dS = \oint \mathbf{a}_r \cdot \mathbf{a}_r\, dS = \int dS = 4\pi R^2$$

Example 1.16

We can consider a field whose divergence increases more rapidly with r than in the previous example:

$$\mathbf{V} = \mathbf{r} = \mathbf{a}_r r \qquad \nabla \cdot \mathbf{V} = 3$$

$$\int \nabla \cdot \mathbf{V}\, dv = 3 \int dv = 3v$$

$$\oint \mathbf{V} \cdot \mathbf{n}\, dS = \oint r \mathbf{a}_r \cdot \mathbf{a}_r\, dS = R \int_{r=R} dS = 4\pi R^3 = 3v$$

Since the divergence is uniform throughout the volume, with value 3, the volume integral is merely this value times the volume.

1.13 THE LAPLACIAN

The divergence of the gradient of a scalar function f occurs very prominently in electromagnetics. In rectangular coordinates, it is

$$\nabla \cdot \nabla f = \frac{\partial^2 f}{\partial x^2} + \frac{\partial^2 f}{\partial y^2} + \frac{\partial^2 f}{\partial z^2} = \nabla^2 f \qquad (1.32)$$

where the notation ∇^2 is appropriate to indicate an operator of the second derivatives. Note that such a simple form is not found in other coordinate systems, where $\nabla \cdot \nabla f$ may even contain first-derivative terms. However, as with divergence and curl, this notation is still used. The correct forms for $\nabla^2 f$, called the *Laplacian* of f, are shown in the table of vector operations which appears on the inside cover of this book, along with a summary of the other operations we have studied.

EXERCISES

Extensions of the Theory

1.1. Trigonometric relations can be simply derived from vector relations. The unit vectors **a** and **b** lie in the xy plane at angles α and β to the x axis, respectively.
   (a) Express each in terms of its components.
   (b) Derive the expression for $\cos(\alpha - \beta)$.
   (c) Derive the expression for $\sin(\alpha - \beta)$.

1.2. Show that $\mathbf{r} \cdot \mathbf{n} = d$ is the equation of a plane that is normal to the constant unit vector **n** and a distance d from the origin.

1.3. Three vectors **A**, **B**, and **C** are the sides of a triangle. Show that $C^2 = A^2 + B^2 - 2AB \cos(\mathbf{A}, \mathbf{B})$. This is the law of cosines.

1.4. In curvilinear coordinates, the basis vectors **a** always have unit magnitude but change in direction. Show that $d\mathbf{a}$ is always perpendicular to **a**. *Hint:* Start from $\mathbf{a} \cdot \mathbf{a} = 1$.

1.5. Prove that the divergence of any curl is identically zero.

1.6. Prove that the curl of any gradient is identically zero.

Problems

*1.7. The coordinate systems most used are (x, y, z), (ρ, ϕ, z), and (r, θ, ϕ). Given the following points in parenthesis, in one coordinate system, find the coordinates in the other two systems:
   (a) $(x, y, z) = (3, 4, 5)$
   (b) $(r, \theta, \phi) = (5, 60°, 90°)$
   (c) $(\rho, \phi, z) = (7, \pi/6, 2)$

*1.8. Find $\mathbf{A} + \mathbf{B}$, $\mathbf{A} - \mathbf{B}$, $\mathbf{A} \cdot \mathbf{B}$, and $\mathbf{A} \times \mathbf{B}$ for each of the following:
   (a) $\mathbf{A} = 7\mathbf{a}_x + 9\mathbf{a}_y$ and $\mathbf{B} = 2\mathbf{a}_x - 5\mathbf{a}_y$
   (b) $\mathbf{A} = \mathbf{a}_x - 3\mathbf{a}_y + 2\mathbf{a}_z$ and $\mathbf{B} = \mathbf{a}_x + 3\mathbf{a}_y - 2\mathbf{a}_z$
   (c) $\mathbf{A} = 2\mathbf{a}_\rho + (\pi/3)\mathbf{a}_\phi + 2\mathbf{a}_z$ and $\mathbf{B} = \mathbf{a}_\rho + (\pi/2)\mathbf{a}_\phi + 2\mathbf{a}_z$
   (d) $\mathbf{A} = 5\mathbf{a}_r + (\pi/2)\mathbf{a}_\theta + (\pi/6)\mathbf{a}_\phi$ and $\mathbf{B} = \mathbf{a}_r - (\pi/4)\mathbf{a}_\theta - (\pi/6)\mathbf{a}_\phi$

*1.9. **(a)** Show that $\mathbf{A} = 7\mathbf{a}_x + \mathbf{a}_y - 3\mathbf{a}_z$ and $\mathbf{B} = \mathbf{a}_x + 2\mathbf{a}_y + 3\mathbf{a}_z$ are orthogonal.
   (b) Show that **B** and $\mathbf{C} = 2\mathbf{a}_x + 4\mathbf{a}_y + 6\mathbf{a}_z$ are parallel.
   (c) Find the angle between **A** and **C**.

*1.10. Repeat Problem 1.9 with the vectors

$$\mathbf{A} = 7\mathbf{a}_\rho + \mathbf{a}_\phi - 3\mathbf{a}_z \qquad \mathbf{B} = \mathbf{a}_\rho + 2\mathbf{a}_\phi + 3\mathbf{a}_z$$
$$\mathbf{C} = 2\mathbf{a}_\rho + 4\mathbf{a}_\phi + 6\mathbf{a}_z$$

*1.11. Determine the volume of the parallelepiped whose edges are the vectors

$$\mathbf{A} = \mathbf{a}_x - \mathbf{a}_y - 6\mathbf{a}_z \qquad \mathbf{B} = \mathbf{a}_x - 3\mathbf{a}_y + 4\mathbf{a}_z$$
$$\mathbf{C} = 2\mathbf{a}_x - 5\mathbf{a}_y + 3\mathbf{a}_z$$

1.12. Prove that $\mathbf{A} \times (\mathbf{B} \times \mathbf{C}) = \mathbf{B}(\mathbf{A} \cdot \mathbf{C}) - \mathbf{C}(\mathbf{A} \cdot \mathbf{B})$ for any vectors $\mathbf{A}$, $\mathbf{B}$, and $\mathbf{C}$. *Hint:* Find the x component.

1.13. Given that $\mathbf{A} = A(x, y, z)\mathbf{a}_z$, show that $\nabla \times \mathbf{A}$ lies in the xy plane.

*1.14. Given $f(x, y, z) = 5(x^2 + y^2) - 2z$, find $\nabla f(0, 0, 0)$ and $\nabla f(2, 1, 5)$. Sketch both the appropriate $f = $ constant surfaces and gradient vectors.

*1.15. A surface is given by $y^2 - yz + zx = 1$. Find $\mathbf{n}$ to the surface.

1.16. If point P is located at $\mathbf{r}' = \mathbf{a}_x x' + \mathbf{a}_y y' + \mathbf{a}_z z'$ and point Q is located at $\mathbf{r}'' = \mathbf{a}_x x'' + \mathbf{a}_y y'' + \mathbf{a}_z z''$. The vector from P to Q is $\mathbf{r} = \mathbf{r}'' - \mathbf{r}'$. Show that $\nabla' r = -\nabla'' r = \mathbf{r}/r$ (the superscript on ∇ gives the variables of differentiation).

1.17. **(a)** Find $\nabla \phi$, where ϕ is the planar coordinate angle. *Hint:* Express ϕ in terms of other variables.
   (b) Repeat part (a) for θ, the spherical polar coordinate angle.

1.18. Perform the line integral $\oint \mathbf{F} \cdot d\boldsymbol{\ell}$ along the path $ABCDA$ in Fig. 1.25, when $\mathbf{F} = 3\mathbf{a}_x - \mathbf{a}_z$.

1.19. Evaluate the line integral of $\mathbf{E} = K\rho^2 z \mathbf{a}_\phi$ over a closed path that lies on a cylinder, centered on the z axis, with radius 2, extending from $z = 0$ to $z = 3$, and between the x and y axes in the first octant. See Fig. 1.26.

1.20. Repeat the evaluation of Example 1.6 for the line integral of $\mathbf{V}$ traveling around the path in a counterclockwise direction. Being careful about sign conventions, show that you arrive at the same result.

1.21. The electrostatic field of a point charge is, in spherical coordinates, $\mathbf{E} = \mathbf{a}_r K/r^2$, where K is a constant. Show that the line integral of $\mathbf{E}$ is zero about any closed path not through the origin.

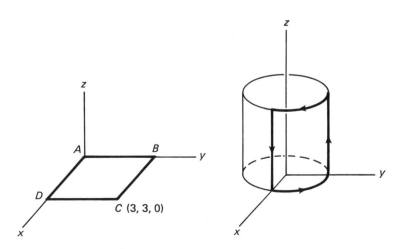

Figure 1.25 For Exercise 1.18.   **Figure 1.26** For Exercise 1.19.

1.22. Given the vector $\mathbf{F} = y\mathbf{a}_x - x\mathbf{a}_y$, evaluate the contour integral of $\mathbf{F}$ for the following:
(a) From $x = 1$ to $x = 4$ along the curve $y = 1/x$.
(b) Around the square with corners at $(0,0)$, $(1,0)$, $(1,1)$, and $(0,1)$.
(c) Around the circle $r = 1$.

1.23. Given the vector field $\mathbf{D} = D_0\mathbf{a}_x$, where D_0 is a constant, find the flux of $\mathbf{D}$ through each of the surface of the volume bounded by the planes $x = 0$, $y = 0$, $z = 0$, and $x + y + z = 1$.

1.24. Determine the total quantity of the field

$$\mathbf{B} = (x + y^2)\mathbf{a}_x + xz\,\mathbf{a}_y + (z^3 + xy)\mathbf{a}_z$$

that flows through the surface in the $z = 2$ plane, bounded by the lines $x = 0$, $x = 3$, $y = 1$, and $y = 2$.

1.25. Find the flux through the hemispherical surface $y > 0$ of the unit sphere centered at the origin for the vector field $\mathbf{A} = \mathbf{a}_z(z - \rho)/(z^2 + \rho^2)^{1/2}$.

1.26. A theorem, which is related to the divergence theorem, states that for a scalar function $f(x, y, z)$, one has

$$\int \nabla f\, dv = \oint f\mathbf{n}\, dS$$

Evaluate the integrals if $f = 1/r$ and v is the hemispherical volume $y > 0$ of the unit sphere centered at the origin. Show, by symmetry and geometric reasoning, that the integrals are zero if extended over the entire sphere.

1.27. Use the simplest method to calculate the circulation of the vector field

$$F = xy^2z\,\mathbf{a}_x + x^2yz\,\mathbf{a}_y + xyz^2\mathbf{a}_z$$

about the closed path $z = 1$, $\rho = (29)^{1/2}$, and $0 \le \phi \le 2\pi$.

1.28. (a) Evaluate the line integral around the circle with $\rho = a$ and $z = 0$ of the field

$$\mathbf{A} = \mathbf{a}_\rho\frac{2}{\rho} + \mathbf{a}_\phi \sin\frac{\phi}{2} + 5\mathbf{a}_z$$

(b) Use the simplest method to calculate the integral of $\nabla \times \mathbf{A}$ over the hemisphere defined by $\rho^2 + z^2 = 9$, $z > 0$.

1.29. A vector force field is given by

$$\mathbf{F} = \mathbf{a}_\rho(\sin^2\phi - z\,\cos^2\phi) + \mathbf{a}_\phi 2(1 + z)\sin\phi\,\cos\phi - \mathbf{a}_z\rho\,\cos^2\phi \text{ newtons}$$

(a) Calculate the work done by this force, $W = \oint \mathbf{F} \cdot \boldsymbol{\tau}d\ell$, around the path $(\rho, \phi, z) = (2,0,0)$, $(2, \pi/2, 0)$, $(2, \pi/2, 3)$, $(0, \pi/2, 3)$, $(0,0,0)$, and $(2,0,0)$ meters.
(b) Show that $\nabla \times \mathbf{F} = 0$.
(c) Determine a function V such that $\mathbf{F} = -\nabla V$. *Hint:* Write the equation for each component of ∇V and integrate separately, adjusting the integration constants.

1.30. Verify Stokes' theorem for $\mathbf{A} = \mathbf{a}_\phi \sin \theta$, in spherical coordinates, over the hemispherical surface $z > 0$, of the unit sphere centered at the origin.

1.31. For the vector field $\mathbf{V} = \mathbf{a}_\phi K r^2$, where K is a constant, evaluate both integrals of Stokes' theorem for the surface segment of the sphere of radius R, centered at the origin and bounded by the $x = 0$, $y = 0$, and $z = 0$ planes in the first octant.

1.32. A vector identity similar to the divergence theorem states that

$$\int \nabla \times \mathbf{A} \, dv = -\oint \mathbf{A} \times \mathbf{n} \, dS$$

Verify this for $\mathbf{A} = \rho \mathbf{a}_\phi$, where S is the cylindrical surface having radius ρ_0 and height z_0, whose axis is $\rho = 0$.

1.33. Determine the net flux (i.e., the total amount) of the vector field

$$\mathbf{F}(\rho, \phi, z) = \rho \mathbf{a}_\rho + \mathbf{a}_\phi + z \mathbf{a}_z$$

leaving a cylindrical *closed* surface defined by $\rho = 1$, $0 \le \phi \le \pi$, $0 \le z \le 1$. Verify your result with the divergence theorem.

1.34. Show that $\oint \mathbf{r} \cdot \mathbf{n} \, dS = 3v$ for any surface S enclosing the volume v.

*1.35. Find $\nabla^2 \mathbf{A}$, where $\mathbf{A} = \mathbf{a}_x(x^2 - y^2) + \mathbf{a}_y(y^2 - z^2) + \mathbf{a}_z(z^2 - x^2)$.

1.36. Find the surface integral of the vector

$$\mathbf{F} = (y + x^2)\mathbf{a}_x + (y^2 - x)\mathbf{a}_y$$

over the surface of a cube with sides $2m$ that are parallel to the coordinate planes, and centered at the origin of coordinates. Check your result using the divergence theorem.

1.37. Verify the divergence theorem for $\mathbf{V} = \phi \mathbf{a}_\rho + \rho \mathbf{a}_\phi - \mathbf{a}_z$ and for the volume bounded by the surfaces $\phi = 0$, $\rho = \rho_0$, $\phi = \pi/2$, $\rho = \rho_0/2$, $z = 0$, and $z = 1$.

1.38. Evaluate $\oint \mathbf{V} \cdot \mathbf{n} \, dS$ over the unit cube, three of whose faces lie in the $x = 0$, $y = 0$, and $z = 0$ planes, if $V = \mathbf{a}_x(x^2 + z^2) + \mathbf{a}_y zy + \mathbf{a}_z(x^2 - y^2)$.

*1.39. Find $\nabla \times \mathbf{A}$ and $\nabla \cdot \mathbf{A}$ at the point $(1, \pi/2, \pi/6)$ for the following:
(a) $\mathbf{A} = \mathbf{a}_x(x^2 + y^2) + \mathbf{a}_y xyz + \mathbf{a}_z(z^2 - x^2)$
(b) $\mathbf{A} = \mathbf{a}_\rho \rho^2 + \mathbf{a}_\phi \rho \cos \phi + \mathbf{a}_z(\rho^2 - z^2)$
(c) $\mathbf{A} = \mathbf{a}_r r(1 - \cos \theta) + \mathbf{a}_\theta r \cos \theta + \mathbf{a}_\phi \sin \theta \cos \phi$

APPENDIX 1

A1.1 DERIVATIONS IN CURVILINEAR COORDINATES

A1.1.1 Curl of a Vector

In this chapter we evaluated the curl of a vector using the rectangular coordinate system. Here we extend that analysis to the case of curvilinear coordinates, denoted by x_1, x_2, and x_3. Let us consider the contour in Fig. A1.1, consisting of four segments

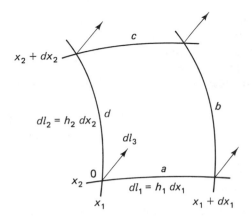

Figure A1.1   Infinitesimal contour path in general coordinates.

along the coordinate axes x_1, x_2, $x_1 + dx_1$, and $x_2 + dx_2$. These are taken to be in a curvilinear system. Moving in a counterclockwise direction from point 0,

$$\oint \mathbf{V} \cdot \boldsymbol{\tau} \, d\ell = [V_1 \, d\ell_1]_{\text{path } a} + [V_2 \, d\ell_2]_{\text{path } b} + [V_1 \, d\ell_1]_{\text{path } c} + [V_2 \, d\ell_2]_{\text{path } d}$$

$$(A1.1)$$

We take the vector at the point 0 to be $\mathbf{V} = V_1 \mathbf{a}_1 + V_2 \mathbf{a}_2 + V_3 \mathbf{a}_3$ and the separation of the path segments are $h_1 \, dx_1$ and $h_2 \, dx_2$. Not only are the vector components functions of x_1 and x_2, but, because the coordinates are curved and the metrics change with position, the path lengths $d\ell_i = h_i \, dx_i$ are also functions of x_1 and x_2. Therefore,

$$[V_1 \, d\ell_1]_{\text{path } a} = V_1 h_1 \, dx_1 \tag{A1.2a}$$

$$[V_2 \, d\ell_2]_{\text{path } b} = \left[V_2 + \frac{\partial V_2}{\partial x_1} dx_1 \right] \left[h_2 + \frac{\partial h_2}{\partial x_1} dx_1 \right] dx_2 \tag{A1.2b}$$

$$[V_1 \, d\ell_1]_{\text{path } c} = \left[V_1 + \frac{\partial V_1}{\partial x_2} dx_2 \right] \left[h_1 + \frac{\partial h_1}{\partial x_2} dx_2 \right] (-dx_1) \tag{A1.2c}$$

$$[V_2 \, d\ell_2]_{\text{path } d} = V_2 h_2 (-dx_2) \tag{A1.2d}$$

Substituting Eqs. (A1.2) into Eq. (A1.1), and canceling terms, yields

$$\oint \mathbf{V} \cdot \boldsymbol{\tau} \, d\ell = V_2 \frac{\partial h_2}{\partial x_1} dx_1 \, dx_2 + \frac{\partial V_2}{\partial x_1} h_2 \, dx_1 \, dx_2$$

$$- V_1 \frac{\partial h_1}{\partial x_2} dx_2 \, dx_1 - \frac{\partial V_1}{\partial x_2} h_1 \, dx_1 \, dx_2 + \cdots$$

$$= \frac{1}{h_1 h_2} \left[\frac{\partial (V_2 h_2)}{\partial x_1} - \frac{\partial (V_1 h_1)}{\partial x_2} \right] d\ell_1 \, d\ell_2 \tag{A1.3}$$

where the first-order small terms have canceled and we have dropped the third-order small terms in dx_i. We have also substituted $h_1 \, dx_1 h_2 \, dx_2 = d\ell_1 \, d\ell_2$, which is the area enclosed by the path in the $x_1 x_2$ surface. This area can be associated with its normal direction, so that $d\ell_1 \, d\ell_2 \to \mathbf{a}_3 \, dS$, since the unit vector normal to the $x_1 x_2$ surface is the $\mathbf{a}_3$ coordinate vector.

We wish to go to the limit that the path reduces to enclose only a point. From Eq. (A1.3), we see that it is dS that must approach zero and that the rotation or curl at the point must be expressed as the circulation per unit area:

$$(\text{curl of } \mathbf{V})_{x_3} = \lim_{dS \to 0} \frac{\oint \mathbf{V} \cdot \boldsymbol{\tau} \, d\ell}{dS} \qquad (\text{A1.4a})$$

$$= \frac{1}{h_1 h_2} \left[\frac{\partial}{\partial x_1} (V_2 h_2) - \frac{\partial}{\partial x_2} (V_1 h_1) \right] \qquad (\text{A1.4b})$$

It should be clear that we could have selected an area $d\ell_2 \, d\ell_3 \to \mathbf{a}_1 \, dS$ or $d\ell_3 \, d\ell_1 \to \mathbf{a}_2 \, dS$ to obtain different components of the curl of $\mathbf{V}$; they can be obtained by cyclically permuting the subscripts $1 \to 2 \to 3 \to 1$. These are components of the vector

$$\begin{aligned}
\text{curl } \mathbf{V} = \; & \frac{\mathbf{a}_1}{h_2 h_3} \left[\frac{\partial}{\partial x_3} (V_2 h_2) - \frac{\partial}{\partial x_2} (V_3 h_3) \right] \\
& + \frac{\mathbf{a}_2}{h_3 h_1} \left[\frac{\partial}{\partial x_1} (V_3 h_3) - \frac{\partial}{\partial x_3} (V_1 h_1) \right] \qquad (\text{A1.5a}) \\
& + \frac{\mathbf{a}_3}{h_1 h_2} \left[\frac{\partial}{\partial x_2} (V_1 h_1) - \frac{\partial}{\partial x_1} (V_2 h_2) \right]
\end{aligned}$$

This form can be obtained by expanding the determinant

$$\text{curl } \mathbf{V} = \frac{1}{h_1 h_2 h_3} \begin{vmatrix} h_1 \mathbf{a}_1 & h_2 \mathbf{a}_2 & h_3 \mathbf{a}_3 \\ \dfrac{\partial}{\partial x_1} & \dfrac{\partial}{\partial x_2} & \dfrac{\partial}{\partial x_3} \\ h_1 V_1 & h_2 V_2 & h_3 V_3 \end{vmatrix} \qquad (\text{A1.5b})$$

A1.1.2 Divergence of a Vector

We wish to calculate the divergence of a vector from the relation

$$\text{div } \mathbf{V} = \lim_{dv \to 0} \frac{\oint \mathbf{V} \cdot \mathbf{n} \, dS}{dv} \qquad (\text{A1.6})$$

For this consider the infinitesimal volume element between coordinate surfaces of a general curvilinear coordinate system, as shown in Fig. A.1.2. We note the convention that $\mathbf{n}$ is positive when taken directed out of an enclosed volume, whereas other directions are positive when taken along increasing x_i.

Therefore, the flux of $\mathbf{V}$ through the left $x_2 x_3$ face (containing the point 0) is

$$\mathbf{V} \cdot \mathbf{n} \, dS_{\text{left face}} = -V_{1(\text{left})} \, d\ell_{2(\text{left})} \, d\ell_{3(\text{left})} = -V_1 h_2 \, dx_2 h_3 \, dx_3 \qquad (\text{A1.7a})$$

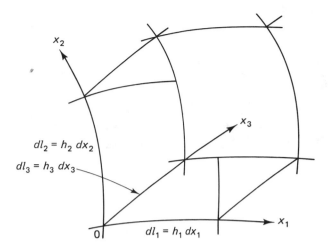

x_2

$dl_2 = h_2\, dx_2$

$dl_3 = h_3\, dx_3$

x_3

Figure A1.2 Infinitesimal volume
element.

0 $dl_1 = h_1\, dx_1$ x_1

Through the right x_2x_3 face, the flux is

$$\mathbf{V} \cdot \mathbf{n}\, dS_{\text{right face}} = V_{1(\text{right})}\, d\ell_{2(\text{right})}\, d\ell_{3(\text{right})}$$

$$= \left[V_1 + \frac{\partial V_1}{\partial x_1}\, dx_1 \right]\left[h_2 + \frac{\partial h_2}{\partial x_1}\, dx_1 \right] dx_2 \left[h_3 + \frac{\partial h_3}{\partial x_1}\, dx_1 \right] dx_3$$

$$= V_1 h_2\, dx_2\, h_3\, dx_3 + V_1 h_2\, dx_2 \frac{\partial h_3}{\partial x_1}\, dx_1\, dx_3$$

$$+ V_1 \frac{\partial h_2}{\partial x_1}\, dx_1\, dx_2 h_3\, dx_3 + \frac{\partial V_1}{\partial x_1}\, dx_1 h_2\, dx_2 h_3\, dx_3$$

$$+ \text{terms in } (dx_i)^4$$

$$= V_1 h_2\, dx_2 h_3\, dx_3 + \frac{\partial}{\partial x_1}(V_1 h_2 h_3)\, dx_1\, dx_2\, dx_3 + \cdots \qquad \text{(A1.7b)}$$

Adding Eqs. (A1.7a) and (A1.7b) gives the net outward flux of V along x_1. Divide
this by $dv = d\ell_1\, d\ell_2\, d\ell_3 = h_1 h_2 h_3\, dx_1\, dx_2\, dx_3$ and let $dv \to 0$ to yield

$$(\text{div } \mathbf{V})_1 = \lim_{dv \to 0} \frac{\left(\oint \mathbf{V} \cdot \mathbf{n}\, dS\right)_1}{dv} = \frac{1}{h_1 h_2 h_3} \frac{\partial}{\partial x_1}(V_1 h_2 h_3) \qquad \text{(A1.7c)}$$

Contributions to the divergence of $\mathbf{V}$ from the other faces of the volume can be
found by permuting the subscripts. The total is

$$\text{div } \mathbf{V} = \frac{1}{h_1 h_2 h_3}\left[\frac{\partial}{\partial x_1}(V_1 h_2 h_3) + \frac{\partial}{\partial x_2}(V_2 h_3 h_1) + \frac{\partial}{\partial x_3}(V_3 h_1 h_2) \right] \qquad \text{(A1.8)}$$

2
MAXWELL'S EQUATIONS

2.1 THE CONCEPTUAL CHANGE

In 1864 James Clerk Maxwell published *A Dynamical Theory of the Electromagnetic Field,* in which he accomplished two formidable tasks: summarizing, in a concise mathematical manner, the then diffuse understanding of electricity and magnetism, and extending that understanding to predict the existence of electromagnetic waves that propagate through space. At the time, electrostatics and magnetostatics were more or less distinct disciplines, each with its own system of measurement and its own definitions and standards. Furthermore, the current teachings held that electric and magnetic changes took place instantaneously at all distances, without the aid of any intervening mechanism. This was referred to as *action at a distance.*

Maxwell had been very impressed by the intuitive approach and experiments of Michael Faraday, who proposed that there existed a mechanism for action of the electric and magnetic interactions. Faraday, one of the foremost experimental physicists of his time, pictured *lines of force* that spread out from electric charges, magnetic poles, and currents, and which then acted on other charges, poles, and currents. It had been shown by Oersted, in 1819, that a current in a wire exerts a magnetic force, that is, it could deflect a magnetic compass needle. This set off a flurry of experimental activity by such men as Sir Humphrey Davy and André Ampère, who showed that iron filings formed into rings about a current-carrying wire and that a current-carrying coil acted like a bar magnet. Following these leads, Faraday reported, in 1831, that *changing* the current in one loop of wire caused an

38

electromotive force to be developed in a second loop, so that current flowed in the second loop. This induction of an electric effect through a change of the magnetic lines about a current, now known as Faraday's law of electromagnetic induction, provided a much sought after connection between the electric and magnetic domains.

Faraday's extensions of Davy's iron filings demonstration had convinced him that the lines they display represent a real change in the space about a magnet or current. In 1846 he published a paper on "Ray Vibrations," proposing that force was transmitted by waves running along these lines. Whereas Faraday was not able to put his results in proper mathematical form, his work captivated Maxwell, who published his first paper on Faraday's lines of force in 1855, at the age of 23. Maxwell set out to incorporate Faraday's ideas and discoveries into the known body of knowledge in a consistent way. Following the methods of his time, he did this through mechanical analogues, by developing models of belts, pulleys, and gears that displayed the properties of electric and magnetic force lines. As he did this, he made two major discoveries:

1. For consistency with the other properties, not only must one have Faraday's law, that moving, or changing, magnetic lines induce an electric effect, but it was also necessary that moving, or changing, electric lines induce a magnetic effect. Thus, the displacement of electric flux acts like a current; he referred to a changing electric flux as a *displacement current*.

2. Once a mathematical description of the mechanical system was at hand, it could be divorced from the structures used to derive it. The *fields* of Faraday's electric and magnetic lines are the means through which electric and magnetic phenomena act. Maxwell then showed that a solution to his equations exists according to which the electromagnetic fields propagate away from their source, like a wave in water.

To determine the electromagnetic velocity of propagation, Maxwell had to carefully examine the basis of the two unit systems that were used for electric (esu units) and magnetic (emu units) quantities. The speed of propagation c was equal to the ratio of the magnitudes of the unit charge in esu and emu. Besides being a superlative applied mathematician, Maxwell was also an extremely capable experimentalist. By 1868 he had set up a torsion balance in which he opposed the force between two charged capacitor plates and the force between two current coils. From this he found that $c = 2.88 \times 10^{10}$ cm/sec. Such a great speed could only correspond to light, which, at the time, was widely believed to travel instantaneously between distant points. While this value is slightly lower than the present value of 2.99×10^{10} cm/sec, it established that light was an electromagnetic phenomenon, and with a theoretically based finite velocity.

Maxwell's theory had a mixed reception. Its promise was tremendous, but its novelty and level of difficulty put off even the most accomplished analysts. And, in an age when science was transmitted through personal contact and correspondence, the reclusive Scotsman, Maxwell, was not well known in the scientific centers of continental Europe. Also, Maxwell's theory did not give a clear account of electric charges, which Faraday's great experiments on electrolytic dissociation and deposition had indicated to be discrete quantities.

Maxwell died in 1879 and it wasn't until nearly 20 years later that Heinrich Hertz was able to reformulate Maxwell's work in a more comprehendable way. In 1888 Hertz was able to demonstrate that electromagnetic waves other than light were possible when he achieved the first radio transmission from the oscillating current of an induction coil. An indication of the impact and enlightenment resulting from Maxwell's work can be gleaned from the following quote:

> I can best give you an idea of the sweeping changes in viewpoint brought about by the theory of Faraday and Maxwell by telling you of the time I spent as a student, 1887–1891.
>
> -
>
> My time of study coincided with the period of Hertz's experiments. At first, however, electrodynamics was still presented to us in the old manner—in addition to Coulomb and Biot–Savart, Ampère's law of the mutual action of two elements of current and its competitors, the laws of Grassmann, Gauss, Riemann, and Clausius, and as a culmination the law of Wilhelm Weber, all of which were based on the Newtonian concept of action at a distance. The total picture of electrodynamics thus presented to us was awkward, incoherent, and by no means self-contained. Teachers and students made great effort to familiarize themselves with Hertz's experiments step by step as they became known and to explain them with the aid of the difficult original presentation in Maxwell's Treatise.
>
> It was as though scales fell from my eyes when I read Hertz's great paper: "Fundamental Equations of Electrodynamics for Bodies at Rest." Here Maxwell's equations, purified by Heaviside and Hertz, were made the axioms and the starting point of the theory. The totality of electromagnetic phenomena is derived from them systematically by deduction. Coulomb's law, which formerly provided the basis, now appears as a necessary consequence of the all-inclusive theory. Electric currents are always closed. Current elements arise only as mathematical increments of line integrals. All effects are transmitted by the electromagnetic field which may be represented by force-line models. *Action at a distance* gives way to field action. . . . (From Arnold Sommerfeld, *Electrodynamics* [New York: Academic Press, Inc., 1952].)

We have dealt only with Maxwell's electrodynamics, and we recognize that its consequences have affected every aspect of our lives. However, this was only part of his great contribution to scientific understanding. From 1859 to 1877 Maxwell published papers and books on the kinetic theory of gasses, the theory of heat, and thermodynamics. His work provides the basis on which the modern understanding of these areas is built. He also ardently pursued experiments on light and color vision. In the course of his life, James Clerk Maxwell changed the way we understand the physical universe. He revolutionized our understanding of electromagnetism, the way we understand thermodynamics, the way we use mechanics, our comprehension of light, optics, and vision. He must be ranked among the great geniuses of all time.

Maxwell was also a reluctant revolutionary of the entire world outlook of modern scientists. The publication of *Principia* by Isaac Newton, in 1687 forever changed the way people regarded nature. He demonstrated that mechanical motion (in particular of astronomical bodies) was based on laws of motion, and could be mathematically and quantitatively understood and predicted. (Newton invented differential and integral calculus to prove this.) As a result, the world view became one in which natural phenomena occur through the mechanical interaction of particles. This was the conceptual origin of the theory that light is transmitted by means of luminiferous corpuscles. When light came to be regarded as a wave, the mechanical outlook demanded the existence of a medium of transmission, the aether, in analogy with the transmission of mechanical waves. With Maxwell's proof that light was electromagnetic, the mechanical nature of aether became unnecessary, and the aether itself eventually lost acceptance.

Maxwell himself was a proponent of the physical existence of aether, though there was great conflict between its lack of mechanical influence and the properties it must have in order to explain its evident electromagnetic behavior. In spite of his own reservations, however, he had initiated a fundamental change of concept. Einstein has said of Maxwell:

> Neglecting the important *individual* results which Clerk Maxwell's life-work produced in important departments of physics, and concentrating on the changes wrought by him in our conception of the nature of physical reality, we may say this: — Before Clerk Maxwell people conceived of physical reality — in so far as it is supposed to represent events in nature — as material points, whose changes consist exclusively of motions, which are subject to partial differential equations. After Maxwell they conceived physical reality as represented by continuous fields, not mechanically explicable, which are subject to partial differential equations. This change in the conception of reality is the most profound and fruitful one that has come to physics since Newton . . . . (Albert Einstein, *Essays in Science* [New York: Philosophical Library Inc., 1934].)

2.2 MAXWELL'S VECTOR EQUATIONS

While vectors were known and used by Maxwell, he did not have the concise vector notation we have used in the previous chapter, so his results had to be written in terms of a large number of partial differential equations. Today, Maxwell's electromagnetic results can be summarized in the following four vector equations:

$$\nabla \times \mathbf{H} = \mathbf{J} + \frac{\partial \mathbf{D}}{\partial t} \tag{2.1}$$

$$\nabla \times \mathbf{E} = -\frac{\partial \mathbf{B}}{\partial t} \tag{2.2}$$

$$\nabla \cdot \mathbf{B} = 0 \tag{2.3}$$

$$\nabla \cdot \mathbf{D} = \rho \tag{2.4}$$

In the following chapters we will study the physical properties of Maxwell's equations in detail. For now, however, let us note some of their general mathematical features in terms of the vector relations we have studied.

As we have seen, the divergence theorem states that the flux of a vector field across a closed surface equals the total divergence of the field in the enclosed volume, that is, the divergence acts as a source of that vector. If $\nabla \cdot \mathbf{D} = \rho$ [Eq (2.4)], then ρ (whatever it may be we have not yet said) is the source of $\mathbf{D}$. Similarly, Stokes' theorem indicates that the strength of a vector field that circulates about a closed path equals the quantity of its curl through the area bounded by that path. The curl of the vector can therefore be considered to act as its source. If $\nabla \times \mathbf{H} = \mathbf{J}$

[Eq. (2.1), when there is no time variation so the term $\partial \mathbf{D}/\partial t$ can be neglected], then **J** (whatever it may be we have not yet said) is a measure of the source of **H**.

Two aspects of the magnetic field are involved, described by **B** and **H**, and two properties of the electric field are involved, given by **D** and **E**. Maxwell's equations give curl and divergence of both the magnetic field and the electric field, in terms of the sources of these fields. In vector calculus, there exists a theorem by which a vector is completely defined if its curl and its divergence are known. From this point of view, Eqs. (2.1) to (2.4) are complete presentations of the electric and magnetic fields.

In addition, mathematicians divide vector fields into two pure types: those which have zero curl (called *irrotational vectors*) and which diverge from their sources; and those which curl about their sources (called *solenoidal vectors*) but that have zero divergence. Under conditions when there is no time variation, the electric and magnetic fields are, respectively, examples of the two types.

The two magnetic and the two electric field quantities are generally taken to be linearly related so that

$$\mathbf{B} = \mu \mathbf{H} \qquad \mathbf{D} = \varepsilon \mathbf{E} \qquad\qquad (2.5\text{a, b})$$

where μ and ε are the magnetic permeability and electric permittivity, respectively. Equations (2.5) are called the constitutive equations. μ and ε are usually scalar quantities (so that vectors **B** and **H** are parallel and vectors **D** and **E** are parallel), which serve to relate the magnitudes and units of the linked fields. μ and ε depend on the medium in which the fields exist, so that the relation between the two electric fields and between the two magnetic fields is complicated by the presence of matter.

3

ELECTROSTATICS

3.1 ELECTRIC FLUX DENSITY

We begin our detailed study of Maxwell's equations with Eq. (2.4):

$$\nabla \cdot \mathbf{D} = \rho \tag{3.1}$$

In this equation, ρ is a volume charge density in units of coulombs per cubic meter (C/m^3). The symbol ρ_v is often used to emphasize that a volume density is intended. Charge, like mass, is a fundamental property of matter that is not thought of as being derived from other properties (as is force or speed). Unlike mass, positive and negative charges are common, existing in all atomic systems that make up matter. Equation (3.1) refers to charge that has been separated from its atomic source, free charge that can be moved about. Later in this chapter, we will refer to conditions in which positive and negative charge in an atomic or molecular system move apart but do not separate. This gives rise to what we will call *bound charge*. Maxwell's equation, Eq. (3.1), refers only to free charge, and throughout this work we will take ρ to mean free charge, unless specifically indicated otherwise. In keeping with convention, the symbol ρ refers to positive charge density; a minus sign $(-)$ is associated with ρ if negative charge is present.

In a charged volume, where the charge density can vary from point to point, the charge contained within an infinitesimal volume dv is $dq = \rho \, dv$. From the previous discussion of vectors, we understand Eq. (3.1) to mean that the electric quantity $\mathbf{D}$ diverges from a positive charge, that is, a positive charge acts as a source of

D. Conversely, it converges onto a negative charge, that is, a negative charge acts as a sink for **D**. Equation (3.1) can be integrated through the charge contained in the volume

$$\int \nabla \cdot \mathbf{D}\, dv = \int \rho\, dv = \int dq = Q\,(\text{total, in volume}) \tag{3.2}$$

Together with the divergence theorem, this yields

$$\oint \mathbf{D} \cdot \mathbf{n}\, dS = Q\,(\text{enclosed}) \tag{3.3}$$

since the total Q in the volume is identical to the Q enclosed by the surrounding surface. Equation (3.3) is known as *Gauss' law*; it is the integral form of the differential equation Eq. (3.1). A dimensional analysis of Eq. (3.3) gives $D \cdot \text{area} = \text{charge}$, so that the dimensions of **D** are C/m^2. From this, it appears that there is a flux (measured in coulombs) of **D** across the closed surface equal to the charge (in coulombs) contained within. The density of this flux, that is, the flow across each unit area, is equal to **D**, in C/m^2; **D** is called the *electric flux density*.

Gauss' law is an important conceptual relation. It is particularly powerful in those cases where symmetry allows us to solve the integral. This is illustrated by the following important example.

Example 3.1

A spherical shell of radius R, centered at the origin of coordinates, has a uniform distribution of *surface charge* ρ_s (C/m^2). Determine **D** at all points.

Note that the charge is distributed on a surface, with area density ρ_s (in C/m^2), rather than through a volume with volume density ρ_v (in C/m^3). The total charge on the surface is found by integrating the charge per infinitesimal area times the area:

$$Q = \int_R \rho_s\, dS = 4\pi R^2 \rho_s \tag{3.4}$$

To apply Gauss' law, we consider **D** on a hypothetical sphere of radius r, concentric with the sphere R. Figure 3.1(a) shows that, on this surface, $\mathbf{n} = \mathbf{a}_r$. The entire geometry being considered has spherical, or point, symmetry. By this is meant that the physical situation does not change if we construct any axis through the origin and rotate the problem about that axis through an arbitrary angle. Such rotations can bring any point on the hypothetical, Gaussian, surface to the position formerly occupied by any other point on that surface. Since the physical situation is the same in the two positions, we conclude that vector **D** must be the same, in magnitude and in direction, at any two points on the surface r. This means that **D** is not a function of angular location, $\mathbf{D} \neq \mathbf{D}(\theta, \phi)$ so that $\mathbf{D} = \mathbf{D}(|\mathbf{r}|)$ only. Furthermore, $|\mathbf{r}|$ is constant on the sphere, so that the argument of **D** is constant and $\mathbf{D}(r) = \text{constant}$ *on that surface.*

Figure 3.1(b) shows a particular situation where **D** is at an arbitrary angle, so that it has both normal and tangential components to the surface. A rotation axis is shown through the point of interest and if the angle of rotation is 180°, then the tangential component is reversed. Since the two situations must be equivalent, the only way that both the tangential component and its negative can exist at the same point is if the tangential component is equal to zero so that **D** is entirely radial and $\mathbf{D} = D(r)\mathbf{a}_r$.

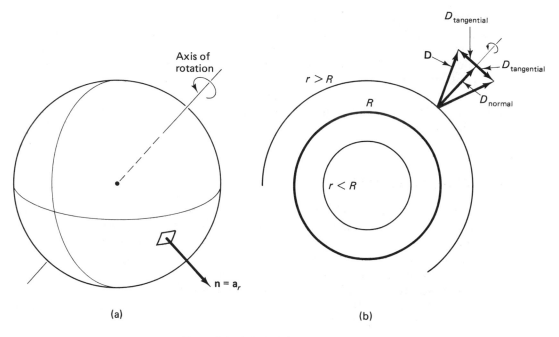

Figure 3.1   Spherical Gaussian surfaces.

Applying Gauss' law, we now integrate $\mathbf{D}$ over the entire surface r:

$$\int \mathbf{D} \cdot \mathbf{n}\, dS = \int_{r=\text{const.}} D(r)\mathbf{a}_r \cdot \mathbf{a}_r\, dS = D\int_r dS = 4\pi r^2 D \qquad (3.5)$$

According to Gauss' law, this must be equal to the charge enclosed by the Gaussian surface. In the case that $r < R$, all the charge is on the greater sphere, so that the enclosed charge is zero and we have

$$\mathbf{D} = 0 \qquad r < R \qquad\qquad (3.6\text{a})$$

When $r > R$, the Q (enclosed) is given by Eq. (3.4), so that $4\pi r^2 D = Q$. Combining this magnitude result with the directional information given previously, we have

$$\mathbf{D} = \mathbf{a}_r \frac{Q}{4\pi r^2} \qquad r > R \qquad\qquad (3.6\text{b})$$

Equation (3.6a) expresses the important result that the field is zero everywhere within a uniform spherical, surface charge distribution. Equation (3.6b) gives its value at points outside of the distribution. Note that this $\mathbf{D}$ is dependent on Q/r^2; it is as though the entire charge were concentrated at the origin. This result includes the case of a point charge Q, if we let $R \rightarrow 0$.

In this proof, some very powerful concepts of symmetry have been applied to resolve most of the properties of $\mathbf{D}$ and to solve the integral of Eq. (3.5).

It is important to point out that Maxwell's equation, Eq. (3.1), and Gauss' law, Eq. (3.3), as well as the results of Example 3.1, relate $\mathbf{D}$ to the charge sources and to the geometry. Nowhere is there any reference to the medium in which the charge and field exist, be it a conductor such as copper, an insulator such as alumina, be it wood or water. In cases where the free-charge distribution is known, the flux density is derived solely from that charge. While this often makes it relatively simple to calculate **D**, it is the other electric field quantity in Maxwell's equations, **E**, called the *electric field strength,* which is of primary importance in studying electric phenomena. Therefore, we turn to an examination of this field strength and its relation to the flux density **D**.

3.2 ELECTRIC FIELD STRENGTH

Let us place a small *test* charge q into a region of space in which there exists an electric field, so that q experiences a force, exerted by the field. This field can arise from various bodies carrying charge distributions, but we will not be concerned with the detailed sources except for the following limitation. Since q is being used to test the field, it is necessary that it be so small that the force *it* exerts on the sources of the field are negligible. Then the sources are not disturbed and the test charge samples the original field. As the force on q is proportional to its magnitude, we define the charge independent field strength as

$$\mathbf{E} = \mathbf{F}/q \qquad (3.7)$$

We see from this that **E** has units of newtons/coulomb.

Note that while **D** is a conceptual emanation of electric flux from charge, **E** is an experimental force exerted by that charge. If our understanding is to be consistent, the electric force exerted by a charge distribution that generates a field, should be proportional to the density of electric flux generated by that distribution. We will examine this relation in more detail in a later section; for now, let us accept that this proportionality exists. In a vacuum, we take the proportionality factor to be $1/\varepsilon_0$, so that $\mathbf{E} = \mathbf{D}/\varepsilon_0$. ε_0 is called the *permittivity of free space.* Then, from Eq. (3.6b), the field strength outside of a spherical charge distribution is

$$\mathbf{E} = \frac{Q}{4\pi\varepsilon_0 r^2}\mathbf{a}_r \qquad (3.8)$$

As mentioned previously, this result includes the important case of an infinitesimal charge. We will see later that ε_0 must be changed when the field medium is not a vacuum.

3.3 ELECTRIC POTENTIAL

Equation (3.7) indicates that the mechanical forces, work done, and energy stored are derived from **E**. Let the charged body q be moved from point a to point b in the field region. If this motion is done in a quasistatic way (i.e., so slowly that it is always in equilibrium and we need not consider its kinetic energy), then the work done by the external mover is given by the externally applied force times its displacement:

$$W = \int_a^b \mathbf{F}_{\text{external}} \cdot \boldsymbol{\tau} d\ell = -q \int_a^b \mathbf{E} \cdot \boldsymbol{\tau} d\ell \tag{3.9}$$

The minus sign before the second integral arises because the external force must act *against,* or balance, the force exerted by the electric field **E**. For example, if *a* is on one charged body and *b* is on another, then Eq. (3.9) expresses the energy required to move a charge *q* between the two.

We can gain some further insight into work and energy in an electric field by referring to Maxwell's equation, Eq. (2.2):

$$\nabla \times \mathbf{E} = -\frac{\partial \mathbf{B}}{\partial t} \tag{3.10}$$

For the present, we shall consider only the condition when there is **no change with time** ($\partial \mathbf{B}/\partial t = 0$), so that

$$\nabla \times \mathbf{E} = 0 \qquad \text{static case} \tag{3.11}$$

If we insert this expression into Stokes' theorem, we have

$$\int \mathbf{n} \cdot \nabla \times \mathbf{E}\, dS = 0 = \oint \mathbf{E} \cdot \boldsymbol{\tau} d\ell \tag{3.12}$$

Comparing Eqs. (3.12) and (3.9), we see that the work per unit charge, W/q, is equal to zero when a closed path is traversed, that is, when the path terminates at its starting point. Whatever energy is lost during one part of a closed path, for example, between *a* and *b* on Fig. 3.2, is returned during the other, for example, from *b* to *a*. Thus, the total energy is conserved; the statement that $\nabla \times \mathbf{E} = 0$ is equivalent to a statement of the law of conservation of energy of the electrostatic field.

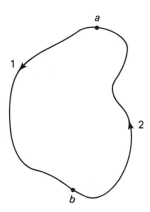

Figure 3.2   Closed-path integration.

If the closed path is divided into two parts, denoted by 1 and 2, as shown in Fig. 3.2, then

$$\oint \mathbf{E} \cdot \boldsymbol{\tau} d\ell = \int_a^b \mathbf{E} \cdot \boldsymbol{\tau}_1 d\ell_1 + \int_b^a \mathbf{E} \cdot \boldsymbol{\tau}_2 d\ell_2 = \int_a^b \mathbf{E} \cdot \boldsymbol{\tau}_1 d\ell_1 - \int_a^b \mathbf{E} \cdot \boldsymbol{\tau}_2 d\ell_2 = 0$$

(3.13)

or

$$\int_1 \mathbf{E} \cdot \boldsymbol{\tau} d\ell = \int_2 \mathbf{E} \cdot \boldsymbol{\tau} d\ell \tag{3.14}$$

This states that the work done in moving between points a and b is the same regardless of whether one moves along path 1 or along path 2. Since these paths are arbitrary, the work done, or the energy expended, is the same along any path connecting the points. As a result, the energy must be a function of the end-point positions alone and not dependent upon how one arrives at that position. That function of position gives the energy that is potentially available at each point; it is a potential energy.

Example 3.2

Determine the potential-energy function due to a point charge Q at the origin.

Consider a point test charge q in the vicinity of Q. From Eq. (3.8), we see that if Q is positive, then $\mathbf{E}$ is directed away from Q, along positive $\mathbf{a}_r$. If Q is negative, the minus sign causes $\mathbf{E}$ to be directed inward, along $-\mathbf{a}_r$. We take Q to be positive. Note now, from Eq. (3.7), that if q is positive, then $\mathbf{F}$ is parallel to $\mathbf{E}$ (outward, along $\mathbf{a}_r$), that is, like charges repel. If q is negative, then $\mathbf{F}$ is opposite from $\mathbf{E}$ (inward), that is, opposite charges attract.

Equation (3.9) gives the work done by the external force as

$$W = -q \int_a^b \mathbf{E} \cdot \boldsymbol{\tau} d\ell \tag{3.15a}$$

From Eq. (3.8) we see that $\mathbf{E}$ is entirely along $\mathbf{a}_r$, so only the $\mathbf{a}_r$ component of $\boldsymbol{\tau} d\ell$ enters the scalar product. We therefore move q along a radial line from point a at distance r_a to point b at distance r_b, as shown in Fig. 3.3(a). The integral is

$$W = -\frac{Qq}{4\pi\varepsilon_0} \int_a^b \frac{\mathbf{a}_r}{r^2} \cdot \mathbf{a}_r \, dr$$

$$= \frac{Qq}{4\pi\varepsilon_0}\left(\frac{1}{r_b} - \frac{1}{r_a}\right) \tag{3.15b}$$

or

$$W = q[V(r_b) - V(r_a)] \tag{3.16a}$$

where

$$V(r) = \frac{Q}{4\pi\varepsilon_0 r} \qquad \text{point charge} \tag{3.16b}$$

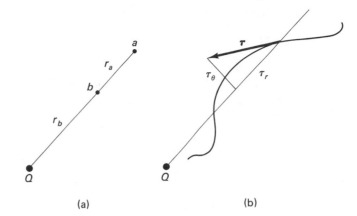

Figure 3.3 Path segments.   (a)   (b)

$V(r)$ is called the *electrostatic potential function* of the point charge Q. $V = W/q$ is the potential energy per unit test charge, in units of joules/coulomb, abbreviated *volts*. Note that V depends only on the charge Q and on the distance r at which the potential is determined. If Q were held in place and q released, it would fly out to infinity, repelled by the force $q\mathbf{E}$, thereby minimizing its potential energy, since $qV(\infty) = 0$. Useful work could be obtained, for example, by having a charge strike a target.

If we take q around a closed path, returning to its starting point, Eq. (3.12) indicates that the electric potential will return to its initial value, that is, no *net* work is done, as demanded by the conservation of energy. In detail we examine this for the path segment shown in Fig. 3.3(b), where the tangent $\boldsymbol{\tau}$ has both $\mathbf{a}_\theta$ and $\mathbf{a}_r$ components. As $\mathbf{E}$ is entirely parallel to $\mathbf{a}_r$, the integral of $\mathbf{E} \cdot \boldsymbol{\tau}$ reduces to only the integral along $\mathbf{a}_r$, giving the same result as before.

3.4 DISTRIBUTED CHARGE

The discussion of Eqs. (3.11) to (3.14) does not involve any specific charge distribution. Consequently, a potential function exists for every charge distribution [though having a different position dependence than given by Eq. (3.16b)], so that Eq. (3.16a) is quite general. This can be illustrated by considering a volume containing a charge density $\rho(\mathbf{r})$ that varies with position. The infinitesimal volume, shown in Fig. 3.4, containing the charge $dQ = \rho\,dv$, is so small that it can be regarded as a point charge. Then Eq. (3.7) gives the field it produces and Eq. (3.16b) gives its potential. The principle of superposition tells us that the field and potential of a distributed charge are the sum of the fields and potentials of the component parts. Therefore, the field and potential of the entire volume of charge, at any point, can be found by integrating Eqs. (3.7) and (3.16b) over the entire charge volume.

$$\mathbf{E} = \int \frac{dQ}{4\pi\varepsilon_0 r^2}\mathbf{r}_0 = \frac{1}{4\pi\varepsilon_0}\int \frac{\rho}{r^2}\mathbf{r}_0\,dv \qquad (3.17)$$

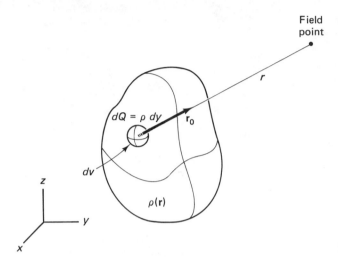

Figure 3.4   Volume charge distribution.

$$V = \int \frac{dQ}{4\pi\varepsilon_0 r} = \frac{1}{4\pi\varepsilon_0} \int \frac{\rho}{r} \, dv \qquad (3.18)$$

where r is the distance and $\mathbf{r}_0$ is the unit vector from the infinitesimal charge dQ to the field point.

If instead of the volume charge distribution, with $dQ = \rho_v \, dv$, we had been dealing with a surface charge distribution, then we would have made the substitution $dQ = \rho_s \, dS$. A third type of distribution that is encountered is a line charge, which has charge along a curve ℓ. In this case, $dQ = \rho_\ell \, d\ell$, and the three-dimensional integration of dv is replaced by a one-dimensional integration along the linear charge.

Note that the integrals of Eqs. (3.17) and (3.18) contain ρ and r, both of which vary with position as one moves within the charge distribution to perform the integrals. As a result, these integrals are not simple, unless symmetry allows reduction of the forms involved. In addition, Eq. (3.17) contains the unit vector $\mathbf{r}_0$, directed from the source point to the field point. This vector is also a function of the source points over which the integral is being performed. In fact, it is generally true that the integral for $\mathbf{E}$ must be divided into three scalar integrals, one for each of its components. The fact that this is not necessary when evaluating V is one reason for using V in complicated field calculations.

Example 3.3

Two long coaxial cylinders, of radii a and b, hold equal and opposite charge per unit length, ρ_ℓ. Find the potential when $a < \rho < b$. The medium is a vacuum.

(We first note that these finite-radius cylinders actually carry surface charge. However, for long cylinders, it is more convenient to express this as a linear charge density. If the inner surface has density ρ_s then $\rho_\ell = 2\pi a \rho_s$.)

It should be clear to the reader that arguments similar to those used in Example 3.1 can be applied to the geometry shown in Fig. 3.5(a), which has rotational symmetry about the cylinder axis. These lead to the conclusion that $\mathbf{D} = \mathbf{a}_\rho D(\rho)$. Gauss' law is applied to the cylindrical surface shown in Fig. 3.5(a), containing only the inner charged surface. Since $Q_{\text{enclosed}} = \ell \rho_\ell$, this gives

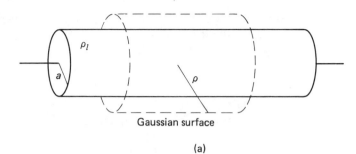

Gaussian surface

(a)

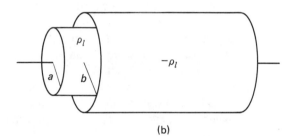

Figure 3.5   Axially symmetric charge
distribution.

(b)

$$\oint \mathbf{D} \cdot \mathbf{a}_\rho \, dS = D \cdot \ell 2\pi\rho = \ell\rho_\ell \qquad \text{when } \rho > a$$

$$= 0 \qquad \text{when } \rho < a$$

which lead to the results

$$\mathbf{E} = \mathbf{a}_\rho \frac{\rho_\ell}{2\pi\varepsilon_0\rho} \qquad \rho > a \qquad (3.19a)$$

$$= 0 \qquad \rho < a \qquad (3.19b)$$

The larger cylinder produces no electric field in the region $\rho < b$, so that Eq. (3.19a) gives
the entire field in that region. From this and Eq. (3.9), the potential at any point between
the surfaces is found by integrating along the radius from a to ρ, where $a < \rho < b$:

$$V(\rho) - V(a) = W/q = -\int_a^\rho \mathbf{E} \cdot \boldsymbol{\tau} \, d\ell$$

$$= -\int_a^\rho \mathbf{a}_\rho \frac{\rho_\ell}{2\pi\varepsilon_0\rho} \cdot \mathbf{a}_\rho \, d\rho$$

$$= -\frac{\rho_\ell}{2\pi\varepsilon_0} \ln \frac{\rho}{a} \qquad (3.20)$$

Finally, the difference of potential between the charged surfaces is

$$V = |V(b) - V(a)| = \frac{\rho_\ell}{2\pi\varepsilon_0} \ln \frac{b}{a} \qquad (3.21)$$

In this discussion, we derived the potential as the integral of the electric field
strength **E**. If follows, then, that the electric field is the derivative of the potential.

Since **E** is a vector and V is scalar, the derivative function that connects them must be the gradient operator

$$\mathbf{E} = -\nabla V \tag{3.22}$$

We should note that this also follows from Eq. (3.11), that $\nabla \times \mathbf{E} = 0$, and the vector identity that the curl of any gradient equals zero. Any real charge distribution is made up of infinitesimal point charges and, since the superposition principle states that the total effect of the distribution is the sum (or integral) of the individual contributions, it follows that Eq. (3.22) is always true. Among the consequences of Eq. (3.22) is that **E** is always perpendicular to the constant-V surfaces. Note that ∇V is a vector directed toward increasing values of V; the negative sign in Eq. (3.22) causes **E** to point toward lower values of V, that is, the driving force on a positive charge is directed toward a state of lower potential energy.

From Eq. (3.22), it is seen that in addition to the units $E = F/q \rightarrow$ newtons/coulomb, E has the units of $dV/d\ell \rightarrow$ volts/meter. We can also now find the units of ε_0. These are $\varepsilon_0 = D/E \rightarrow$ (coulomb/m^2)/(volts/meter) = C/V–m = farads/m, where we have used the definition of capacitance: 1 farad = 1 C/V. Capacitance (in farads) is studied in a later chapter, but the reference is made here to that circuit parameter to reflect the practical utility of the SI units.

3.5 POLARIZATION

Since $\mathbf{D} = \varepsilon_0 \mathbf{E}$ in a vacuum, we inquire into the physical difference between **D** and **E**, and how this relation changes when the fields are in other media.

In equilibrium, the positive and negative charges in matter are in balance so as to produce zero net field. For example, a simple model of a neutral atom has a symmetrical cloud of negative charge surrounding a positive nucleus, so that the centers of the (+) and the (−) charge coincide, as shown in Fig. 3.6(a). However, when an electric field is applied to a dielectric body, a force $q\mathbf{E}$ is exerted on the positive charge while a force $-q\mathbf{E}$ is experienced by the negative charge. These opposite forces cause the two charges to move apart a distance δ to a new equilibrium position, where their attractive interaction balances the applied electric force. This separation of (+) and (−) electric *poles* creates an electric *dipole* whose dipole moment is defined as $\mathbf{p} = q\boldsymbol{\delta}$; its magnitude equals the charge times the separation distance and its vector direction is from the negative to the positive charge, as shown in Fig. 3.6(b). Of course, the real situation is not so simple. Molecules are not so symmetrical and may even have permanent dipole moments, and charge displacement and rotation is limited by chemical bonding and intra- and interatomic forces, not to mention the randomizing effects of thermal vibrations. Nevertheless, the result is that most materials polarize to some degree when placed in an electric field, that is, they generate electric dipoles. A measure of the ability of the dielectric body to produce dipoles is its *polarization*, $\mathbf{P} = \Sigma \mathbf{p}/$volume, the (vector) dipole moment per unit volume.

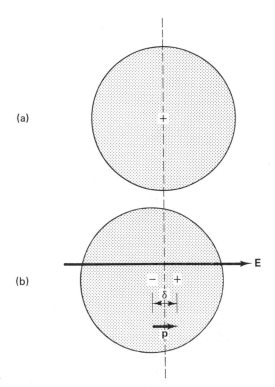

(a)

(b)

E

Figure 3.6 Atomic model of polarization.

From the simple atomic description, it can be seen that the directed distance $\boldsymbol{\delta}$, and, therefore, $\mathbf{p}$ and $\mathbf{P}$, are proportional to $\mathbf{E}$, the force field that generates them. This is expressed by

$$\mathbf{P} = \varepsilon_0 \chi_e \mathbf{E} \qquad (3.23)$$

where the proportionality constant is written in the form $\varepsilon_0 \chi_e$; χ_e is called the electric susceptibility of the particular medium.

Early physicists did not have our developed concept of atomic and molecular structure, but they appreciated the separation of charge in a volume. This separation moves positive charge to one side of the volume (in the direction of the applied force $\mathbf{F} = q\mathbf{E}$), and moves negative charge to the opposite side. As a result, equal and opposite charges appear on the opposite surfaces of the dielectric volume. These charges are said to be *induced* by the applied field that polarizes the medium. Figure 3.7 shows a dielectric slab between the plates of a large capacitor. The plates have been charged by an external source and contain equal and opposite charge densities ρ_s, which produce a flux density $\mathbf{D}$ in the volume occupied by the dielectric. Imagine that the dielectric is divided into thin sections perpendicular to the applied field, as shown. Each of these sections develops surface charge as a result of its own internal charge displacement. Since the volumes are in contact and the material is uniform, the $(+)$ charge on the surface of one section cancels the equal $(-)$ charge from its neighbor, as shown in the figure-insert. Only on the outer surface of the

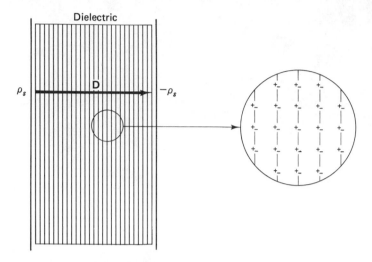

Figure 3.7   Dielectric slab between charged capacitor plates.

body is there no cancellation, so the net effect of the applied field is to induce a surface charge; this is also shown in Fig. 3.8(a). The induced charge is also referred to as a *bound* charge, since it cannot be removed from the surface like any *free* charge that we can manipulate, such as the charge which was originally placed on the capacitor plates. On Fig. 3.8 a prime is used to denote the bound charge.

Once the surface charges are induced, they generate an additional electric flux density from positive to negative, shown by $\mathbf{D}'$ on Fig. 3.8(a). This induced flux is indistinguishable from the applied flux except that it is oppositely directed, so that the magnitude of the total internal flux density is $|\mathbf{D} + \mathbf{D}'| = |\mathbf{D}| - |\mathbf{D}'|$. Note that the flux density $\mathbf{D}$ is generated by the free external charge and the flux density $\mathbf{D}'$ is generated by the induced charge that is bound to the body. They are equivalent fields and we use the prime to distinguish them as a matter of convenience.

Figure 3.8(b) shows the sections separated. Each has a volume $v = \ell_1\ell_2\ell_3$, and is taken to have a charge Q' induced on its surfaces. The dipole moment of the volume is $p = Q'\ell_1$ and its dipole moment per volume is $P = p/v = Q'\ell_1/\ell_1\ell_2\ell_3 = Q'/\ell_2\ell_3 = Q'/A = \rho'_s$, where A is the area of the charged face, and ρ'_s is the induced-surface charge density. It is shown in an exercise at the end of this chapter that two parallel sheets with surface charge density ρ_s produce an electric flux density between them, which is given by $D = \rho_s$. Therefore, the magnitude of the induced flux density is given by $D' = \rho'_s = P$ and, from their directions in Fig. 3.8(a), $\mathbf{D}' = -\mathbf{P}$. The total flux density is the vector sum of the two partial densities: $\mathbf{D} + \mathbf{D}' = \mathbf{D} - \mathbf{P}$.

It was pointed out previously that there must be a proportionality between the force quantity, $\mathbf{E}$, and the flux density, which now consists of the induced plus free flux densities. This proportionality is through a universal constant ε_0:

$$\mathbf{E} = (\mathbf{D} - \mathbf{P})/\varepsilon_0 \tag{3.24a}$$

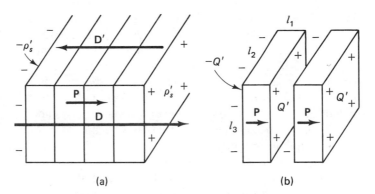

(a)   (b)

Figure 3.8   Volume polarization and induced surface charge.

Combining Eqs. (3.23) and (3.24a), we have

$$\mathbf{D} = \varepsilon_0(1 + \chi_e)\mathbf{E} = \varepsilon\mathbf{E} \qquad (3.24b)$$

Note that in the absence of polarization, $\chi_e = 0$, so that $\varepsilon = \varepsilon_0$; this is the case for a vacuum. ε is the permittivity of the medium with susceptibility χ_e. Generally, $\varepsilon \neq \varepsilon_0$, reflecting the fact that both polarization flux and free-charge flux must be accounted for when determining the electric field intensity **E**. A material figure of merit is its relative permittivity:

$$\varepsilon_r = \varepsilon/\varepsilon_0 = 1 + \chi_e \qquad (3.25)$$

From Eq. (3.24b), we see that ε and ε_0 relate **D** and **E**, the two vector electric fields. This involves a change of magnitude and a change of units. The different units imply a relation between the two different physical aspects of the fields, as a force generator and as a flux emanation. While we will not consider it here, in some cases, for example, crystalline materials, **D** and **E** are not parallel, so that ε must also effect a change in direction, that is, a rotation.

The fundamental parameter ε_0, the vacuum permittivity, has a measured value:

$$\varepsilon_0 = 8.85419 \times 10^{-12} \text{ C/V–m} \qquad \text{(or F/m)} \qquad (3.26a)$$

so that, for estimation purposes,

$$1/4\pi\varepsilon_0 = 8.988 \times 10^9 \approx 9 \times 10^9 \text{ m/F} \qquad (3.26b)$$

3.6 FRICTION AS A CAUSE OF CHARGE SEPARATION

A variety of techniques can be used to generate free electric charge of the sort we have been discussing. Most mechanisms that produce direct current are applicable to charging conductors, for example, the chemical forces in a battery cause charge separation and this force can be used to provide opposite charge on the plates of a capacitor that is connected to the battery terminals. Similarly, any mechanism that causes atomic and molecular ionization has separated positive and negative charges.

Atoms and molecules that are subject to a sufficiently high electric field can be ionized by that field. Atoms and molecules at high temperatures have an increased probability of spontaneously ionizing, and molecular ionization can sometimes be initiated by light and ultraviolet radiation or by ambient high-energy radiation such as cosmic rays and radioactive decay products emitted from the earth. It is known that heating certain materials cause them to emit electrons that can be directed and gathered; this is the source of the beams of modern cathode-ray tubes.

The occurrence of static electricity has been known for so long that its first observation is lost in antiquity. The very word *electric* derives from the Greek word for amber, since ancient Greeks knew of generating static charge by rubbing an amber, or other resin, rod with a soft cloth. The generation of electric charge through friction is therefore one of mankind's oldest experimental scientific phenomena. It is one that we have all experienced by walking in a carpeted room on a cool dry day.

And it is more than an idle, interesting occurrence. Extremely high voltages can be generated by accumulating charge that is separated through friction. Many, if not most, industries take precautions against electrostatic discharge (ESD) that can damage equipment, cause personnel accidents, and generate fires and explosions. An example of the industrial cost of ESD is the damage of many semiconductor junctions by mishandled integrated circuit chips. On a much grander scale, it is known that electrical displays frequently accompany volcanic eruptions: frictional motion of molten magma against the volcanic walls seems to separate charge, which is carried upward when the magma is ejected.

Yet, despite its long history, its commonplace occurrence, and its cost to industry and commerce, the nature of the mechanism by which frictional charge separation occurs is poorly understood. Frequently, it is not even certain what sign of charge will be generated on an object by a particular frictional history. Theories have been proposed, some of which fit some observations, but, because of the vastly diverse types of materials, surfaces, and conditions, there are likely different explanations for what appears to be a single phenomenon. Quantitative relations tend to be phenomenological in nature, though some refer back to more fundamental properties of matter. Research is underway, but progress is surprisingly slow in this area of electrical behavior.

Fortunately, engineers are resourceful and means have been found to avoid the problem, even while recognizing that its basic nature is unknown. Thus, libraries of stored information on large tape reels are run through channels with conducting walls to draw off the charge that accumulates through friction of the tape with those walls; workers in many industries routinely wear loosely grounded wristlets, or sit or stand on noninsulating pads that prevent their accumulating dangerous levels of charge and high voltages; hospital operating rooms must have floors of reasonable conductivity in order to dissipate charge that might cause explosions in the various gasses that are commonly in use; atomized and ionized water droplets are sprayed into air-conditioning and heating ducts to reduce the frictional charge generated by moving air.

The nature of this frictional charge generation is, of course, related to the very nature of friction itself, which is similarly understood in an inadequate way. These are problems of a very fundamental nature that constitute an open area for research.

EXERCISES

Extensions of the Analysis

*3.1.   We will show, in the next chapter, that $\mathbf{E} = 0$ inside a good conductor. With this, apply Gauss' law to determine the flux density outside of a large conducting sheet that carries a surface charge density of ρ_s C/m^2.

3.2.   An infinite plane contains a charge density of ρ_s C/m^2. Using Gauss' law and symmetry arguments, determine the direction of the electric flux density due to the charge and show that its magnitude is $|D| = \rho_s/2$.

*3.3.   Apply the superposition principle to the results of Exercise 3.2 to find the electric flux density in the region between and outside of two planar charge distributions with opposite signs: $\pm\rho_s$.

3.4.   At great distances, any charge distribution appears to be a point charge, where Q is the net charge of the distribution. Establish a criterion for the acceptability of this approximation and estimate the distance at which one might expect it to fail for a distribution consisting of a charge of $+2.0 \times 10^{-6}$ C and a charge of -2.1×10^{-6} C separated by 0.01 cm.

*3.5.   We have discussed the fact that the electric field from a continuous charge distribution can be obtained by integrating the vector contributions of the individual infinitesimal charges. Of course, we also recognize that the field due to discrete charges can be obtained by adding the vector contributions of the individual charges.

   (a) Consider that four point charges are located at the corners of a square of side length 2, centered about the z axis, as shown in Fig. 3.9(a). Find $\mathbf{D}$ at points on the z axis.

   (b) The charge arrangement is changed to that shown in Fig. 3.9(b). Determine the direction of $\mathbf{D}$ at points on the z axis.

Problems

3.6.   Example 3.3 is for a geometry with line, or axial, symmetry. Explain why $\mathbf{E}$ must be independent of z. By considering arbitrary rotations about that axis, show that $\mathbf{E} = \mathbf{E}(\rho)$ only. Take a 180° rotation about an axis perpendicular to the symmetry line and show that $E_z = 0$ and that $\mathbf{E} = \mathbf{a}_\rho E$.

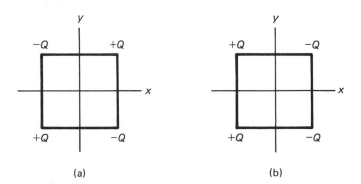

Figure 3.9   For Exercise 3.5.   (a)   (b)

3.7. A system of charges in free space consists of a point charge $Q = +4 \times 10^{-3}$ C at the origin, surrounded by a spherical shell with the volume charge density $\rho_v = -10^{-3}$ C/m^3, in the region 1 m $< r <$ 1.25 m. Calculate and plot the electric field strength as a function of r.

*3.8. Given a very long cylindrical surface of radius a carrying a surface charge density of ρ_s C/m^2. Derive an expression for the flux density for $\rho > a$ and for $\rho < a$.

*3.9. Given the electric field in air:

$$\mathbf{E}(\rho, \phi, z) = \rho_0[\rho\mathbf{a}_\rho + \mathbf{a}_\phi + z\mathbf{a}_z] \text{ V/m}$$

(a) Determine the charge density.
(b) Verify Gauss' law in integral form by taking the volume $\rho \leq 1, 0 \leq \phi \leq \pi, 0 \leq z \leq 1$.

3.10. A positive spherical charge core extends from the origin to a distance R_0 and has a uniform charge density ρ_0 C/m^3. Negative charge lies in a shell between $2R_0$ and $4R_0$ and has a charge density $\rho_v = \rho_0 R_0^2/r^2$. Find the static electric flux density for all $r > 0$. What is the charge unbalance?

3.11. Charge is distributed through a long cylindrical volume of space with a volume distribution about the axis given by

$$\rho_v = 3 - \frac{\rho}{\rho_0} \qquad \rho \leq \rho_0$$

$$= 0 \qquad \rho \geq \rho_0$$

(a) What is the charge per unit axial length (i.e., ρ_ℓ along the z axis)?
(b) Find the electric flux density D at points for which $\rho > \rho_0$. Briefly mention any symmetry arguments you need.

3.12. As shown in Fig. 3.10, a spherical charge distribution has a neutral inner core with radius a, surrounded by a spherical volume with outer radius $2a$, containing a positive charge with density $\rho_v = A/r$, surrounded by a thin spherical shell with radius $3a$, on which is a negative surface charge density ρ_s.
(a) Find an expression for ρ_s if the total charge is zero.

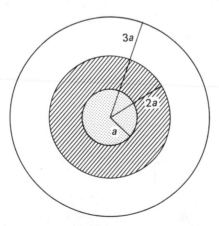

Figure 3.10 For Exercise 3.12.

(b) Under the conditions of part (a), find the electric flux density at *all* points of space.

3.13. (a) Given the electric vector flux density $\mathbf{D} = \mathbf{a}_y D_0 y^3$, determine the charge density in space that produces this field.

(b) Find the total charge in the volume enclosed by the planes $x = -2$, $x = 0$, $y = -2$, $y = 0$, $z = -2$, and $z = 0$.

3.14. Positive charge is placed within a spherical volume of radius a about the origin, with volume density $\rho_v = Ar$ C/m^3 (A = constant). In addition, negative charge is distributed over a spherical shell of radius $2a$, with a surface density $\rho_s = -Aa^2/12$ C/m^2. Derive the expressions for the electric field strength and for the electric potential at all points.

*3.15. Given the vector field $\mathbf{A} = \mathbf{a}_x y^2 + \mathbf{a}_y x^2$.

(a) Find the circulation of $\mathbf{A}$ about the square with corners at $(0,0,0)$, $(1,0,0)$, $(1,1,0)$, and $(0,1,0)$.

(b) Can this vector represent an electric field? Why?

*3.16. (a) The circle $\rho = a$ and $z = 0$ has an equal but opposite charge placed on its two halves. The dividing line between the halves is $y = 0$. Determine the *direction* of the electric field at points on the z axis.

(b) Determine the *direction* of the electric field at points on the z axis if both halves of the circle carry charge of the same sign.

*3.17. A drop of oil is suspended in space between two large metal plates that are 10 cm apart and that have a difference of potential of 10^4 V. Determine the charge on the drop if its mass is 0.1 gm.

*3.18. Determine the ratio of the gravitational and electrostatic forces between two protons. (The gravitational force is $\mathbf{F} = \mathbf{a}_r Gm_1m_2/r^2$, where $G = 6.67 \times 10^{-11}$ m^3/kg–sec^2.)

3.19. Two spheres, each with 0.1 gm mass, are suspended from a common point by weightless strings that are 10 cm long. Determine the charge on each sphere if each string is at an angle of $15°$ to the vertical.

3.20. (a) Show that only one of following vector fields can be an electric field.

$$\mathbf{M} = \mathbf{a}_x M_0 y^3 \qquad \mathbf{N} = \mathbf{a}_x N_0 x^3$$

(b) For the electric field, find the flux out of the volume enclosed by the planes $x = -2$, $x = 0$, $y = -1$, $y = +1$, $z = -1$, and $z = +1$.

*3.21. A region of space contains an electric field $\mathbf{E} = \mathbf{a}_x E$. What can you conclude about the change of E with position if the region is filled with a constant charge density?

3.22. An infinitely long cylinder of radius a is filled with a positive charge of constant volume density ρ_v. The medium is air.

(a) Calculate the vector electric flux density at all points.

(b) Recall that energies, and therefore potentials, have arbitrary reference and can be shifted by any amount. Therefore, if the potential is zero at $\rho = a$, find it at all points. What is the potential at the origin?

*3.23. Find the amount of electric charge associated with the scalar potential $V = xy^2z$ in air in the .olume $0 \le x, y, z \le 1$.

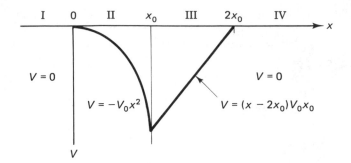

Figure 3.11   For Exercise 3.24.

3.24. In a region of space, the electric potential is constant on planes perpendicular to the x axis. Its variation along the x axis is shown in Fig. 3.11.
   (a) Determine the vector electric field intensity in regions I through IV.
   (b) Find the volume density of charge in regions I through IV.
   (c) Calculate the surface charge densities on the planes between the regions. *Hint:* Using the result of part (a), apply Gauss' law to a *thin* pill box whose axis is parallel to the x axis.
   (d) Note that $V = 0$ for $x \leq 0$ and for $x \geq 2x_0$, that is, the charge between these planes has no effect outside of the planes. Use your solutions to parts (b) and (c) to show that the net charge is zero between the planes.

3.25. An electrostatic potential in volts is given by

$$V = \frac{1}{r^2} - \frac{2}{r} \qquad r \text{ in meters}$$

If a particle with charge $q = +2$ C is released from rest at $r = 3$ m, find
   (a) The force acting on it at the moment of release.
   (b) Its kinetic energy at $x = 1$ m. (Work done by the field = K.E.)
   (c) Where would you place the positive charge to have it remain stationary, that is, in equilibrium? Is this equilibrium stable? (Stable equilibrium exists at a point if the particle will return to that point when given an infinitesimal displacement.)
   (d) Answer part (c) for a negative charge.

3.26 . The coaxial system of Example 3.3 is air-filled. Find the maximum difference of potential of the two cylinders if breakdown of the air dielectric is to be avoided. Spark discharge breakdown in dry air occurs at $\sim 10^8$ V/m.

3.27. It is said that the electrostatic field is *conservative*.
   (a) What is conserved?
   (b) Prove that this statement is true. Explain how your proof establishes or demonstrates the criteria of your answer to part (a).

3.28. Two long, thin concentric cylinders have radii of a and b meters, where $b > a$. The region between a and b is filled with a dielectric with $\varepsilon = 3\varepsilon_0$. All other regions contain air. The cylindrical surfaces carry opposite charge, $\pm \rho_\ell$ coulombs per unit length.

(a) Determine the electric field strength at all points.

(b) What is the polarization of the dielectric (with units)?

3.29. A very long metal cylinder of 3 cm radius carries a positive surface charge which has 10^{-9} C/m of length. It is surrounded by a hollow dielectric cylinder that has $\varepsilon = 3\varepsilon_0$, and inner and outer radii of 6 and 9 cm, respectively.

(a) Find D at points 2, 5, 7, and 10 cm from the cylinder axis.

(b) Find E at the same points as part (a).

(c) What is the induced charge per meter of length on the inner surface of the dielectric cylinder? On the outer surface?

3.30. A cleaved diamond is found to develop an induced polarization of 10^{-7} C/m^2. What is the average atomic dipole moment? What is the separation of the nucleus and the center of the electron cloud of the carbon atoms? Take the density of a diamond to be 3.5 g/cm^3 and the effective atomic separation to be 2.8×10^{-10} m.

3.31. A parallel-plate capacitor consists of two metal plates between which there are two layers of dielectrics, as shown in Fig. 3.12. Equal and opposite charges of 10^{-7} coulomb are placed on the metal plates.

(a) Find D, E, and P in each dielectric.

(b) Find the charge density on the surface of each of the dielectrics and the *net* induced charge at their interface.

3.32. A point charge Q is at the center of a hollow dielectric sphere whose inner radius is r_i and whose outer radius is r_o. The dielectric has relative permittivity ε_r. Find D, E, V, and P for $0 < r \le \infty$.

3.33. Two concentric cylindrical conductors have the space between them filled by two dielectrics, as shown in Fig. 3.13 (next page). Equal and opposite charges of ρ_ℓ C/m of length are on the conductors.

(a) Show that

$$\mathbf{D} = 0 \qquad\qquad \rho < a, \qquad \rho > c$$

$$\mathbf{D} = \mathbf{a}_\rho \rho_\ell / 2\pi\rho \qquad a < \rho < c$$

(b) Find $\mathbf{E}$ in each medium.

(c) Find the polarization in each medium.

(d) What is the difference of potential between the conductors?

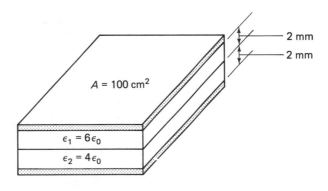

2 mm

2 mm

$A = 100$ cm^2

$\varepsilon_1 = 6\varepsilon_0$

$\varepsilon_2 = 4\varepsilon_0$

Figure 3.12 For Exercise 3.31.

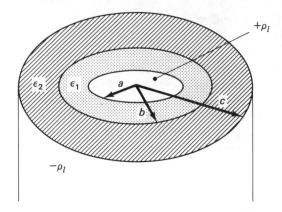

Figure 3.13   For Exercise 3.33.

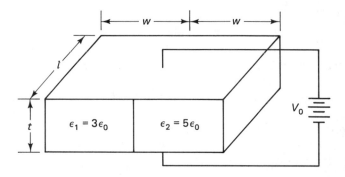

Figure 3.14   For Exercise 3.34.

3.34. A parallel-plate capacitor has two dielectric materials, side by side, as shown in Fig. 3.14. The capacitor is connected to a battery with voltage V_0.
   (a) What is the electric field strength between the plates?
   (b) What is the polarization P and flux density D in each dielectric?
   (c) Find the ratio of the charges on the plates above the two dielectrics.

3.35. A spherical region about the origin contains a volume charge density:

$$\rho = \rho_0\left(1 - \frac{r^2}{a^2}\right) \qquad r < a$$

$$= 0 \qquad r > a$$

where a is the radius of the sphere, and ρ_0 is a constant. Solve for the electric field strength at all points of space.

4
MAGNETOSTATICS

4.1 MAGNETIC FLUX

We begin the discussion of magnetic fields with Maxwell's equation, Eq. (2.3):

$$\nabla \cdot \mathbf{B} = 0 \qquad \text{or} \qquad \oint \mathbf{B} \cdot \mathbf{n} \, dS = 0 \qquad (4.1a, b)$$

where the integral form has been derived from the differential equation through application of the divergence theorem. Unlike the electric flux, which leaves positive charge and enters negative charge, Eq. (4.1a) indicates that there are no sources or sinks from which the magnetic flux density $\mathbf{B}$ diverges; it is continuous. The integral form makes the equivalent statement that the same magnetic flux enters and leaves each closed surface, so that the net flux across the surface is zero. Figure 4.1 shows this for a curved surface, where lines of $\mathbf{B}$ enter and leave several times. Note that the number of times each line enters is equal to the number of times it leaves. $\mathbf{B}$, the magnetic flux density, is continuous; lines of $\mathbf{B}$ have no beginning or end.

4.2 CURRENTS

Maxwell's equation, Eq. (2.1), gives the curl of the magnetic field:

$$\nabla \times \mathbf{H} = \mathbf{J} + \frac{\partial \mathbf{D}}{\partial t} \qquad (4.2)$$

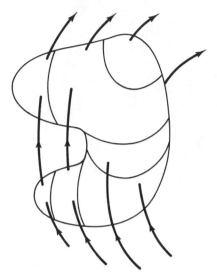

Figure 4.1 Conservation of magnetic flux.

As in the discussion of electricity, we shall, for now, consider only the case where there is *no time variation* of the fields. Then

$$\nabla \times \mathbf{H} = \mathbf{J} \qquad (4.3a)$$

where

$$\mathbf{J} = \frac{d\mathbf{I}}{dS} \qquad |\mathbf{I}| = \frac{dQ}{dt} \qquad (4.3b, c)$$

Here **I** is a filamentary current, defined in direction and magnitude as the charge per second (1 coulomb/sec = 1 ampere) flowing past a point on a thin filament (or wire). We observe the usual convention of taking the current to be a movement of positive charge; since the mobile electrons are negative, a flow of electrons along one direction constitutes a current in the opposite direction. While the electric fields studied in Chapter 3 were generated by stationary charge, Eq. (4.3a) indicates that a magnetic field is generated by current, that is, by charge in motion. The statement that there is no time variation means, in this context, that the currents are steady. **J** is the current density, in units of A/m². As shown in Fig. 4.2, a current flowing across an area is made up of many filaments of area dS, and $d\mathbf{I}$ is the current that crosses each dS. The density **J** may depend upon position.

Current is generated by a force exerted on charge, generally, by an electric field, which causes it to move. In most common situations and materials, the stronger the field, the greater is the mean velocity of the charge carriers and the greater the current. This linear relation is expressed by *Ohm's law*:

$$\mathbf{J} = \sigma \mathbf{E} \qquad (4.4)$$

where σ is the conductivity of the material, describing the ease with which charge moves and current flows. σ has the units of $J/E \rightarrow (A/m^2)/(V/m) = A/V\text{–m} = 1/\text{ohm–m} = \text{mho/m} = \text{siemens/m}$. [By international agreement, the *siemen* (S) replaces the older unit *mho*.] The reader may not recognize Eq. (4.4) as a form of

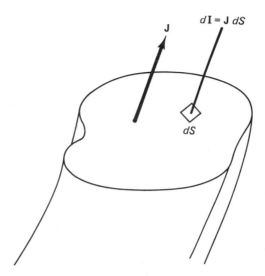

Figure 4.2   Current density.

Ohm's law. The correspondence can be seen by examining a section of conductor shown in Fig. 4.3. Here an electric field is applied across the conductor by the battery with terminal voltage V. Then, for the uniform field: $\mathbf{E} = -\nabla V = -dV/d\ell = \Delta V/\Delta\ell = V/\ell$. As current is related to the uniform current density by $J = I/A$, Eq. (4.4) then becomes

$$I/A = \sigma V/\ell \qquad \text{or} \qquad V = I(\ell/\sigma A) = IR \qquad (4.5)$$

where $R = \ell/\sigma A$ is called the *resistance* of the sample.

Equation (4.4) implies that the charge carriers are able to move when an electric field is applied. This is true to some extent in most materials, whether by occasional electronic hopping between molecular sites in insulators, where σ is very small, or by the free migration of detached electrons in metals, where σ is very large. Conductivity is reputed to have one of the greatest ranges of variation of any material physical parameter, from 10^{-10} S/in to 10^{10} S/m. In insulating materials, where the electron motion is negligible, polarization phenomena manifest themselves as discussed in the previous chapter.

A metal is a good conductor because it has a large amount of charge that is free to move. This has an interesting consequence in the electrostatic case. Figure 4.4

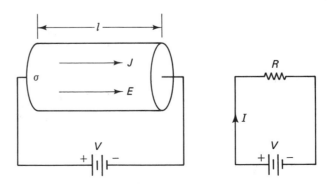

Figure 4.3   Relations in Ohm's law.

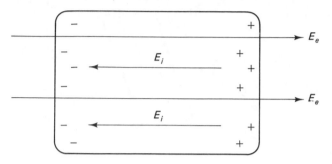

Figure 4.4   Conductor in electrical equilibrium.

shows a good conductor, insulated from its surroundings and placed in an externally applied electric field of magnitude E_e. As a result of the force being exerted on it, the mobile charge moves to one side of the body. (If the mobile charge is negative, then it moves opposite to the field.) This leaves an imbalance of charge with opposite polarity. The charge thus separated produces an internal field of magnitude E_i, which is oppositely directed from E_e; the total internal field is $E_t = E_e - E_i$. Charge migration continues until field E_i becomes great enough so that $E_t = 0$, at which time the force on the free charge vanishes, so the charge flow, that is, the internal current, ceases. This occurs very quickly in a good conductor, on the order of 10^{-18} sec. We then have the situation, in electrostatic equilibrium, that a good conductor has zero internal electric field.

A distinction must be made between the situations of Figs. 4.3 and 4.4. Figure 4.4 displays a case of eventual static equilibrium, that is, the lack of charge motion is the result of a balance of electrostatic field forces. In Fig. 4.3, however, that equilibrium is absent. Because the sample is not isolated — it is wired to the battery — the charge that would accumulate at the ends of the sample, as in Fig. 4.4, is drawn off at one end and returned at the other, preventing the establishment of electrostatic equilibrium. A dynamic equilibrium — a steady state — is established when V is applied to the conductor in Fig. 4.3, but it is, strictly speaking, incorrect to speak of *magnetostatics*. Perhaps *equilibrium magnetics* would be appropriate, but we shall conform to the customary usage of magnetostatics to denote the situation in which currents are unchanging.

4.3 MAGNETIC FIELDS

Equation (4.3a) states that **H** curls about the current; the lines of **H** circle their source. Figure 4.5 shows several current lines within a closed curve. Such a curve can be considered to be the boundary of an infinite number of open surfaces, that is, surfaces which sit on the curve and do not enclose a volume. For simplicity the surface shown is planar, and the plane surface has unit normal **n**. If we integrate $\nabla \times \mathbf{H}$ over that surface, we have

$$\int \mathbf{n} \cdot \nabla \times \mathbf{H}\, dS = \int \mathbf{n} \cdot \mathbf{J}\, dS \qquad (4.6a)$$

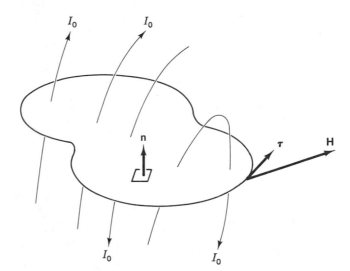

Figure 4.5 Net current through the area is I_0.

Applying Stokes' theorem to the left-hand integral, and applying Eq. (4.3b) to the right-hand integral gives

$$\oint \mathbf{H} \cdot \boldsymbol{\tau} \, d\ell = I(\text{total crossing area}) \tag{4.6b}$$

Figure 4.5 shows $\mathbf{H}$ and $\boldsymbol{\tau}$ at a point on the contour. Equation 4.6b is known as *Ampère's law*, stating that the integral of $\mathbf{H}$ about a closed path is equal to the total, or net, current that crosses an area bounded by that path. Here the meaning of the word *net* is that an upward current through the area is balanced by an equal downward current through that area, as shown in Fig. 4.5. If each arrow in this figure represents a current I_0, then its component crossing the surface is $I_0 \cos(\mathbf{n}, \mathbf{I}_0) = I_n$, where $(\mathbf{n}, \mathbf{I}_0)$ denotes the angle between $\mathbf{I}_0$ and the surface normal, and I_n is the component of $\mathbf{I}_0$ that is normal to the surface. The tangential component of $\mathbf{I}_0$ does not carry current across the surface, so it is not considered. From Eq. (4.6b), we can determine the units of $\mathbf{H}$ to be amperes/meter.

Ampère's law is very useful in those situations where there is sufficient symmetry to allow evaluation of the integral. We illustrate this by the following two examples. Because some of the symmetry arguments are the same for both, and because comparison of their differences is instructional, we will solve both examples together.

Example 4.1

Determine the magnetic field produced by the very long, straight wire filament carrying current I, shown in Fig. 4.6(a).

Example 4.2

Determine the magnetic field produced by the very long, straight cylindrical solenoid, shown in Fig. 4.6(b), which is wound with n turns/meter and which carries the current I. For this problem we assume the pitch of the turns to be so small that the current can be con-

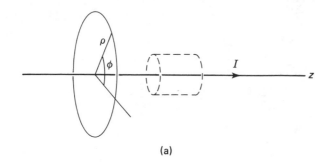

(a)

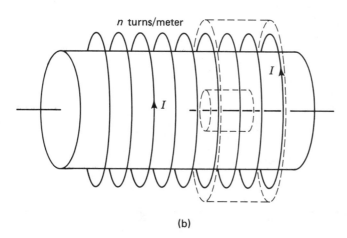

n turns/meter

(b)

Figure 4.6   (a) Long straight conductor.
(b) Section of a long solenoidal winding.

sidered to be entirely circumferential, and the wires to be so fine that the circumferential current flow is continuous about the coil, with surface density nI A/m.

The wire and solenoid are considered to be sufficiently long that they are, effectively, infinite. This means that the portion that we examine is so far from the ends that every plane perpendicular to the axis is equivalent to every other; since all are so far from the ends, they can all be assumed to be at the center. (The limits of this assumption are explored in problems at the end of the chapter.) As a result, the field cannot be a function of z, the position along the axis.

Both configurations have axial symmetry, meaning that any rotation about the z axis leaves the physical situation unchanged. As a result, any point in a plane perpendicular to the axis can be made to coincide with any other that is at the same radial distance in the plane. Since this rotation does not change the physical situation, the points must be equivalent, so the field cannot be a function of ϕ, the angle about the axis.

Having eliminated z and ϕ, we are left with $\mathbf{H} = \mathbf{H}(\rho)$ only. We now examine the individual components of $\mathbf{H}$.

The property of flux closure (that is, the lines form closed loops) guides us with regard to the radial component. We have not yet established a relation between the two magnetic field aspects, $\mathbf{H}$ and $\mathbf{B}$, but we can, in analogy with the electric case, at least assume that as long as we stay in a single medium, they are parallel, so that closure of one ($\mathbf{B}$) implies closure of the other ($\mathbf{H}$). Since the field lines close on themselves, each component must reverse, that is, change its sign, in some region. However, neither configuration we are studying has a mechani··· for radial reversal (with the solenoid, the radial field inside and outside might be reversed, but this still does not lead to a closed loop), so we must conclude that $H_\rho = 0$. This can be made more convincing by applying Eq. (4.1b) to any of the dashed cylindrical surfaces in Fig. 4.6. For clarity the surface is also shown in Fig. 4.7.

$$\oint \mathbf{H} \cdot \mathbf{n}\, dS = \int_{\text{end 1}} H_z\, dS - \int_{\text{end 2}} H_z\, dS + \int_{\text{side}} H_\rho\, dS = 0$$

The integral over end 2 carries a minus sign because its $\mathbf{n}$ is along $-\mathbf{a}_z$. Since $\mathbf{H}$ does not change with z, the two end integrals cancel. We are therefore left with $\int H_\rho\, dS_{\text{side}} = 2\pi\rho\ell H_\rho = 0$, meaning that $H_\rho = 0$.

The expression for $\mathbf{H}$ has now been reduced to

$$\mathbf{H} = \mathbf{a}_\phi H_\phi(\rho) + \mathbf{a}_z H_z(\rho)$$

At this point, the properties of the two problems diverge, so we treat them separately.

Example 4.1 (cont.)

For the straight wire, there is no J_ϕ current, so that the ϕ component of Eq. (4.3a) is

$$0 = J_\phi = (\nabla \times H)_\phi = -\partial H_z/\partial\rho$$

indicating that H_z is constant. However, this conflicts with the closure property that H_z must reverse in some region, unless the constant value is zero.

We therefore have $\mathbf{H} = \mathbf{a}_\phi H(\rho)$, so we can now evaluate Ampère's integral, $\oint \mathbf{H} \cdot \boldsymbol{\tau}\, d\ell$, around the circular path shown in Fig. 4.6(a). As the circle has constant ρ, the integrand $H(\rho)$ is constant and can be removed from the integral, giving

$$\oint \mathbf{H} \cdot \boldsymbol{\tau}\, d\ell = \oint \mathbf{a}_\phi H(\rho) \cdot \mathbf{a}_\phi \rho\, d\phi = \rho H \oint d\phi = 2\pi\rho H$$

This is equal to current I through the integration loop, so that $H = I/2\pi\rho$, or, in complete form,

$$\mathbf{H} = \mathbf{a}_\phi \frac{I}{2\pi\rho} \tag{4.7}$$

Example 4.2 (cont.)

To find H_ϕ for the solenoid, we apply Ampère's law to either of the concentric paths 1 and 2, shown in Fig. 4.8, for which $\boldsymbol{\tau} = \mathbf{a}_\phi$. Since both paths are at constant ρ, we have that $H(\rho)$ is constant, which can be removed from the integral. Thus,

$$\oint \mathbf{H} \cdot \boldsymbol{\tau}\, d\ell = \oint H_\phi \mathbf{a}_\phi \cdot \mathbf{a}_\phi \rho\, d\phi + \oint H_z \mathbf{a}_z \cdot \mathbf{a}_\phi \rho\, d\phi = 2\pi\rho H_\phi = 0$$

where the last equality follows from the fact that there is no current through the path of integration. We therefore have $H_\phi = 0$.

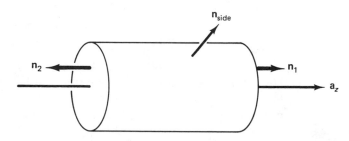

Figure 4.7 Cylindrical surface for flux closure.

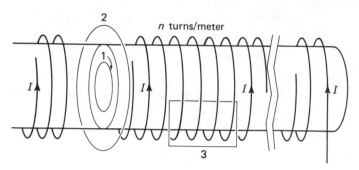

Figure 4.8   Solenoidal windings.

This argument is correct for path 1, inside the solenoid. However, because of the finite pitch of real helical windings, the current crossing path 2 is actually I, so that, *outside* of the solenoid, it looks like a single current conductor and H_ϕ is given by Eq. (4.7). As applications of solenoidal windings only use the interior field, H_z, and since H_ϕ(outside) $\ll H_z$(inside), this exterior field is generally ignored.

We can learn something about H_z from both the differential and integral forms of Ampère's equation. Since we have only the single component $H_z(\rho)$, the differential form gives $(\nabla \times \mathbf{H})_\phi = -\mathbf{a}_\phi \partial H_z / \partial \rho = \mathbf{J}$. Both within and outside of the solenoid $\mathbf{J} = 0$, so that the only place H_z can change is *at* the windings, where $\mathbf{J} \neq 0$. H_z must be constant inside and (possibly with a different value) outside. Now we have already established that $H_\rho = 0$, so that the continuous solenoid had no field directed from inside to out. Consequently, the internal flux doesn't leak through the windings to be continuous with the external flux. The external flux must connect with the internal flux, which leaves the solenoid at one end and closes on itself at the other end. For a short solenoid, the flux returns through the region around the solenoid. For a long solenoid, this returning flux must pass through a larger external volume of space, making its density small at any point. This is illustrated in Fig. 4.9. In the limit that the length becomes infinite, the external flux density becomes zero, so that

$$H_z(\text{out}) \rightarrow 0 \qquad \text{length/radius} \gg 1$$

With this result, we can apply Ampère's law, in integral form, to path 3 in Fig. 4.8, to find

$$\oint \mathbf{H} \cdot \boldsymbol{\tau} d\ell = H_z(\text{in})\ell + H_z(\text{out})\ell = I_{\text{through path}}$$

For the path of length ℓ, there are $n\ell$ turns, each with current I, crossing the path so the enclosed current is $n I \ell$. Since $H_z(\text{out}) = 0$,

$$H_z(\text{in}) = nI \tag{4.8}$$

For each of the fields found in these examples, it is easy to show that $\nabla \times \mathbf{H} = 0$, even though $\mathbf{H} \neq 0$ and $\mathbf{H}$ may be a function of position. This emphasizes that curl $\mathbf{H}$ is a *point* function, meaning that in each region (actually, at each point), it gives the current density at the point where it is being evaluated. We can have $\nabla \times \mathbf{H} = 0$ at one place, even though there is current at another that generates that $\mathbf{H}$.

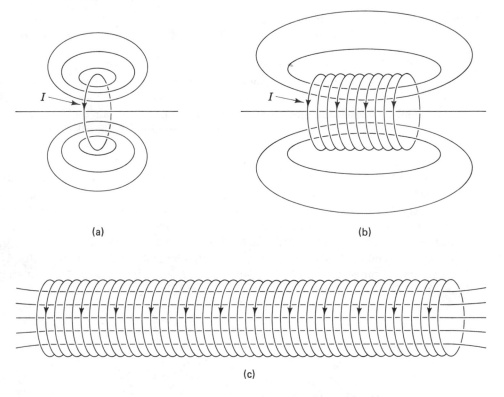

(a)   (b)

(c)

Figure 4.9   Field-line closure for short and long solenoids.

It was pointed out earlier that nI is the surface current on the solenoid, in A/m of length. We will later use this result of Eq. (4.8): that the internal field of a solenoidal surface current is equal in magnitude to that of the surface current density.

4.4 FORCES AND TORQUES

We saw in Chapter 3 that static charge experiences a force proportional to an electric field $\mathbf{E}$ in which it resides. In addition, charge in motion, that is, current, experiences a force proportional to a magnetic field $\mathbf{B}$ through which it passes. This is generally expressed in two equivalent forms:

1. A charge q moving with velocity $\mathbf{v}$ through a region containing a magnetic flux density $\mathbf{B}$ experiences the Lorentz force:

$$\mathbf{F} = q(\mathbf{v} \times \mathbf{B}) \tag{4.9a}$$

2. A current element $\mathbf{I}\, d\ell$ placed in an externally applied magnetic field is subject to the force

$$d\mathbf{F} = \mathbf{I}\, d\ell \times \mathbf{B} \tag{4.9b}$$

Note that $\mathbf{I}\,d\ell$ in Eq. (4.9b) is a current element that is placed in a preexisting field. This field is not to be confused with any field that the current element itself may generate.

The force law can be illustrated by considering the example of a square current loop in an applied magnetic field, that is, a field that is generated by some external means, as shown in Fig. 4.10(a). For simplicity, $\mathbf{B}$ is taken with its planar component parallel to one of the square edges; this does not alter our conclusions. The forces on the sides of the square are found by using the right-hand rule in Eq. (4.9b), their magnitudes and directions are as shown in Figs. 4.10(b) and (c).

One component of the force on each side is equal to $B_n IL$, where B_n is the component of $\mathbf{B}$ along $\mathbf{n}$, the normal to the loop area. These tend to pull the loop apart (or to collapse it if $\mathbf{I}$ or $\mathbf{B}$ are oppositely directed), but do not result in any motion of a rigid loop. On the top and bottom sides, however, are the additional forces $B_p IL$, due to the planar component of $\mathbf{B}$. These constitute a couple that exerts a torque on the loop and causes it to rotate. The torque acting on the loop is the force times the perpendicular separation of the couple elements: $\tau = (B_p IL)L = B_p IA = B_p m$, where m, called the *magnetic dipole moment* of the loop, is equal to the circulating current times its area: $m = IA$ $(A = L^2)$. When equilibrium is reached, as

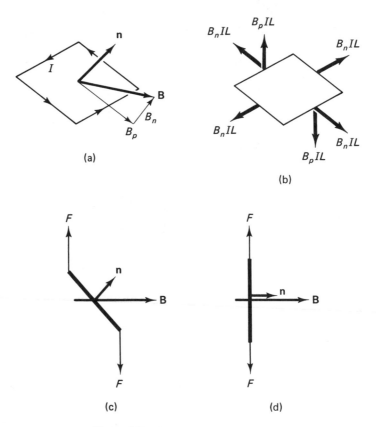

Figure 4.10 Current loop in a magnetic field.

shown in Fig. 4.10(d), the loop has its area perpendicular to **B** and its normal **n** is along **B**.

The magnetic dipole moment of a current loop is generally expressed as a vector along the area normal, according to the right-hand rule with the fingers along I: $m = IA\mathbf{n}$. In terms of this dipole moment, the torque on a dipole in an applied field is

$$\boldsymbol{\tau} = \mathbf{m} \times \mathbf{B} \tag{4.10}$$

4.5 PHYSICAL REALITY AND FIELD THEORY

We have been developing the theory of electricity and magnetism in a logical way, by deduction from Maxwell's equations. It was pointed out in Chapter 2 that the historical development of electromagnetism began with a build-up of experimental observations which appeared to be only distantly related. In this, the experimental basis of magnetostatics is the force that exists between two current elements $\mathbf{I}_1 d\boldsymbol{\ell}_1$ and $\mathbf{I}_2 d\boldsymbol{\ell}_2$. The force on element 2 is given by Eq. (4.9), now written as

$$d\mathbf{F}_2 = \mathbf{I}_2 d\boldsymbol{\ell}_2 \times d\mathbf{B}_1 \tag{4.11a}$$

where $d\mathbf{B}_1$ is the flux density (at element 2) arising from element 1. It will be seen later that the magnetic flux density from element 1 is

$$d\mathbf{B}_1 = \mu_0 \frac{\mathbf{I}_1 d\boldsymbol{\ell}_1 \times \mathbf{r}_0}{4\pi r^2} \tag{4.11b}$$

where $\mathbf{r}_0 r$ is the vector distance from source point 1 to field point 2. Taken together, these give the Biot–Savart law for the force between two current elements in air:

$$d\mathbf{F}_2 = \frac{\mu_0}{4\pi r^2} \mathbf{I}_2 d\boldsymbol{\ell}_2 \times (\mathbf{I}_1 d\boldsymbol{\ell}_1 \times \mathbf{r}_0) \tag{4.11c}$$

Physically, only the currents and forces are observable and can be measured, not the fields. It is only for convenience of understanding, and in analyzing complex examples, that the observable force interaction is separated into one part describing the field generation and another where the field acts to exert a force.

The same situation exists with regard to electrostatics, where the force interaction between two charges dq_1 and dq_2 in air is given by Coulomb's law.

$$d\mathbf{F} = \mathbf{a}_r \frac{dq_1 \, dq_2}{4\pi\varepsilon_0 r^2}$$

For convenience, this is separated into the generation of an electric field **E** and the action of that field to produce the force.

These considerations, in both the magnetic and electric cases, raise questions of the reality of the fields. Classical physicists, following Faraday and Maxwell, held to physical existence of the fields and had detailed theories of changes in the medium through which they act, the aether. With experimental evidence that the aether does not exist, and development of the Theory of Relativity, the question of existence of the fields has faded from prominence, being regarded as irrelevant to determining the ultimate interactions. Modern field theorists interpret electromagnetic interactions through the virtual particles and quasiparticles of quantum field theory and these, in turn, have characteristics which are verifiable only in terms of their observable actions. Such considerations raise philosophical questions about the significance of scientific truths, about human comprehension of the universe, and about the ultimate meaning of nature and reality. As was mentioned in Chapter 2, these issues constitute a fascinating outgrowth of Maxwell's work, and one of great importance to the philosophically minded; in the context of this text it is left for the reader to draw his or her own conclusions about their significance.

4.6 MAGNETIZATION

In the previous section, it was shown that a magnetic dipole rotates so that its magnetic moment lies parallel to an applied field. Figure 4.11(a) shows the fields *generated* by the currents in the four sides of the loop of such a dipole; each side acts like a straight current element of the type analyzed in Example 4.1; its field curls about its respective side according to the right-hand rule. Outside of the square, the fields due to opposite sides tend to cancel. However, the fluxes from all four sides are parallel and additive through the area of the loop, producing a magnetic field normal to the loop and directed according to the right-hand rule, as shown in Fig. 4.11(b). Alternatively, the field of this current loop can be viewed as that from a very short solenoid of the type discussed in Example 4.2; while we have pictured a square loop for ease of demonstration, the same result is to be expected regardless of the loop shape. In Chapter 6, it will be shown that the field produced by the current loop is proportional to its magnetic moment m.

Simplistically, we can visualize electrons moving about an atomic nucleus or a molecular core as being current loops of the sort we have been discussing. There are on the order of 10^{22} atoms per cm^3 in a solid, so that, even though each magnetic moment is minute, it is not unreasonable to expect their combined effect to generate a large field strength if they all align. Figure 4.12 shows a material body placed in a field **H** produced by external currents. The associated flux density **B** acts on the atomic loops and tends to align them with the field. (The degree of alignment is limited because the loops are held in place by interatomic bonding forces, and align-

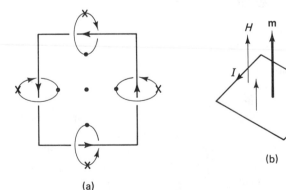

(a)

(b)

Figure 4.11 Magnetic field of a current loop.

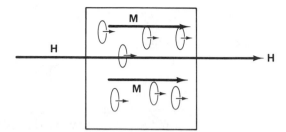

Figure 4.12   Magnetization induced by an
external field.

ment is limited by thermal randomizing effects.) These internal current loops then produce an additional magnetic field, which we denote by the vector symbol **M**, called the *magnetization*. This new field is indistinguishable from the **H** field produced by the external currents, and the difference of notation is merely a convenience to indicate different origins.

   Early physicists who studied magnetic behavior had no concept of quantum interactions that produce magnetism or of atomic structure. However, they did know the results of Example 4.2, that a solenoidal surface current density **K** generates an internal field of magnitude equal to $|\mathbf{K}|$. Therefore, Ampère proposed that the response of a magnetic body to an applied field is the result of *induced currents* that are caused to flow around the perimeter of each infinitesimal element. Figure 4.13(a) shows a magnetic material enclosed in solenoidal windings which carry current I.

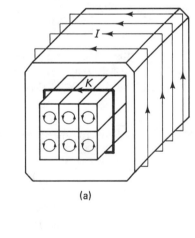

(a)

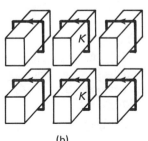

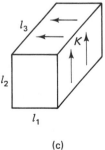

Figure 4.13   Amperian currents in a
magnetic material.   (b)   (c)

Also shown are a few very small elements of the material. In Fig. 4.13(b), these elements are separated to show their induced Ampèrian currents. Referring to Fig. 4.13(a), we see that when the elements are in contact, each side carries equal and opposite induced currents from the contacting neighbors. As a result, the internal currents cancel and one is left only with the outermost surface current density **K**, called the *bound* current density. Consequently, the field we have called **M** is equal in magnitude to $|\mathbf{K}|$, or, turning the relationship around, the Ampèrian currents **K** and the hypothetical induced surface current density are equal to $|\mathbf{M}|$.

Since the field produced by each small element of material is proportional to its magnetic dipole moment, a measure of the response of a magnetic material is the dipole moment of a unit volume of the material. Figure 4.12(c) shows a representative element that has

$$\text{dipole moment/volume} = \frac{IA}{\text{volume}} = \frac{K\ell_3 \times \ell_1 \ell_2}{\ell_1 \ell_2 \ell_3} = K$$

From this, we see that the quantity denoted by **M** is not only the induced magnetic field and the Ampèrian induced surface current density, but it is equal to the induced magnetic dipole moment per unit volume. **M** is called the *magnetization* of the sample.

Recall from Fig. 4.12, that the total internal field of a magnetizable body is **H** + **M**. As in the electrostatic case, the field property that produces force (here it is **B**) must be proportional to this total field. The proportionality constant is denoted by the symbol μ_0, so that

$$\mathbf{B} = \mu_0(\mathbf{H} + \mathbf{M}) \tag{4.12}$$

The strength of magnetization **M** is dependent on the aligning or driving force **B**. Historical development, however, has us express it as proportional to the externally generated field **H**:

$$\mathbf{M} = \chi_m \mathbf{H} \tag{4.13}$$

where χ_m is called the magnetic susceptibility. With this, Eq. (4.12) can be written as

$$\mathbf{B} = \mu_0(1 + \chi_m)\mathbf{H} \tag{4.14a}$$

or

$$\mathbf{B} = \mu\mathbf{H} \qquad \mu = \mu_0(1 + \chi_m) \tag{4.14b, c}$$

A figure of merit for magnetic material is the relative magnetic permeability:

$$\mu_r = \mu/\mu_0 = 1 + \chi_m \tag{4.15}$$

In the presence of magnetizable material, $\mu \neq \mu_0$ because an additional field, generated by the Ampèrian or bound currents, adds to the field from the free currents in contributing to the total flux density **B**.

The units of **B** are webers/m^2 or teslas (1 T = 1 Wb/m^2). The constant μ_0 is

$$\mu_0 = 4\pi \times 10^{-7} \text{ Wb/A–m} \qquad (\text{Wb/A–m} = \text{henrys/m}) \tag{4.16}$$

For most materials, the value of χ_m is constant, so that a graph of B, or of M, versus H is a straight line. Also, for most materials, $\chi_m \ll 1$, so that $\mu \approx \mu_0$ and the slope of this line is small. There exists a particularly important class of materials, ferromagnetics, typified by iron, nickel, and cobalt, for which $\chi_m \gg 1$, so that $\mathbf{M}$ generated by the bound currents is many times greater than $\mathbf{H}$ from the external currents. This makes $\mu \gg \mu_0$, so that $\mathbf{B}$ is large and the useful magnetic properties are greatly enhanced, for example, the ability to exert forces is greatly increased. The large value of $\mathbf{M}$ in these materials is due to a new type of physical behavior and, as a result, Eqs. (4.13) to (4.15) are no longer applicable. Instead, the observed response is represented by Fig. 4.14. Here we see that $\mathbf{M}$ and $\mathbf{B}$ are no longer proportional to $\mathbf{H}$, that is, the relation is nonlinear. Furthermore, the relation is not single-valued, since for each $\mathbf{H}$, the magnetization $\mathbf{M}$ and the flux density $\mathbf{B}$ can have a range of values. As $\mathbf{H}$ changes, $\mathbf{M}$ and $\mathbf{B}$ follow a different path when $\mathbf{H}$ is decreased from that which they take when $\mathbf{H}$ increases; this is called *hysteresis*. The various nested hysteresis loops shown on Fig. 4.14 are obtained by cycling $\mathbf{H}$ between the maximum values for that loop.

During each cycle of an exciting current in a transformer or motor, the magnetization of the ferromagnetic material traverses a hysteresis loop and, in each cycle, the energy required to turn the atomic moments is dissipated as heat. This energy per unit volume of the magnetic material is equal to the area of the B–H hysteresis loop. In addition to these energy losses, the nonlinear responses of $\mathbf{M}$ and $\mathbf{B}$ to the variations of $\mathbf{H}$ result in distortion of signals that pass through a ferromagnetic device, such as a signal transformer. Consequently, the designers of many devices strive to select materials that have narrow, more nearly linear loops, such as loop 1 in

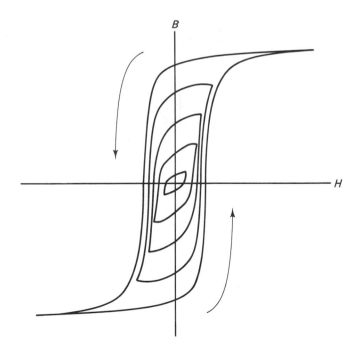

Figure 4.14   Magnetic hysteresis.

Fig. 4.15. In contrast, many applications, such as permanent magnets and magnetic computer memories, depend on having relatively square, open loops, such as loop 2 in Fig. 4.15. For these, large values of **M** and **B** remain when the applied **H** is removed, and a strong reverse **H** is required to reduce **M** to zero, making the magnetization very stable, that is, permanent. This will be considered again in Chapter 6.

Figures 4.14 and 4.15 show hysteresis loops for materials cycled through large fields. It is important to note that even these materials respond linearly, that is, they behave according to Eqs. (4.13) and (4.14), when very small variations of **H** are involved, for example, for waves in the microwave and optical range, where the field variations are generally quite small. Figure 4.16 is half of a hysteresis loop showing two regions where the full field variation was stopped and cycled through a smaller change. As the change decreases, we first find a *minor* loop, as at (a), and then a linear variation, as at (b).

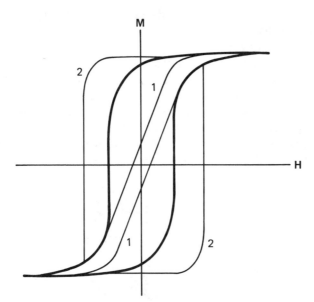

Figure 4.15   Types of hysteresis loops.

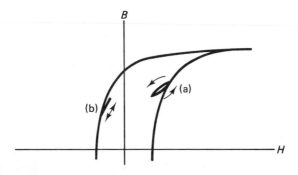

Figure 4.16   Changes for small field variations.

4.7 UNITS, STANDARDS, AND MEASUREMENTS

Historically, electrostatics and magnetostatics developed separately; they were independent disciplines and each had its own system of units, though both the electrostatic units (esu) and the electromagnetic units (emu) are cgs systems. When interrelations were discovered between electric and magnetic phenomena, it was found that descriptions of common events involved quantities, such as current and charge, which were measured in vastly different units in esu and in emu. The two areas of study could only be reconciled by introducing a large numerical adjustment factor into their joint equations. The adjusted system of units, called the *Gaussian system*, carries this factor: $c = 3 \times 10^{10}$ cm/sec.

Of more recent introduction is the SI, or International System of Units, based on an older Giorgian system. This has been adopted for almost all engineering applications because of its internal consistency, simplicity, and direct relation to the practical units and measures encountered by the engineer. SI units are used in this book.

It is useful to know how the electrical and magnetic quantities are systematically defined and determined in the SI system. The basic unit is defined in terms of the magnetic force law. By using very precise geometry and force measurements, *the standard ampere* is determined to be the current that, when flowing in parallel wires 1 meter apart in a vacuum, causes them to exert a mutual interactive force of 2×10^{-7} newtons per meter of their length. This is illustrated in Fig. 4.17. From Eqs. (4.7) and (4.9b), we see that for this arrangement,

$$F/\ell \text{ (newtons/m)} = \mu_0 I^2/2\pi r = \mu_0/2\pi \qquad (4.17)$$

Since F/ℓ is exactly 2×10^{-7}, the value of μ_0 is precisely given by Eq. (4.16).

Only a current of exactly 1 A will satisfy Eq. (4.17). Once the exact ampere of current has been found in this way, the unit of charge is determined from

$$I = \frac{dQ}{dt} \qquad (4.18)$$

that is, a current of 1 ampere equals a flow of 1 coulomb of charge per second. Having a precisely obtainable unit of charge allows, in turn, determination of the electric permeability of free space from Coulomb's law. The force between two charges of 1 coulomb each, separated by 1 meter in vacuum, is found from Eqs. (3.7) and (3.8):

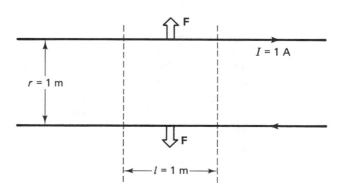

Figure 4.17 Schematic force bridge to determine the standard ampere.

$$F \text{ (newtons)} = Q^2/4\pi\varepsilon_0 r^2 = 1/4\pi\varepsilon_0 \tag{4.19}$$

Measurement of this force gives the value of ε_0 in Eq. (3.26a).

This outline of the relation between fundamental units of electricity and magnetism is the conceptual basis for determining all measurement standards in the field
of electromagnetics. Because of the practical difficulty of carrying out the standards
measurements described above, it was, for a long time, the practice to use standard
chemical cells as voltage reference, and a specified mercury column as the standard
ohm. Here, however, reproducibility and reliability were problems.

Recently, there have developed exquisitely precise and accurate techniques for
determining some of the fundamental constants and standards of electricity and magnetism. While one rarely uses an Ampèrian force bridge or Coulomb balance today,
all of the more sophisticated methods must be shown to be equivalent to those described, or the unit redefinition must have a known relation to the units based on
these methods.

The area of metrology, that is, precise and accurate measurements and standards determinations, has lead to important advances in techniques used in many
areas. For example, the time is past when the standard of time was taken from the
unsteady rotation rate of the earth. Fantastic precision in the measurement of time
and frequency have allowed the second to be defined in terms of the period of the
cesium atom. Recently, it has been found that a quantum mechanical phenomenon of
superconductivity, Josephson tunneling, can be used to define voltage in terms of a
frequency measurement that can be made with great accuracy. With increasingly accurate and precise primary standards, it is easier to calibrate secondary standards for
use in research laboratories and in industrial field work, so that the improved standards allow greater accuracy overall.

This is but one example of modern metrology. Modern science depends criti-
cally upon reliable measurements, and can only advance as fast as the techniques of
measurement allow. With these refined methods, new phenomena have been found
and new fields of scientific effort have been opened.

EXERCISES

Extensions of the Theory

4.1. We recognize that charge q moving with velocity $\mathbf{v}$ constitutes a current (if
the charge is negative, the current is in the opposite direction), so that
$\mathbf{I}\,d\ell \to q\mathbf{v}$ (check the units). More specifically, consider the cylindrical conductor shown in Fig. 4.18 (next page), which has n free electrons per unit
volume, moving with an average speed v. Show that the current density is
$J = nqv$. *Hint:* Take $|L|$ (in m) $= |v|$ (in m/sec), and note that the current is
equal to the charge moving past a point each second.

4.2. Each of the electrons considered in Exercise 4.1 starts from rest and is accel-
erated by an applied electric field $\mathbf{E}$. After a mean time τ, it collides with an
atom and comes to rest, after which the process is repeated. Derive an ex-

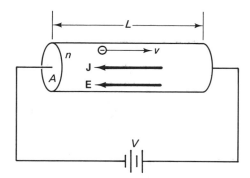

Figure 4.18   For Exercise 4.1.

pression for the mean velocity of the electron and obtain an expression for the conductivity from Ohm's law $\sigma = J/E$.

4.3.   Since the divergence of any curl is zero, Maxwell's equation $\nabla \cdot \mathbf{B} = 0$ is automatically satisfied if $\mathbf{B} = \nabla \times \mathbf{A}$, where $\mathbf{A}$ is called the *magnetic vector potential*. For a line current element $\mathbf{I}\,d\ell$, it can be shown that

$$\mathbf{A} = \frac{\mu}{4\pi} \int \frac{\mathbf{I}\,d\ell}{r}$$

where r is the distance from the current element to the field point, and the integral is over all the current elements.
(a) Determine $\mathbf{A}$ at any point, due to an infinite, constant current-line along the positive z axis.
(b) Calculate $\mathbf{B}$ and compare your result with Eq. (4.7).

4.4.   The plane $x = 0$ carries a uniform surface current density $\mathbf{K} = \mathbf{a}_z K_0$ A/m. Using symmetry arguments and Ampère's law, determine the magnetic field strength at all points $x \neq 0$.

4.5.   Equation (4.11b) gives the magnetic field due to an infinitesimal current element. An infinitesimal square current loop is shown in Fig. 4.19. By adding the contributions of the four sides, show that the magnetic field at points on the z axis is $\mathbf{H} = \mathbf{m}/2\pi r^3$, where $\mathbf{m}$ is the dipole moment of the loop.

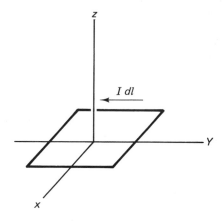

Figure 4.19   For Exercise 4.5.

Point of information: The magnetic field H from the dipole decreases as $1/r^3$, whereas according to the Biot–Savart law, H from a current element decreases as $1/r^2$, and, from Example 4.1, H from a long straight wire decreases as $1/r$.

*4.6. A cylindrical rod of length ℓ and circular cross-sectional area A is permanently magnetized along its length with the dipole moment per unit volume given by $\mathbf{M}$. An external field $\mathbf{B}$ is applied perpendicular to the axis of the rod.

Considering the torques acting on a unit volume of the magnetic dipoles that make up $\mathbf{M}$, show that the total torque on the rod is $BA\ell M$.

Point of Information: This result can be interpreted by saying that a pole of strength $p = MA$ exists at each end of the rod. These poles (north where $\mathbf{M}$ terminates and south where $\mathbf{M}$ originates in the rod) respond to the field $\mathbf{B}$ in the same way as positive and negative charges, respectively, respond to an electric field; they each experience a force pB (in opposite directions). The torque is, therefore, force $\times$ lever arm $= pB\ell = BAM\ell$.

Problems

*4.7. Using Eq. (4.11b), find the magnetic field at the center of a single circle of radius R, and carrying a current I.

*4.8. An electron beam along the z axis constitutes a current with densities:

$$\mathbf{J} = \mathbf{a}_z J_0 \left(1 - \frac{\rho^2}{a^2}\right) \qquad \rho < a$$

$$= 0 \qquad\qquad\qquad \rho > a$$

(a) What is the total current flowing along the z axis?
(b) Calculate the magnetic field at all points of space.

4.9. Imagine that a closer examination of the current of Problem 4.8 indicates that there is *also* a small leakage component given by $\mathbf{J} = \mathbf{a}_\rho \beta a/\rho$ for all ρ, where $\beta \ll J_0$. Discuss how this changes the components of $\mathbf{H}$.

4.10. Two hollow tubes, both of radius a, are separated by a center-to-center distance d, as shown in Fig. 4.20 (next page). Each carries a surface current density of $K = I/2\pi a$ A/m.
(a) Calculate the magnetic field due to one of the cylindrical currents at all points.
(b) What is the total magnetic flux, per meter of tube length, in the region between the cylinders. Carefully define the region you are using.

*4.11. Given the magnetic field $\mathbf{H} = \mathbf{a}_z H_0 \, e^{-ky}$,
(a) determine the current density,
(b) verify Ampère's law by taking the circulation of $\mathbf{H}$ about the path bounding the square $x = 0$, $0 \le y \le a$, and $0 \le z \le a$.

4.12. Two long current-carrying wires are shown in Fig. 4.21. Their separation is s. A field point lies in their plane, a distance ρ from their midline.
(a) Find the field intensity H from *each* wire at point ρ.
(b) If $\rho \gg s$, show that the total field is $\mathbf{H} \approx \mathbf{a}_\phi (Is/2\pi\rho^2)$.

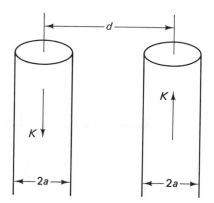

Figure 4.20 For Exercise 4.10.

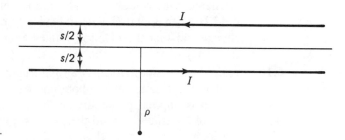

Figure 4.21 For Exercise 4.12.

*4.13. A rectangular loop, in air, is placed in the field of a very long straight conductor carrying a current I, as shown in Fig. 4.22 (see below). What is the total magnetic flux passing through the loop?

*4.14. The following vector fields are in free space:

$$G_1 = a_x y^2 x + a_y x^2 y + a_z \frac{z^3}{3}$$

$$G_2 = a_x x^2 yz + a_y xy^2 z - a_z 2xyz^2$$

Determine whether these might be static electric or magnetic fields, and, if so, find the distribution of field sources in each case.

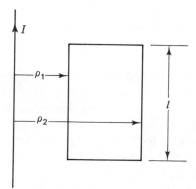

Figure 4.22 For Exercise 4.13.

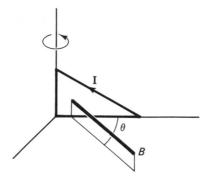

Figure 4.23   For Exercise 4.16.

4.15. A charge with velocity **v** and mass m enters a region containing a uniform field **B**, with **v** and **B** perpendicular to each other. Show that $|\mathbf{v}|$ does not change and that the electron rotates around the field direction with angular frequency $\omega = qB/m$. [See Eq. (4.9a).]

4.16. A rigid frame in the form of a right triangle and carrying a current I is hinged so that it is free to rotate about one leg, as shown in Fig. 4.23. A magnetic field **B** is applied perpendicular to the hinged leg and at an angle θ to the plane of the triangle. Find the force on each side and show that the torque on the frame is $\boldsymbol{\tau} = \mathbf{m} \times \mathbf{B}$.

4.17. Determine the current distributions producing the magnetic fields:

$$\mathbf{H} = H_0(\mathbf{a}_x + \mathbf{a}_y) \qquad\qquad x > s, \qquad\qquad y > s$$

$$= H_0\left(\mathbf{a}_x\frac{y}{s} + \mathbf{a}_y\right) \qquad x > s, \qquad\qquad 0 < y < s$$

$$= H_0\left(\mathbf{a}_x + \mathbf{a}_y\frac{x}{s}\right) \qquad 0 < x < s, \qquad y > s$$

$$= H_0\left(\mathbf{a}_x\frac{y}{s} + \mathbf{a}_y\frac{x}{s}\right) \qquad 0 < x < s, \qquad 0 < y < s$$

where s and H_0 are constants. Draw the current geometry.

4.18. In a transformer core, the magnetization reverses its direction twice during each cycle, as the transformer current causes the flux to vary. This causes energy from the current source to be dissipated by the material in the form of heat.

   (a) If the magnetic material has a maximum magnetization (dipole moment per unit volume) of 1.6×10^6 A/m and a density of approximately 6.7×10^{28} dipoles/m³, determine **m** of an atom.

   (b) Calculate the work done in rotating each magnetic dipole from a position normal to the field to a position parallel to the field. Take $H = 159$ A/m.

   (c) What is the power (energy/second) dissipated at 50 Hz in a unit volume?

*4.19. The solenoid shown in Fig. 4.24(a) has 100 turns, is 10 cm long, and carries $I = 10$ A. It is filled with a material whose magnetization curve is shown in

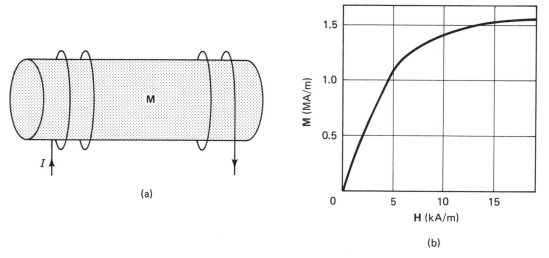

Figure 4.24   For Exercise 4.19.

Fig. 4.24(b) (note that the abscissa is in units of 10^3 A/m and the ordinate is in units of 10^6 A/m). What is the internal flux density (with units)?

4.20. A very long wire of 1-mm radius, and carrying 20 A of current, passes through a snugly fitting ferrite jacket of 5-mm outer radius. The ferrite has $\mu = 700\mu_0$. Surrounding them both is a shield of low-grade iron ($\mu = 3\mu_0$) with inner radius of 7 mm and outer radius of 10 mm.
(a) Sketch the magnitude of B and of M for $0 < \rho < 20$ mm.
(b) Find the Ampèrian current on the outer surface of the ferrite.

*4.21. We have shown that a cylinder of magnetic material that is magnetized along its length can be considered to have hypothetical surface currents, the bound or Ampèrian currents, with surface current density K.

If a solenoid has 20 turns/cm, it is found that a current of 0.1 A is sufficient to saturate a cylindrical sample of iron placed in the solenoid, so that the induced flux density is 0.08 T. Determine the ratio of the induced-surface-current density to the free-current density.

4.22. A solenoid carrying 10 A is 1 meter long and has six layers of wire, each with 120 turns. It is wound over a hollow carbon steel tube that has an outer diameter of 1 cm and an inner diameter of 0.953 cm. The magnetic suscepti-bility of the steel is $\chi = 9$. Find the total flux through the solenoid (i.e., air and steel).

4.23. A 2-m long solenoid having 2000 turns/m is wound on an iron core 4 cm in radius, and carries a steady current of 10 A. The relative permeability of the iron (assumed constant) is 1000. Neglecting end effects,
(a) calculate the intensity of magnetization and the total flux in the iron.
(b) If the solenoid were air-cored, find the number of turns per meter that would be necessary if it were to have the same value of B.

5

TIME-VARYING FIELDS

5.1 QUASISTATICS

In studying thermodynamics, one encounters the conflicting concepts of stability and change. Thermodynamic-state variables (e.g., temperature, pressure, magnetization, etc.) are defined only in uniform systems, but for them to evolve, for example, under outside influences, these systems must be changing and therefore the variables are nonuniform. The resolution of this contradiction lies in the concept of an idealized *quasistatic process*. In this process, change takes place so slowly that at any time the system departure from uniformity and equilibrium is infinitesimal, so that static conditions can still be said to exist.

The same considerations are frequently encountered in electromagnetics. If changes take place slowly enough, then at any instant the fields can be considered static. This has also been referred to as the *quasistationary approximation*. For the approximation to be valid, the fields must be *effectively* constant throughout the system being studied. Since electric and magnetic changes propagate with the velocity of light, c, the effect of a change at one end of a system will be felt at the other end, a distance ℓ away, at a time $t = \ell/c$ later. If the change occurs over an interval τ, and $\tau \gg t$, then the distant point ℓ will appear to sense the change as quickly as it happens, and the fields throughout the system can be said to be uniform at all times; they appear to be static fields, even though they are changing. Because c is so very large, this is true of most systems, except for those undergoing very rapid changes.

Thus, for a system that is 30 cm long, changes are quasistatic (or quasistationary) as long as $\tau \gg 0.3/c = 10^{-9}$ sec. This approximation is quite generally used in circuit analysis and other considerations of *local* systems. Clearly, very large or open systems, that is, where radiation and propagation are important, or systems operating at very high frequencies, require more careful treatment.

It is useful to consider situations where electric and magnetic changes are an inherent part of the behavior, but the fields are still regarded as static. The understanding gained can then be extended to the full time-varying case. For example, consider an infinitesimal charge dq moving in an electric field. As in the static case, it is assumed to experience the forced $d\mathbf{F} = -dq\,\nabla V$, and the work done by the field is equal to the charge times its change of potential, $dW = dq\,\Delta V$. If this occurs in a time dt, the rate at which the field does work, its power expenditure, is $P = dW/dt = (dq/dt)\,\Delta V = I\,\Delta V$, where the moving charge constitutes a current $I = dq/dt$.

5.2 ELECTROSTATIC ENERGY

To study the energy in an electrostatic field, let us perform an idealized conceptual experiment. We consider a small region of space in which such a field exists, and calculate the energy required to eliminate it from that region by generating an equal and opposite field. Figure 5.1 shows a very small volume of a field space, marked by the circle in Fig. 5.1(a), and greatly enlarged in Fig. 5.1(b); if the region is small enough, the field lines will be locally parallel. Now imagine having sections of two parallel planes, which are perpendicular to the field lines, as shown in Fig. 5.1(c). Note that since these planes are normal to the field lines, they are each at a constant potential and their potential difference is denoted by V.

The plan is to determine the energy required to place just the right amount of charge on the planes so as to generate new field lines that will cancel the original lines. The exercises in Chapter 3 show that this can be done throughout the volume between the planar sections, but at the edges of those sections some of the field lines will inevitably leak out of that volume. To minimize this *fringing effect*, we select the area A of the planes such that its length and width (which are each proportional to $\sqrt{A}$) are very large compared to ℓ, the separation of the planes. Then the field volume between them $(A\ell)$ will be very large compared to the volume occupied by the troublesome edge fringing fields $(\sim\sqrt{A}\,\ell)$. The ratio of these field volumes is $v_{\text{fringe}}/v_{\text{direct}} \approx 1/\sqrt{A}$, which can be made arbitrarily small, so the fringing can be ignored.

Figure 5.1(d) has a voltage source connected to the planar sections. Two features of this source require mentioning. (1) The potential of the source is shown as V, but we must understand that its magnitude is increased so it is an infinitesimal amount greater than V, allowing charge to transfer between the planes. (2) The actual introduction of such a source would impermissibly disturb the field pattern, so, in keeping with the idealized conceptual construction, it is understood that the source merely represents the fact that charge can be moved from one plane to the other, and in doing so, an amount of work is done equal to V joules/coulomb.

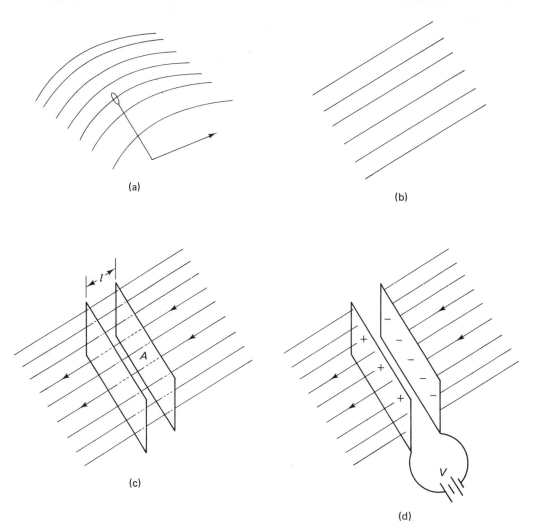

Figure 5.1   Spatial region containing an electric field.

Once charge is transferred, it moves freely on the planes, since they are at equipotential, and the charge distributes itself uniformly over the areas A. Recall from Chapter 3 that the flux density between parallel planes of charge is equal to the charge density on the planes, $\rho_s = D$. Also, for the small separation of concern, the uniform electric field between the plates is $E = V/\ell$. The charging (or charge transfer) continues until the fields generated between the plates are equal and opposite to the original D and E. The net fields are then zero.

In charging the planar sections, current flows under the influence of the applied voltage (or, equivalently, charge is moved by the hypothetical source V). The power and work expended by the source have been derived previously, and are related by

$$dW/dt = P = VI = (E\ell)\,dQ/dt = (E\ell)\,d(\rho_s A)/dt$$

$$= A\ell E\,d\rho_s/dt = vE\,dD/dt \qquad (5.1a)$$

Equation (5.1a) has used the facts that, at every instant, charge is equal to the charge density times the area, $Q = \rho_s A$, that $V = E\ell$, that $\rho_s = D$, and that $A\ell = v$ is the volume between the planes. The electric energy density, that is, the electric energy per unit volume, is defined as $U_e = W/v$, so that Eq. (5.1a) becomes

$$\frac{dU_e}{dt} = \mathbf{E} \cdot \frac{d\mathbf{D}}{dt} \qquad (5.1b)$$

where the vector properties of the fields have been restored. Note that for a linear dielectric,

$$\frac{dU_e}{dt} = \mathbf{E} \cdot \frac{d\mathbf{D}}{dt} = \frac{d}{dt}\left(\frac{1}{2}\varepsilon E^2\right) \qquad (5.2a)$$

so that

$$U_e = \varepsilon\frac{\mathbf{E}^2}{2} = \frac{1}{2}\mathbf{E} \cdot \mathbf{D} \qquad (5.2b)$$

Integrating this electric energy density over an enclosed volume gives the total energy in the volume.

5.3 CAPACITANCE

In circuits and power systems, it is common to have devices that store electrical energy. A basic figure of merit of such a device, its capacitance, is defined as the charge that it can store divided by the energy per unit charge, that is, Q per unit difference of potential, needed to store it:

$$C = \frac{Q}{V} \qquad (\text{coulomb/volt} = \text{farad}) \qquad (5.3)$$

Example 5.1

Calculate C for the parallel-plate capacitor shown in Fig. 5.2.

In this capacitor, we have seen that

$$V = E\ell = \frac{D}{\varepsilon}\ell = \frac{\rho_s}{\varepsilon}\ell = \frac{Q}{A\varepsilon}\ell = \frac{Q\ell}{\varepsilon A}$$

from which

$$C = \frac{Q}{V} = \varepsilon\frac{A}{\ell} \qquad (\text{farads}) \qquad (5.4)$$

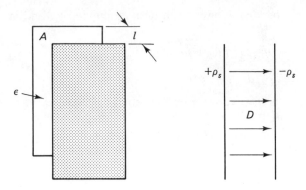

Figure 5.2 Parallel-plate capacitor.

The same limitation applies to this derivation as in deriving Eqs. (5.1) and (5.2), namely, the neglect of the edge fringing fields. As a result, Eq. (5.4) is most accurate when ℓ is small and A is large. In Chapter 8, we will learn how to calculate the fringe fields.

From Eq. (5.4), it can be seen that ε has the units farads/meter. Although it is also defined by $\varepsilon = D/E \rightarrow (C/m^2)/(N/C) = C^2/N\text{-}m^2$, this latter expression is rather awkward and F/m is usually employed. Thus, $1/4\pi\varepsilon_0 = 9 \times 10^9$ m/F.

Example 5.2

Determine the capacitance per unit length of a coaxial cable.

Place equal and opposite charge on each length of the cable center conductor and outer shield, as shown in Fig. 5.3. This charge per unit length is denoted as ρ_ℓ. In this case, the result of Example 3.3 is given by Eq. (3.21).

$$V = \frac{\rho_\ell}{2\pi\varepsilon} \ln \frac{b}{a}$$

The capacitance per unit length is

$$\frac{C}{\ell} = \frac{Q/V}{\ell} = \frac{Q/\ell}{V} = \frac{\rho_\ell}{V}$$

$$= \frac{2\pi\varepsilon}{\ln \dfrac{b}{a}} \qquad \text{(F/m)} \tag{5.5}$$

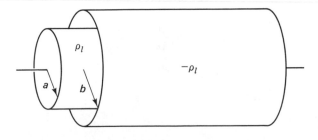

Figure 5.3 Coaxial capacitor.

Equations (5.4) and (5.5) show that the capacitance is a function of the dimensions and materials used. Thus, once a device is fabricated, its capacitance is a constant physical property. For the parallel-plate capacitor, the capacitance can be enhanced by reducing the interelectrode spacing, increasing the electrode areas, and using a dielectric with a high value of ε.

It is often useful to express the energy stored in a capacitor in terms of circuit parameters. Thus, in charging the capacitor, the power is

$$\frac{dW}{dt} = VI = V\frac{dQ}{dt} = V\frac{d(CV)}{dt} = \frac{d(CV^2/2)}{dt} \tag{5.6a}$$

so that

$$W = \frac{CV^2}{2} = \frac{QV}{2} = \frac{Q^2}{2C} \tag{5.6b}$$

Example 5.3

For the case of a parallel-plate capacitor, show that Eq. (5.2b) for the energy density in terms of fields leads to Eq. (5.6b) for the total energy in terms of circuit parameters.

$$W = U_e v = \frac{DE}{2}A\ell = \frac{(Q/A)(V/\ell)}{2}A\ell = \frac{1}{2}QV$$

In deriving Eq. (5.6b), $W = QV/2$, all the charge on each capacitor plate is at the same potential. If this is not the case, that is, if elements of charge dq are separated to reside at different potentials, then for each element this result is $dW = V\,dq/2$. For example, in the case of a volume charge distribution for which V is a function of position, the charge element is $dq = \rho\,dv$, and the total energy of the configuration is

$$W = \frac{1}{2}\int V\,dq = \frac{1}{2}\int V\rho\,dv \tag{5.7}$$

The integration is over the volume charge distribution. Equation (5.7) satisfies our intuitive feeling that the energy generated by charge separation, as in a capacitor, is associated with the charge which exists in its own difference of potential, that is, which is established by separating the charge $\pm Q$. On the other hand, Eq. (5.2b) clearly specifies that there is an energy in each volume of space in which an electric field is found, that is, the energy is associated with the fields themselves, rather than their sources. In Chapter 9, where we study traveling waves and fields that move away from, and become independent of, their sources, this latter interpretation becomes vital. We have all experienced such field energy; for example, we can sense it in the radiant light and heat from the sun, or from a light bulb.

It was pointed out in Chapter 4 that, in spite of having energy, in many respects, electric fields can be regarded as mathematical constructs that are used for convenience in conceptualizing interactions. However, Eqs. (5.1b) and (5.2b) introduce the concept that the field itself has energy. This is remarkable because it implies that the fields should be considered as real physical quantities. It was also pointed out in Chapter 4 that while these interpretations, and others, seem contradictory, this fact does not impede our ability to use them to solve problems.

5.4 FARADAY'S LAW

Until now, the consequences of Maxwell's equations have been examined under static conditions, that is, when charge is stationary and when currents are unchanging, or when they are changing so slowly that the static laws can be assumed to apply at every instant (quasistatic conditions). To begin a more complete discussion of time-varying fields, we first examine the full form of Maxwell's equation for the curl of **E**:

$$\nabla \times \mathbf{E} = -\frac{\partial \mathbf{B}}{\partial t} \tag{5.8}$$

In the static case, **E** had no curl, but the complete statement shows that an electric field does curl around a changing magnetic field. Since it was shown earlier that $\nabla \times \mathbf{E} = 0$ is equivalent to the law of the conservation of energy for the electrostatic field, it now appears that electrical energy is no longer conserved. It will be seen later in this chapter that, with the introduction of time variation, there is a coupling of electric and magnetic fields, so that it is the energy of the combined electromagnetic fields that is conserved.

Integrating Eq. (5.8) over an arbitrary surface area, we have

$$\int \mathbf{n} \cdot \nabla \times \mathbf{E} \, dS = -\int \mathbf{n} \cdot \frac{\partial \mathbf{B}}{\partial t} \, dS \tag{5.9}$$

$$\oint \mathbf{E} \cdot \boldsymbol{\tau} \, d\ell = -\frac{\partial}{\partial t} \int \mathbf{n} \cdot \mathbf{B} \, dS \tag{5.10}$$

where Stokes' theorem has been used to change the integral on the left and, on the right, the order of integration (over the space variables) and differentiation (with respect to the time variable) have been interchanged. This is permissible providing that the geometry of the integral changes slowly compared to **B**. Recall that **B** is the magnetic flux density, so that the last integral in Eq. (5.10) is *the total magnetic flux* crossing the surface of integration:

$$\Phi = \int \mathbf{n} \cdot \mathbf{B} \, dS \tag{5.11}$$

The term on the left of Eq. (5.10) is the line integral of E. From Eqs. (3.15) and (3.16) $\int \mathbf{E} \cdot \boldsymbol{\tau} \, d\ell$ is the voltage, or energy per unit charge, or the potential change between the limits of the integral. Here, however, the integral is over a closed path. In the electro*static* case, this was zero but now, in the *dynamic* case, that is, in the presence of a changing magnetic flux, a nonzero voltage is developed around the closed path. Equation (5.10) can be written in the form

$$V = -\frac{d\Phi}{dt} \tag{5.12}$$

Eqs. (5.8), (5.10), and (5.12) are known as *Faraday's law*. This law constitutes the basis for electromagnetic induction, which includes the operation of electrical transformers and the operation of motors and generators. It can truly be said that this is the foundation of modern technological society.

Michael Faraday (1791–1867), who first reported the discovery of induction in 1831–1832, was a most unique scientist. Born into poverty, he had almost no formal education, and had almost no knowledge of mathematics. However, he possessed such fantastic physical intuition and was such a superb experimentalist that he was eventually acclaimed universally. He disliked the then accepted *action at a distance* as an explanation for electric and magnetic phenomena, and proposed and promoted the concept of *lines of force* as the mechanism of interaction. As discussed in Chapter 2, his insights impressed James Clerk Maxwell, who put them into mathematical form and used them as the basis for his treatise, changing the lines of force into the fields we study here.

Example 5.4

To gain a fuller understanding of Faraday's law, consider two analyses of the simple induction situation shown in Fig. 5.4. A conducting U frame lies in the xy plane, normal to a z-directed magnetic field. A bar, with resistance R, is moving with speed v on the frame in the y direction. Determine the current magnitude and direction in the bar.

First Analysis. The magnetic flux through the area of the U-bar frame is $\Phi = \int \mathbf{B} \cdot \mathbf{n}\, dA = \int B\mathbf{a}_z \cdot (-\mathbf{a}_z)\, dA = -BA = -B\ell y$, where $\mathbf{n}$ is taken, according to convention, by the right-hand rule when circling the perimeter in a counterclockwise manner. By Faraday's law, the induced voltage in the frame is $V = -d\Phi/dt = +B\ell\, dy/dt = B\ell v$. Again, V is in the right-hand, positive, direction with respect to $\mathbf{n}$, so the current flows as shown. The current is $I = V/R = vB\ell/R$.

Second Analysis. Each positive charge carrier in the moving bar will experience the Lorentz force of Eq. (4.9a): $\mathbf{F} = q\mathbf{v} \times \mathbf{B} = q v \mathbf{a}_y \times \mathbf{a}_z B = q v B \mathbf{a}_x$ directed along the length of the bar, and the motion of these carriers constitutes a current in the direction of $\mathbf{a}_x$. In moving the carriers along the bar, the work done by the force is $W = F\ell = q v B \ell$. As a result, a difference of potential exists between the ends of the bar, given by $V = W/q = v B \ell$, as found before.

The point of Example 5.4 is that Faraday's law can be understood as a consequence of the force law, or the force law can be obtained from Faraday's law; it is the relative change, in this case motion, between the charge carriers and flux lines, that is important.

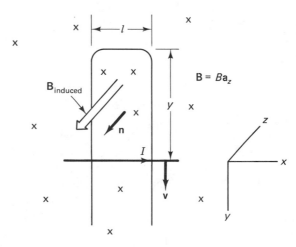

Figure 5.4 Moving bar in a magnetic field.

With the addition of Faraday's law, it can be considered that there are two types of electric fields, both of which result in force on charged bodies and both of which are capable of doing work. One of these diverges from electric charge and has zero curl; this is the field considered in Chapter 3 that satisfies the relations

$$\nabla \cdot \mathbf{E}_1 = \rho/\varepsilon \qquad \nabla \times \mathbf{E}_1 = 0 \qquad (5.13a)$$

The other curls about a changing magnetic flux and has zero divergence; this is the field from Faraday's law that satisfies the relations

$$\nabla \cdot \mathbf{E}_2 = 0 \qquad \nabla \times \mathbf{E}_2 = -\partial\mathbf{B}/\partial t \qquad (5.13b)$$

Clearly, the total field $\mathbf{E} = \mathbf{E}_1 + \mathbf{E}_2$ satisfies Maxwell's equations.

5.5 LENZ'S LAW

The physical significance of the minus signs in Eqs. (5.10) to (5.12) must be pointed out: it implies that $\mathbf{E}$ curls in the *opposite* direction from that along which $\mathbf{B}$ is changing. Thus, for a *change* of $\mathbf{B}$ along the *positive* z axis (if $\mathbf{B}$ is directed along $+z$, then $|\mathbf{B}|$ is increasing, or if $\mathbf{B}$ is along $-z$, then $|\mathbf{B}|$ is decreasing), $\mathbf{E}$ curls about the *negative* z direction. The induced electric field is in such a direction that if the induced current were allowed to flow, it would produce its own magnetic flux to oppose the *change* that generated it. With this additional understanding, Eqs. (5.8), (5.10), and (5.12) are often called the *Faraday–Lenz law*. Lenz's contribution acts as a restraint in that it prevents the buildup of successive changes; the induced effect opposes its generating mechanism. This is illustrated by the following analysis.

Example 5.4 (continued)

As the bar slides down, the flux through the area increases. This flux is along $\mathbf{a}_z$. As expected from Lenz's law, the induced current flows in such a direction that its flux opposes that increase, that is, it is in a direction to reduce $\mathbf{B}$. By the right-hand rule, the magnetic field generated by the induced current, B_{induced} in Fig. 5.4, is along $-\mathbf{a}_z$.

Alternatively, note that the induced velocity of the charge carriers is $v\mathbf{a}_x$ and the force due to this component of $\mathbf{v}$ is $F = q\mathbf{v} \times \mathbf{B} = qv\mathbf{a}_x \times \mathbf{a}_z B = -qvB\mathbf{a}_y$. This force opposes the initial motion of the bar (which is its cause).

5.6 MAGNETOSTATIC ENERGY

To examine magnetic energy, we proceed as in the electrostatic case, by calculating the energy required to develop, or eliminate, the field in a volume of space. Consider flux lines in a region of space, as shown in Fig. 5.5(a). In a very small volume, the lines of $\mathbf{B}$ are locally parallel, as shown in Fig. 5.5(b). Imagine that solenoidal windings are placed over a tube enclosing a certain flux, and that current is passed through the windings to generate an equal field. Recall from Chapter 4 that the field strength within a long solenoid is given by

$$H = nI = \frac{N}{\ell}I \qquad (5.14)$$

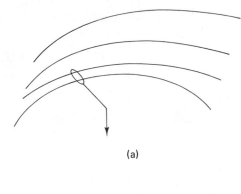

(a)

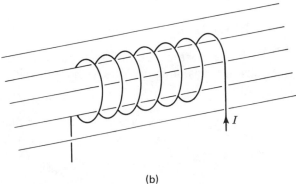

Figure 5.5   Magnetic field region.   (b)

In increasing this current from zero to its final value necessary for the specific H field, the changing flux induces an opposing emf, according to the Faraday–Lenz law, which must be overcome. By considering only the magnitude, from Eqs. (5.10) and (5.12),

$$V = N\frac{d\Phi}{dt} = N\frac{d}{dt}(BA) = NA\frac{dB}{dt} \tag{5.15}$$

The factor N in this equation is due to the fact that Faraday's law describes the voltage generated in a single loop. If N turns are in series, the total voltage is the sum of their individual voltages.

From Eqs. (5.14) and (5.15), the power expended by the current source is

$$P = \frac{dW}{dt} = VI = NA\frac{dB}{dt}\cdot\frac{H\ell}{N} = vH\frac{dB}{dt} \tag{5.16}$$

since the volume enclosed by the windings is $v = A\ell$. It follows that the rate of change of magnetic energy density, $U_m = W/v$, is

$$\frac{dU_m}{dt} = \mathbf{H}\cdot\frac{d\mathbf{B}}{dt} \tag{5.17a}$$

where the vector identities of the fields have been restored. In the linear case, $\mathbf{B} = \mu\mathbf{H}$, so this result can be written in the form

$$\frac{dU_m}{dt} = \mu\mathbf{H} \cdot \frac{d\mathbf{H}}{dt} = \frac{d}{dt}\left(\frac{1}{2}\mu H^2\right) \qquad (5.17b)$$

so that

$$U_m = \frac{1}{2}\mu H^2 = \frac{1}{2}\mathbf{H} \cdot \mathbf{B} \qquad (5.17c)$$

Integrating this magnetic energy density over a volume containing a magnetic field gives the total magnetic energy in the volume.

5.7 INDUCTANCE

As in the electric case, for applications of induction it is important to cast these results in terms of device parameters. For this, write

$$\mathbf{H} = \mathbf{H}(I[t], \mathbf{r}) \qquad (5.18)$$

expressing that, in general, $\mathbf{H}$ is a vector function of position and of its generating current, which has a time-dependence. Then the flux crossing the area of the solenoid in Fig. 5.6 is

$$\Phi = \int \mathbf{B} \cdot \mathbf{n}\, dS = \int \mu\mathbf{H} \cdot \mathbf{n}\, dS \qquad (5.19)$$

and the voltage induced by the changing flux in a single loop is $d\Phi/dt$. For a magnetic device with N turns,

$$V = -N\frac{d\Phi}{dt} = -N\frac{d\Phi}{dI}\frac{dI}{dt} = -L\frac{dI}{dt} \qquad (5.20a)$$

where

$$L = N\frac{d\Phi}{dI} \qquad (Wb/m = \text{henries}) \qquad (5.20b)$$

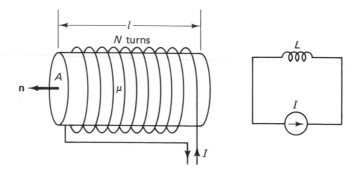

Figure 5.6   Induction coil.

is the inductance of the magnetic device. Since the voltage generated through this inductance, $V = -L\,dI/dt$, is developed in the same circuit as the current that generates the flux, L is also called the coefficient of *self-induction*. If the flux through this
circuit arises from a current in another circuit, then we speak of a coefficient of *mutual induction*.

Example 5.5

Calculate the inductance of the solenoid of Fig. 5.6, using the long-solenoid approximation
described in Example 4.2.

From that example, $\mathbf{H} = \mathbf{a}_z NI/\ell$, so that Eqs. (5.21) give

$$L = N\frac{d}{dI}\left(\mu \int \mathbf{H} \cdot \mathbf{n}\,dS\right) = \frac{N^2\mu A}{\ell} \qquad \text{(henries)}$$

Since the inductance varies as N^2, increasing the number of windings is a very effective way of enhancing L. Increasing N, for a fixed-length coil, implies the use of finer wire,
thereby limiting the current-carrying capability and imposing a practical limit in some
cases. If an iron core is placed within the windings, the value of μ can be increased many
times over that of air, explaining why iron-core inductors, or chokes, are used at power
frequencies.

In the expression for L in Example 5.5, note that N is the number of turns,
which has no units, so that μ has the units henries/meter. Since $\mu = B/H$, it also
has the units $T/(A/m) = T{-}m/A$, or Wb/m$-$A, but these are rather awkward combinations, so that μ is generally referred to in h/m.

In general, the hysteresis properties of a strongly magnetic material, as dis-
cussed in Chapter 4, leads to a complicated dependence of $\mathbf{H}$ and $\mathbf{B}$ on I, and, therefore, on t, so that L is not unique or well-defined. However, for small changes, or by
using average quantities, a linear dependence can often be assumed, that is, B and
$H \sim I$, so that the above considerations can be applied.

At high frequencies, where $\nabla \times \mathbf{E} = -\partial \mathbf{B}/\partial t$ is high, significant electric fields
curl about the changing magnetic field. In relatively good conductors, such as iron,
$J = \sigma E$, so that the curling electric field is accompanied by currents that curl like
eddies in a stream. These *eddy currents* cause energy loss through ohmic heating,
which can be damaging to the component. To reduce these currents, materials with
very low values of σ are used, while still maintaining a high value of μ. The class of
materials called *ferrites* satisfy these criteria. In some ferrites σ may be smaller than
in iron by more than ten orders of magnitude.

The energy necessary to establish a current I in the solenoid can be found from
Eq. (5.20a):

$$\frac{dW}{dt} = IV = IL\frac{dI}{dt} = \frac{d}{dt}\left(\frac{1}{2}LI^2\right) \tag{5.21a}$$

giving, for the stored magnetic energy in an inductor,

$$W = \frac{1}{2}LI^2 \tag{5.21b}$$

Comparing Eqs. (5.17b) and (5.21b) gives the same contrast in the magnetic case as in the electric case, that of associating the energy with either the source currents or the magnetic fields in each volume.

Example 5.6

The coaxial cable shown in Fig. 5.7 carries equal and opposite currents along the inner and outer conducting cylinders. Determine the self-inductance per meter of cable.

In the region between conductors, the inner current generates a magnetic field given by Eq. (4.7):

$$\mathbf{H} = \mathbf{a}_\phi \frac{I}{2\pi\rho}$$

The flux density due to this, $\mathbf{B} = \mu\mathbf{H}$, crosses the circuit composed of the two coaxial elements. To find the resulting self-inductance, take the length ℓ, shown on the figure, and find the flux crossing the area it demarks:

$$\Phi = \int \mathbf{B} \cdot \mathbf{n}\, dS = \int_{z=0}^{\ell} \int_{\rho=a}^{b} \mu \frac{I}{2\pi\rho} \mathbf{a}_\phi \cdot \mathbf{a}_\phi\, d\rho\, dz$$

where $dS = d\rho\, dz$, and $\mathbf{n} = \mathbf{a}_\phi$. As there is only a single intersecting area, $N = 1$ in Eq. (5.20b), giving

$$L = \frac{\mu\ell}{2\pi} \ln \frac{b}{a}$$

and the coefficient of self-induction, per unit length is

$$\frac{L}{\ell} = \frac{\mu}{2\pi} \ln \frac{b}{a} \qquad \text{(h/m)} \tag{5.22}$$

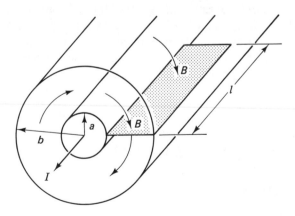

Figure 5.7   Coaxial inductor.

5.8 CONSERVATION OF CHARGE

Continuing the consideration of time variation, consider the full expression of Maxwell's equation for curl **H**:

$$\nabla \times \mathbf{H} = \mathbf{J} + \frac{\partial \mathbf{D}}{\partial t} \tag{5.23}$$

One implication of this can be examined by recalling that the electric flux density **D** diverges from the charge density: $\nabla \cdot \mathbf{D} = \rho$. Therefore, taking the divergence of Eq. (5.23) and using the theorem that, for any vector, the divergence of the curl is identically zero, so that $\nabla \cdot (\nabla \times \mathbf{H}) = 0$,

$$0 = \nabla \cdot \mathbf{J} + \frac{\partial}{\partial t}(\nabla \cdot \mathbf{D}) \tag{5.24}$$

The last term has interchanged the order of the derivatives with respect to time and space variables. Since $\nabla \cdot \mathbf{D} = \rho$, this result becomes

$$\nabla \cdot \mathbf{J} = -\frac{\partial \rho}{\partial t} \tag{5.25}$$

The interpretation of this equation is found by integrating it over a volume:

$$\int \nabla \cdot \mathbf{J}\, dv = -\frac{\partial}{\partial t} \int \rho\, dv \tag{5.26a}$$

$$\oint \mathbf{J} \cdot \mathbf{n}\, dS = -\frac{\partial Q}{\partial t} \tag{5.26b}$$

The integral of Eq. (5.26b) follows from the first integral of Eq. (5.26a) by application of the divergence theorem. In the second integral of Eq. (5.26a), the time derivative and the space integration have been interchanged. This is allowed if the volume geometry changes slowly compared to Q and **D**. Then $\int \rho\, dv = Q$, the total charge within the volume enclosed by the surface S.

Since the left-hand side of Eq. (5.26b) is the total current crossing that surface, the right-hand side indicates that this current, that is, the charge per second, must derive from the change of charge within the volume. The minus sign is necessary to correct for the convention that when current crossing the surface is positive (current out of the volume), Q within the volume must be decreasing, so that dQ/dt is negative. Equation (5.25) then is the law of conservation of charge, stating that whatever current crosses a surface must be equal to the rate of change of charge in the enclosed volume.

Equation (5.23) states that **H** curls about both a real current (density = **J**) and a changing electric flux. From the form of Eq. (5.23), the last term is seen to act like a current density. In Maxwell's original development, this term came from a displacement of his mechanical analog of the electric field, and he called it *displacement current*. Later, it was thought to arise from the electrically induced movement of charge. In a dielectric, this was taken as displacement of the polarization, and, in a vacuum, its origin was conjectured to be a displacement in the aether (the hypothetical medium that was supposed to carry the electric and magnetic fields). In spite of the fact that these rationales are no longer necessary, the term $\partial \mathbf{D}/\partial t = \mathbf{J}_D$ is still called the *displacement current* and $\mathbf{D}$ is sometimes referred to as the *electric displacement*. Because of this, Eq. (5.23) is frequently written

$$\nabla \times \mathbf{H} = \mathbf{J} + \mathbf{J}_D \tag{5.27}$$

At the time that Maxwell developed his theory of electrodynamics, only the static form of Eq. (5.23) was known, $\nabla \times \mathbf{H} = \mathbf{J}$. Maxwell found that consistency of field description could only be obtained by adding the displacement current. This addition imparted an important symmetry to the electrodynamic equations and, as we will see in Chapter 9, led directly to the prediction of electromagnetic propagating waves.

5.9 POYNTING'S THEOREM

It was pointed out earlier, in connection with Eq. (5.8), that when electric and magnetic fields change with time, there is no longer conservation of electric energy or magnetic energy. Where both fields coexist, the total energy density is the sum of the separate contributions:

$$U = U_M + U_E = \frac{1}{2}(\mu H^2 + \varepsilon E^2) \tag{5.28}$$

As the fields change in time, the rate of change of U is

$$\frac{\partial U}{\partial t} = \varepsilon \mathbf{E} \cdot \frac{\partial \mathbf{E}}{\partial t} + \mu \mathbf{H} \cdot \frac{\partial \mathbf{H}}{\partial t} = \mathbf{E} \cdot \frac{\partial \mathbf{D}}{\partial t} + \mathbf{H} \cdot \frac{\partial \mathbf{B}}{\partial t}$$

$$= \mathbf{E} \cdot \nabla \times \mathbf{H} - \mathbf{E} \cdot \mathbf{J} - \mathbf{H} \cdot \nabla \times \mathbf{E} \tag{5.29a}$$

or

$$-\frac{\partial U}{\partial t} = \nabla \cdot (\mathbf{E} \times \mathbf{H}) + \mathbf{E} \cdot \mathbf{J} \tag{5.29b}$$

where Maxwell's curl equations have been used along with the vector identity

$$\nabla \cdot (\mathbf{E} \times \mathbf{H}) = \mathbf{H} \cdot \nabla \times \mathbf{E} - \mathbf{E} \cdot \nabla \times \mathbf{H}$$

Equation (5.29b) is called *Poynting's theorem;* its interpretation can be given if we multiply by dv, an infinitesimal element of volume, and integrate over a volume containing the fields. Applying the divergence theorem to the first term on the right-hand side yields

$$-\frac{\partial}{\partial t} \int U \, dv = \oint \mathscr{P} \cdot \mathbf{n} \, dS + \int \mathbf{E} \cdot \mathbf{J} \, dv \tag{5.30a}$$

where

$$\mathscr{P} = \mathbf{E} \times \mathbf{H} \qquad\qquad (5.30\text{b})$$

is known as Poynting's vector, and the integral of $\mathscr{P}$ is over the surface enclosing the volume v.

The integral on the left-hand side of Eq. (5.30a) is the total energy within the volume. To be specific, assume that this energy is decreasing. Then the time derivative is negative and its preceding minus sign causes the left-hand side to be positive. If the field energy is decreasing, the energy must be going someplace. If, for the moment, we ignore the integral of $\mathbf{E} \cdot \mathbf{J}$, then the energy movement is expressed by the surface integral of Poynting's vector $\mathscr{P}$.

From its units, $\mathscr{P} \sim EH \sim (\text{V/m})(\text{A/m}) = \text{watts/meter}^2 = \text{power/area}$, so that $\mathscr{P}$ is the energy that crosses a unit area each second. The integral of $\mathscr{P}$ gives the total energy/second crossing the surface that encloses the volume within which the energy is decreasing. Thus, the energy (and at the specified rate) that leaves the volume must cross the surface. The exception to this equality is the $\mathbf{E} \cdot \mathbf{J}$ term, which we show is equal to the ohmic power.

For the simple conductor in Fig. 5.8, this term is

$$\int \mathbf{E} \cdot \mathbf{J}\, dv = \int J^2/\sigma\, dv = (I/A)^2(A\ell/\sigma) = I^2(\ell/\sigma A) = I^2 R \qquad (5.31)$$

where the component resistance $R = \ell/\sigma A$. The $\mathbf{E} \cdot \mathbf{J}$ integral gives the loss of electrical energy from the field, by conversion to heat in the conductor. Equation (5.28a), therefore, states that the loss of field energy from within the volume equals the field energy crossing the surface plus that part which stays in the volume but is converted to heat. Poynting's theorem is a statement of the law of conservation of energy of the combined electric and magnetic fields. It is important to note two points. (1) Poynting's theorem is an electromagnetic result that describes the change and flow of electromagnetic energy. The term $\mathbf{E} \cdot \mathbf{J}$ merely describes the fact that a part of the energy is no longer in electromagnetic form; in the illustration, we were able to show that it has been converted to heat. (2) Neither the electric energy nor the magnetic energy is conserved solely; it is their joint energy that is of concern. We saw earlier that the electrostatic relation $\nabla \times \mathbf{E} = 0$ was equivalent to a statement of the conservation of electrostatic energy. In the time-varying case, where $\nabla \times \mathbf{E} = -\partial\mathbf{B}/\partial t$, this is no longer true and Poynting's theorem gives the correct statement.

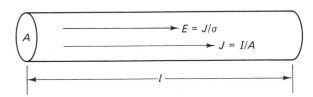

Figure 5.8   Electrical conductor.

5.10 NORMAL BOUNDARY CONDITIONS

In Chapters 3 and 4 and in this chapter, we examined the fields generated by charge
and current distributions, and considered the effects of those fields when they exist
in polarizable and magnetizable matter. No consideration has been given to how
the fields change when they pass between different media, and since they are
usually generated in air but used to influence material bodies, this is a matter of
some importance.

Figure 5.9 shows an interface between medium 1 and medium 2. We assume
our view is so greatly enlarged that the fields do not change within the infinitesimal
region shown, although they can change from point to point on a more macroscopic
scale. On this enlarged scale, the interface is smooth and flat; this is the classical approximation that neglects atomic irregularities. Straddling the interface is a hypothetical closed cylindrical surface on which to apply Gauss' law:

$$\oint \mathbf{D} \cdot \mathbf{n}\, dS = \int \nabla \cdot \mathbf{D}\, dv = \int \rho\, dv = Q_{\text{enclosed}} \tag{5.32}$$

To study the field behavior precisely *at* the interface, the box thickness t will eventually go to zero.

The figure shows positively directed vectors $\mathbf{D}_1$ and $\mathbf{D}_2$ in the two regions.
Note that $\mathbf{D}_2 \cdot \mathbf{n}_2 = D_{2n}$, the normal component of D_2, and $\mathbf{D}_1 \cdot \mathbf{n}_1 = -D_{1n}$, which
is negative because n_1, the outward normal from the enclosed volume, is oppositely
directed from D_{1n}. The surface integral is then

$$\oint \mathbf{D} \cdot \mathbf{n}\, dS = \int -D_{1n}\, dS_1 + \int D_{2n}\, dS_2 + \int \mathbf{D} \cdot \mathbf{n}_3\, dS_3 \tag{5.33}$$

where S_3 is the side cylindrical area. In the limit when t goes to zero, this area and
the last integral vanish. The scale of Fig. 5.9 is so small that $\mathbf{D}$ does not change if

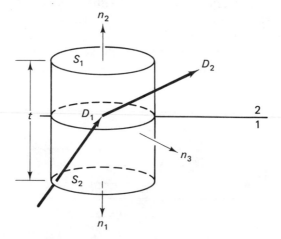

Figure 5.9 Gaussian pillbox at an
interface.

one moves parallel to the surface within the figure, so that D_{1n} and D_{2n} are constant over the areas $S_1 = S_2 = S$, and can be removed from their integrals, giving

$$\oint \mathbf{D} \cdot \mathbf{n}\, dS = D_{2n} S_2 - D_{1n} S_1 = (D_{2n} - D_{1n})S \tag{5.34}$$

When $t \to 0$, the enclosed volume vanishes, so that any volume charge densities will not contribute to Q_{enclosed}, leaving only the free charge Q_s, which lies *on* the surface. Combining Eqs. (5.32) and (5.34) yields

$$D_{2n} - D_{1n} = Q_s/S = \rho_s \tag{5.35}$$

Thus, in crossing between two regions, the normal component of $\mathbf{D}$ changes by an amount equal to the surface density of free charge on the boundary.

Example 5.7

It was pointed out in Chapter 4 that the application of an electric field to a conductor results in a steady state in which free charge has moved to the conductor surface and the internal field is zero. The induced surface-charge density can be found from Eq. (5.35) by taking $D_{\text{internal}} = D_1 = 0$, so that

$$\rho_s = D_{2n} \quad \text{(conductor)}$$

where this $\mathbf{D}$ is the external field. Note that since the internal field is zero, and since the next section will show that $E_{t2} = E_{t1}$, there is no tangential field component, so an external field is entirely normal to a conductor surface.

Equation (5.35) derives from Maxwell's equation $\nabla \cdot \mathbf{D} = \rho$. Using vector $\mathbf{B}$, we can proceed with the same derivation except that now $\nabla \cdot \mathbf{B} = 0$. Since there is no charge density term, we set $\rho = 0$ in Eqs. (5.32) and (5.35), so that the normal component of the magnetic flux density is the same on both sides of the interface:

$$B_{2n} = B_{1n} \tag{5.36}$$

In this regard, note that if there is no free electric charge density on the surface between two regions, then $D_{2n} = D_{1n}$ also.

5.11 TANGENTIAL BOUNDARY CONDITIONS

Figure 5.10 shows the same region of the material interface as in Fig. 5.9, but now with a planar closed path in place of the pillbox. Along this path, we perform the line integral of $\mathbf{H}$, in Ampère's law, using Stokes' theorem and Maxwell's relation:

$$\oint \mathbf{H} \cdot \boldsymbol{\tau}\, d\ell = \int \mathbf{n} \cdot \nabla \times \mathbf{H}\, dS = \int \mathbf{n} \cdot \left(\mathbf{J} + \frac{\partial \mathbf{D}}{\partial t} \right) dS = I + I_D$$

$$\tag{5.37a}$$

where

$$I = \int \mathbf{n} \cdot \mathbf{J}\, dS \quad \text{is the real current crossing the area} \tag{5.37b}$$

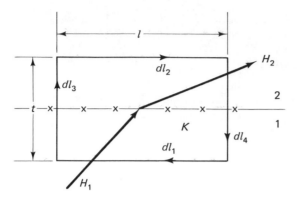

Figure 5.10 Amperian path at an interface.

$$I_D = \int \mathbf{n} \cdot \frac{\partial \mathbf{D}}{\partial t} \, dS \quad \text{is the corresponding displacement current}$$

$$(5.37c)$$

Note that $\mathbf{H}_2 \cdot \boldsymbol{\tau}_2 = H_{2t}$, the tangential component of H_2, and $\mathbf{H}_1 \cdot \boldsymbol{\tau}_1 = -H_{1t}$, the tangential component of H_1. This last term is negative because $\boldsymbol{\tau}_1$ is opposite from the positively directed H_{1t}. Therefore,

$$\oint \mathbf{H} \cdot \boldsymbol{\tau} \, d\ell = \int -H_{1t} d\ell_1 + \int H_{2t} d\ell_2 + \int \mathbf{H} \cdot d\ell_3 + \int \mathbf{H} \cdot d\ell_4$$

$$(5.38)$$

In the limit that $t \to 0$, the path lengths ℓ_3 and ℓ_4 vanish and so do the last two integrals. As before, $\mathbf{H}_1$ and $\mathbf{H}_2$ are constant over the small scale of ℓ, so that

$$\oint \mathbf{H} \cdot \boldsymbol{\tau} \, d\ell = -H_{1t} \int d\ell_1 + H_{2t} \int d\ell_2 = (H_{2t} - H_{1t})\ell \qquad (5.39)$$

When t becomes infinitesimal, so do the area integrals of Eqs. (5.37). The only contribution that may remain occurs in the event that a current flows *on the interface,* through the path. If K amperes of this real surface current are normal to the area, per meter of its width, then $I_{\text{surface}} = K\ell$ and

$$H_{2t} - H_{1t} = I_s/\ell = K \qquad (5.40)$$

This result follows from Maxwell's equation, $\nabla \times \mathbf{H} = \mathbf{J} + \partial \mathbf{D}/\partial t$. If we had used the vector $\mathbf{E}$ rather than $\mathbf{H}$, Maxwell's equation, $\nabla \times \mathbf{E} = -\partial \mathbf{B}/\partial t$ has no term for $\mathbf{J}$ and replaces I_D with the area integral of $\partial \mathbf{B}/\partial t$. Again, when $t \to 0$, this integral vanishes, leaving

$$E_{2t} = E_{1t} \qquad (5.41)$$

so the tangential component of the electric field strength is continuous across an interface.

Example 5.8

A magnetic field is incident from air onto the surface of a ferromagnetic body with $\mu_r = 100$; see Fig. 5.11. Find θ, the minimum normal incident angle from air, such that $\phi > 80°$, where ϕ is the field angle to the surface normal in the magnetic material.

Since there is no real current on the surface, $K = 0$ and the boundary conditions on the fields are

$$H_{2t} = H_{1t} \qquad \frac{B_2}{\mu_2} \sin \phi = \frac{B_1}{\mu_1} \sin \theta$$

$$B_{2n} = B_{1n} \qquad B_2 \cos \phi = B_1 \cos \theta$$

Dividing gives

$$\tan \theta = \frac{\mu_1}{\mu_2} \tan \phi = \frac{\tan \phi}{\mu_r}$$

For $\phi > 80°$, $\tan \phi > 5.67$ and $\theta > 3.25°$. $\mu_r = 100$ is not an exceptional value; materials are available that have μ_r values of several hundreds of thousands. Thus, the magnetic field within ferromagnetic material is nearly parallel to the surface for practically all incident angles.

This example explains the mechanism of magnetic shielding. Modern electronic equipment is often extremely sensitive to the presence of dc and low-frequency magnetic fields, and such fields exist in almost every industrial, commercial, laboratory, and home environment where high currents are flowing in nearby power-transmission systems or in local conduit in building walls. If the volume of concern is surrounded by a shield of highly permeable magnetic material, the fields enter that material and, remaining tangent to its surface, are conducted around the volume instead of passing through it. This greatly reduces the flux density, and magnetic field strength, within the volume. Whereas the exact geometry of the shield and the ambient field strength are important in determining shielding effectiveness,

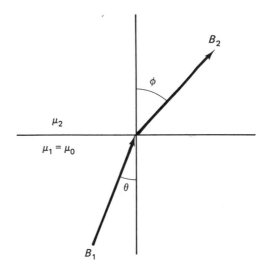

Figure 5.11   Magnetic field angles at an interface.

it is frequently possible to obtain values of $\mu \geq 10^5$, so that field reductions are obtainable by factors of many hundreds. This is discussed in more detail in Chapter 7.

EXERCISES

Extensions of the Analysis

5.1. From the discussion of Exercise 4.3, it follows that when **B** changes, so does **A**. Use the differential form of Faraday's law and the fact that the curl of any gradient is zero to derive the relation

$$\mathbf{E} = -\nabla V - \frac{\partial \mathbf{A}}{\partial t}$$

where V is a scalar function.

5.2. Two coils with N_1 and N_2 turns, respectively, are wound on a common form so that they are effectively superimposed though remaining electrically distinct. The fluxes from the currents in each coil cut both coil circuits, inducing voltages V_1 and V_2. Show that the relation $V_2/V_1 = N_2/N_1$ always applies. This is the transformer equation and the construction described constitutes an electrical transformer.

*5.3. Faraday's law, that a changing magnetic flux induces an emf in a circuit through which it passes, has been used to define the induction of a circuit to itself: its self-induction L. If the changing flux lines from a circuit thread a second circuit, then an emf is induced in this second circuit by means of mutual induction, and the circuit law of induction can be written as $V_2 = -M \, dI_1/dt$, where M is the coefficient of mutual induction.

A long circular-cylindrical air-filled coil of radius r_1 has n_1 turns per meter of length. A second air-filled coil of radius $r_2 < r_1$, with n_2 turns per meter, is inserted coaxially into the first coil. Determine the coefficient of mutual induction per meter of length.

5.4. A ferromagnetic slug with a cross-sectional area of 0.01 m² is wrapped with seven turns of wire, as shown in Fig. 5.12(a). The wrapped slug is then placed in a region where the longitudinal magnetic field has the sawtooth variation shown in Fig. 5.12(b). The magnetic behavior of the material is shown in Fig. 5.12(c).

Carefully draw a diagram showing the voltage induced in the coil over two cycles of the sawtooth wave.

5.5. **(a)** A long solenoidal coil with n turns/meter is filled with material for which $M = \chi H$, where $\chi \gg 1$. Find L, the coefficient of self-inductance per meter.

(b) Rather than being a constant, the material in part (a) has

$$\chi = \chi_0 e^{-H/H_0}$$

where $\chi_0 \gg 1$ and H_0 and χ_0 are constants. Sketch L as a function of the current in the coil. Discuss what is meant by a *nonlinear inductance*.

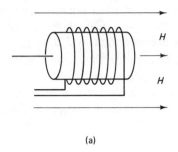

(a)

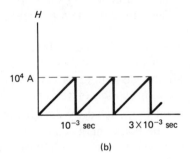

(b)

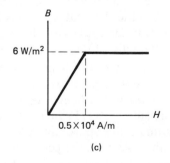

(c)

Figure 5.12   For Exercise 5.4.

*5.6. The energy expended by a force F moving through a displacement $d\ell$ is $dW = F \, d\ell$. This can be generalized to define the thermodynamic force holding a system in equilibrium against a change $d\ell$ as $F = -dW/d\ell$, where W is the energy of the system.

Consider an electromagnet that has pole faces 10 cm in diameter and separated by 10 cm. Calculate the force between the poles if the flux density in the air between them is 2.5 T.

5.7. Due to a momentary fluctuation of electron motion, a volume charge density ρ_0 suddenly appears in a small region of a good conductor. The charge components repel each other and the density rapidly dissipates. Starting from the law of conservation of charge (in differential form), show that $\rho = \rho_0 e^{-t/\tau}$, where $\tau = \varepsilon/\sigma$.

Problems

*5.8. (a) Write an expression for the electrostatic energy density at every point of space due to a charge Q on a spherical surface of radius r_0 about the origin. Determine W, the *total* field energy of the charge, that is, over *all* space.

(b) Calculate the work done to bring another charge Q from infinity to r_0 in the field of the original charge.

(c) Why are the answers to parts (a) and (b) different? Check more advanced texts for a discussion of *self-energy*, which always involves a factor $\frac{1}{2}$.

5.9. If the electron is modeled by the charge distribution of Problem 5.8 then it cannot be a true point charge since $W \to \infty$ when $r_0 \to 0$. If we take its total field energy to be equal to its relativistic mass-energy mc^2, then determine the electron radius.

*5.10. A parallel-plate air capacitor with a plate separation of 2 mm is connected to a 10-volt source.

(a) Find E and D in the capacitor.

(b) The source circuit is now opened, trapping the charge on the plates, and a 2-mm dielectric slab with $\chi_e = 3$ is inserted between the plates. Now find D and E between the plates, that is, in the dielectric.

(c) What is the ratio of the capacitances with and without the dielectric?

*5.11. (a) Find the capacitance of a spherical air-dielectric capacitor that has an inner electrode with radius r_1 and an outer electrode with radius $r_2 > r_1$.

(b) *The capacitance of a body* is defined as its capacitance with all other relatively distant objects. Estimate this capacitance by letting $r_2 \to \infty$ in part (a). Estimate the approximate numerical value for an average person (with units). Assume the person has the density of salt water.

5.12. A parallel-plate air capacitor has its plates 2 mm apart and has a plate area of $72\pi \text{ cm}^2$. It is connected to a 100-V battery and the plates are then separated until they are 1 cm apart. Discuss the change of **D**, **E**, C, the energy density, and the total energy in the capacitor. Neglect fringing.

5.13. Answer Problem 5.12 if the battery is disconnected before separating the plates.

*5.14. Find the capacitance for Problems 3.31 and 3.34.

5.15. Two identical parallel-plate capacitors C are each charged to the same difference of potential of V_0 volts and disconnected from the battery.
   (a) Find the charge and energy stored in each capacitor.
   (b) The positive plate of each capacitor is now connected to the negative plate of the other through resistanceless wires. Now find the charge and energy stored in each capacitor. In view of this result, discuss the nature of the law of conservation of energy.

5.16. Two concentric cylindrical conductors have the space between them filled by two dielectrics, as shown in Fig. 5.13. Equal and opposite charges of 7 coulombs per meter of length are placed on the conductors. Find D, E, and P in each medium and calculate the capacitance per meter length. For the conductors, $a = 0.74$ m, $b = 1.00$ m, $c = 1.65$ m, $\varepsilon_1 = 3\varepsilon_0$, and $\varepsilon_2 = 5\varepsilon_0$.

5.17. A parallel-plate air capacitor has circular plates of 3-cm radius and 1-cm separation. It is charged from a source whose voltage varies as $V = 100(1 - e^{-t/100})$. Find the magnetic field at a point that is in the center plane of the capacitor and that is 6 cm from its axis when $t = 0$, 100, and 10^{10} seconds. *Hint:* Use Maxwell's displacement equation and Ampère's law.

5.18 A very long wire carries a current I_0 along the positive y axis. The medium surrounding the wire has $\mu = \mu_0 (1 + \rho/a)$, where ρ is the distance from the wire. A conducting U frame is placed near the wire in the xy plane, as shown in the Fig 5.14, and a sliding bar is moved along the frame with uniform velocity v. Determine the current i in the frame if the bar has resistance R, as the bar moves from $x = a$. Indicate the current direction. *Note: i* will be a function of time.

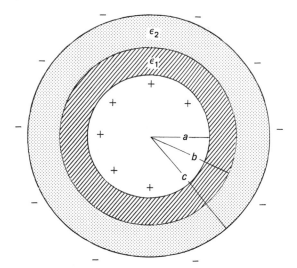

Figure 5.13 For Exercise 5.16.

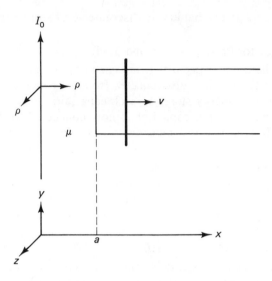

Figure 5.14   For Exercise 5.18.

*5.19. A plasma discharge carries a current of density

$$\mathbf{J} = \mathbf{a}_z 100(1 - \rho) \text{ A/m}^2 \qquad \rho \le 1$$
$$= 0 \qquad\qquad\qquad\qquad \rho > 1$$

A square wire loop is placed as shown in Fig. 5.15. The coil is turned 90°
about the y axis in 1 microsecond.
(a) Calculate the flux through the coil oriented as shown.
(b) Find the voltage across the ends of the coil when it is rotated.
(c) Indicate the polarity of ends a and b. Does it matter in which direction the
coil is turned?

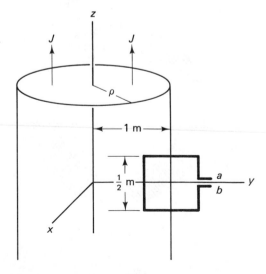

Figure 5.15   For Exercise 5.19.

5.20. Faraday's law states that the induced voltage in a circuit is equal to the rate of change of magnetic flux through the area of that circuit. In this chapter, we have primarily considered the case where the flux change is due to a changing field. It is also possible to change the flux by changing the geometry of the circuit, since $\Phi = \int \mathbf{B} \cdot \mathbf{n}\,dS$. (For example, in Problem 5.18, the geometry is changed by rotating the loop so as to change the area it presents to the flux.) Consider the following:

(a) A conducting ring is painted around the circumference of a balloon with a gap equal to $1/1000$ of the circumference. The balloon is inflated such that its volume increases at a constant rate, $\alpha = dv/dt$, and with the ring area normal to a constant magnetic field B. Since the area through which the flux passes is growing, so must the total flux through the ring. From this, write an expression for the time-dependence of the electric field intensity in the gap.

(b) A movable rod with a weight of 1000 newtons rests on a horizontal track with a coefficient of static friction of 0.2, as shown in Fig. 5.16. The rails and leads have no electrical resistance, but the bar has $0.3\ \Omega$ resistance. Find the field magnitude and direction to initiate motion of the rod toward the battery, which has $V = 30$ volts.

*5.21. Two coplanar loops are shown in Fig. 5.17. As I_1 increases, in what direction does I_2 flow (clockwise or counterclockwise)? Explain.

*5.22. A uniform sheet current of 0.125 A/m flows from $z = -\infty$ to $z = +\infty$ on the cylindrical surface with $\rho = 4$ cm. A small loop with area 0.01 cm² has its center at $\rho = 1$ m and its axis is normal to z. In 1/10 millisecond, its axis is turned to be along z. Estimate the voltage generated in the loop. *Note:* There are two possible answers.

5.23. A wire, with 1 ohm/km resistance, is wound into a circular loop with 1000 turns and a radius of $1/3$ m. The ends of the wire are separated by a gap of 0.1 mm. A field $B = 30t$ tesla (t in sec) is normal to the loop. Find the voltage across and the electric field intensity in the gap. If the air in the gap breaks down (its resistance $\to 0$), find the current that flows.

*5.24. A very thin wire carries the current $I = I_0 \sin \omega t$ along the x axis from $-\infty$ to $+\infty$. A second wire carries the same current from $-\infty$ to $+\infty$ along the y axis. A square loop has its sides parallel to the x and y axes and has two of its corners at $(x, y) = (3, 1)$ and $(5, 3)$. Find the emf induced in the loop.

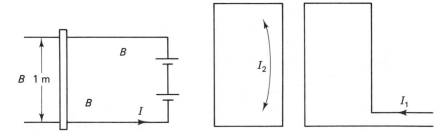

Figure 5.16 For Exercise 5.20.   **Figure 5.17** For Exercise 5.21.

*5.25. A 5-turn coil is wrapped around a 10-cm-radius circular cylindrical bar magnet with $M = 10^6$ A/m, as shown in Fig. 5.18.
   (a) Find the average emf generated in the coil if the magnet is pulled out completely in 1 millisecond.
   (b) Recognizing that this system acts like a generator, which of the terminals is positive with respect to the other?

5.26. In the circuit shown in Fig. 5.19, the end wires each have a resistance of 3 ohms; the resistances of the horizontal wires are negligible. If the magnetic field passing through the circuit is $\mathbf{B} = 2 \cos(60\pi t - 2\pi x)\mathbf{a}_z$ tesla, find the current in the circuit and the voltage across a 3-ohm resistor.

5.27. A rectangular wire loop with a resistance of 0.02 ohm rotates about the side of the loop that lies along the z axis, as shown in Fig. 5.20. There is a constant

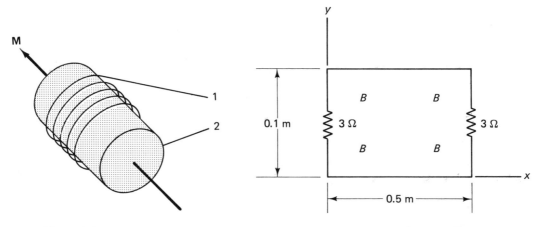

Figure 5.18 For Exercise 5.25.   **Figure 5.19** For Exercise 5.26.

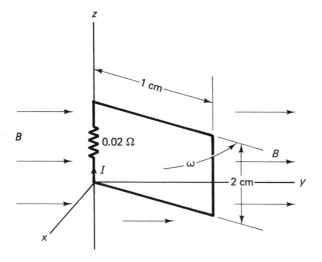

Figure 5.20 For Exercise 5.27.

magnetic field, $\mathbf{B} = 0.01\mathbf{a}_y$ T, and the angular rotation speed is $\omega = 2$ rad/sec. Write an expression for the induced current, with the direction shown, if the loop lies in the xz plane at $t = 0$.

*5.28. Two identical coils AB and CD are wound on the hollow form shown in Fig. 5.21. Each has N turns and a cross-sectional area A. An axial field $H = H_0 \cos\omega t$ is applied.

(a) Find the voltage across terminals 1–2 if the windings are connected, in order: (i) 1–AB–CD–2, and (ii) 1–AB–DC–2.

(b) An iron slug of area A is inserted in coil AB. The iron has $\mu = 10\mu_0$. What are the voltages across terminals 1–2 for the two configurations of part (a)?

5.29. Radio-frequency interference (RFI) is a phenomenon whereby radiation from a source is unintentionally detected by another circuit and processed as though it were part of the intended signal, thereby leading to errors, misinterpretations, and failures. This problem illustrates an analysis of this behavior.

An omnidirectional antenna is one that radiates equally in all directions. If a small antenna of this sort radiates 50 watts, its magnetic field intensity at a distance of 200 m is approximately 5×10^{-4} A/m. If the magnetic field from this radiation is normal to a circuit board on which are printed two parallel lines 1 cm apart and 10 cm long, determine the emf induced in this circuit if $f = 10^8$ Hz.

5.30. A long current-carrying loop has its near sides separated by $4s$ and its far ends are so distant that their effects are negligible. A small square loop, of side $2s$ and carrying a current I, is centered in the large loop, as shown in Fig. 5.22(a).

(a) Calculate the coefficient of mutual induction between the circuits. *Hint:* See Problem 4.13 and note that $M_{12} = M_{21}$.

(b) A magnetic slug, with permeability $\mu_r = 10,000$ and dimensions $2s \times s/8$, is centered in the small loop, as shown in Fig. 5.22(b). Estimate the new mutual inductance. *Hint:* Assume $s/16 \ll s$).

(c) Repeat part (b) if the slug is turned, as shown in Fig. 5.22(c).

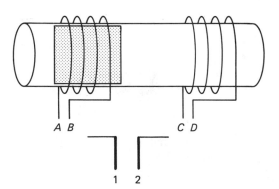

Figure 5.21   For Exercise 5.28.   1   2

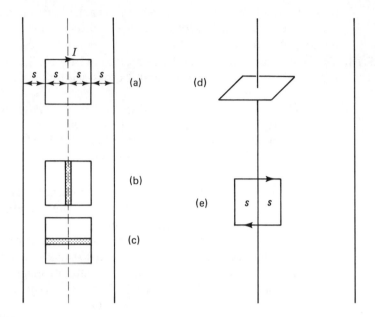

Figure 5.25   For Exercises 5.30 and 5.31.

*5.31. Repeat part (a) of Problem 5.30 if the empty loop is placed normal to and centered about one of the long wires, as shown in Fig. 5.22(d). Repeat for the empty loop still centered on the loop wire, but now in the plane of the large loop, as shown in Fig. 5.22(e).

*5.32. Current $I = I_0 \cos \omega t$ flows in the toroidal coil of Fig. 5.23. Find its coefficient of mutual induction with loops 1, 2, 3, and 4.

5.33. Two very long parallel pairs of transmission lines run as shown in Fig. 5.24. Find the mutual induction of a 1-m length of the lines.

5.34. An air-dielectric parallel-plate capacitor with circular plates of radius a and separation s is connected to a voltage source $V = V_0 \sin \omega t$. Neglect fringing.
   (a) Find **E** and **D** in the capacitor.
   (b) Find the magnetic field strength in the capacitor at distance a from its axis. *Hint:* Use Ampère's law.
   (c) Calculate Poynting's vector at $\rho = a$.
   (d) Discuss whether the interelectrode region radiates or absorbs energy.

5.35. Calculate Poynting's vector at the surface of the resistive element of Fig. 5.8 (use the fact that the tangential component of E is continuous across a material interface) and show that the power entering a length of the element is equal to the ohmic loss in that length.

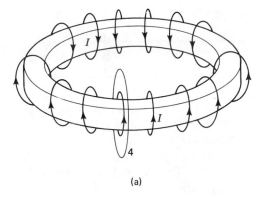

(a)

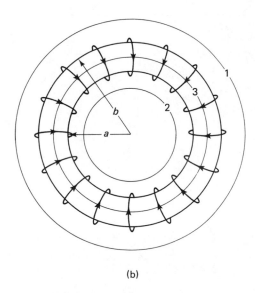

(b)

Figure 5.23   For Exercise 5.32.

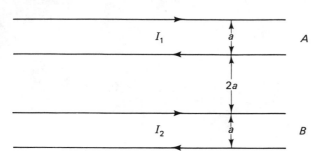

Figure 5.24 For Exercise 5.33.

5.36. An electric field is incident from air, at an angle of 30°, onto a dielectric with $\varepsilon_1 = \sqrt{3}\varepsilon_0$, as shown in Fig. 5.25.
   (a) Determine the refraction angle, ϕ if the surface is uncharged.
   (b) A free charge is placed on the dielectric surface and it is found that $\phi = 76°$. The air field magnitude is $E_0 = 1.54$ V/m. What is the surface charge density?

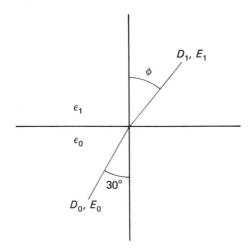

Figure 5.25 For Exercise 5.36.

6

TECHNIQUES IN ELECTROSTATICS

The major conceptual features of electrostatic fields were discussed in Chapters 3 and 5, including their symmetry properties, their energies, the forces they exert, and their interactions with matter. And using Gauss' law, we were able to calculate the fields from charge distributions that have a significant amount of symmetry. However, many problems arise in which the sources are unknown or symmetry is not sufficient to allow solution from Gauss' law alone, so that other techniques are necessary to carry out detailed calculations in these cases.

6.1 CALCULATIONS FROM CHARGE

The electric field strength and the electric potential of a point charge Q at the origin have already been shown to be

$$\mathbf{E} = \mathbf{a}_r \frac{Q}{4\pi\varepsilon r^2} \qquad V = \frac{Q}{4\pi\varepsilon r} \qquad \mathbf{E} = -\nabla V \qquad (6.1)$$

For a continuous distribution of charge, Eqs. (6.1) can be applied to each infinitesimal charge element dQ by taking the equivalence $Q_{\text{point charge}} \rightarrow dQ$, where

$$dQ_v = \rho_v \, dv \qquad dQ_s = \rho_s \, dS \qquad dQ_\ell = \rho_\ell \, d\ell \qquad (6.2)$$

Here ρ_v (frequently written without the subscript), ρ_s, ρ_ℓ are the volume, surface, and line charge densities, respectively. The effect of all the charge of a distribution is then found by integrating over the entire distribution. This procedure is illustrated in the following examples.

Example 6.1

Calculate the electric field of a charge Q distributed uniformly on a straight line of length $2L$. Compare the result with the field from an infinite line charge.

Figure 6.1(a) shows the finite line charge along the z axis, and the field point (ρ, z). As this example has axial symmetry, we employ cylindrical coordinates. Also, since the total charge is given, we have a linear charge density $\rho_\ell = Q/2L$. Adding the contributions of elements $dQ = \rho_\ell dz'$, where z' is the dummy integration variable along the wire, gives

$$\mathbf{E} = \int \frac{dQ}{4\pi\varepsilon_0} \frac{\mathbf{r}_0}{r^2} = \frac{\rho_\ell}{4\pi\varepsilon_0} \int_{-L}^{L} \frac{dz'}{r^2} \left(\mathbf{a}_\rho \frac{\rho}{r} + \mathbf{a}_z \frac{z - z'}{r} \right)$$

where $\mathbf{r}_0 = \mathbf{a}_\rho \rho/r + \mathbf{a}_z(z - z')/r$ is the unit vector directed from the source element $d\ell$ to the field point at (ρ, z), and $r = [\rho^2 + (z - z')^2]^{1/2}$. $\mathbf{r}_0$ has been expressed in terms of its radial and axial components (note that because of axial symmetry, there is no component $\mathbf{a}_\phi$), so that the vector integration has been reduced to two scalar integrations. These are straightforward, leading to

$$\mathbf{E}_{\text{line}} = \frac{\rho_\ell}{4\pi\varepsilon_0} \left[\mathbf{a}_\rho \frac{\cos\theta_2 - \cos\theta_1}{\rho} + \mathbf{a}_z \left(\frac{1}{R_1} - \frac{1}{R_2} \right) \right]$$

where the R_i and θ_i are shown in Fig. 6.1(b).

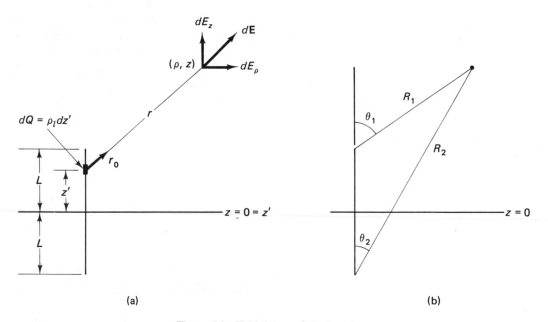

(a)   (b)

Figure 6.1   Field due to a finite line charge.

From Eq. (3.19a), for an infinite line charge,

$$\mathbf{E}_\infty = \mathbf{a}_\rho \frac{\rho_\ell}{2\pi\varepsilon_0\rho}$$

For comparison of $\mathbf{E}_{\text{line}}$ and E_∞, let us take the symmetry plane $z = 0$. Then $R_1 = R_2$, making $E_z = 0$. We also have

$$-\cos\theta_1 = \cos\theta_2 = \cos\theta = \frac{L}{(L^2 + \rho^2)^{1/2}} \qquad z = 0$$

where the θ_i are angles from the line ends to the field point, measured from the positive z axis. Then

$$E_\rho = \frac{\rho_\ell}{2\pi\varepsilon_0\rho}\frac{L}{(L^2 + \rho^2)^{1/2}} = E_\infty \cos\theta$$

This result demonstrates that if we are willing to accept an error in the field calculated, we can use the much simpler infinite-line charge result. For example, we will have an error of 10 percent or less as long as we stay within the region marked by $\theta \leq 24°$. If we can only tolerate a 1-percent error, then we are restricted to $\theta < 8°$. It is valuable to recognize that a real finite line can be treated as being infinite as long as one stays close to its center, and that the difference can be estimated.

Farther from the center of the finite line, departures from the infinite-line case become greater. In the extreme, this can be seen by placing the expression for E_ρ into an alternative form. Since $\rho_\ell = Q/2L$, at a point in the medial plane ($z = 0$),

$$E_\rho = \frac{Q}{4\pi\varepsilon_0\rho^2}\frac{1}{[1 + (L/\rho)^2]^{1/2}}$$

indicating that when ρ becomes large, $E_\rho \rightarrow Q/4\pi\varepsilon_0\rho^2$, the field of a point charge. When $\rho = 2L$, the difference of E_ρ from the point-charge result is 10 percent; when $\rho = 7L$, the error is only 1 percent. Thus, while there is a range over which we cannot use one or another of the simple engineering approximations (i.e., infinite-line or point-charge) for the exact solution, it can be regarded as a relatively narrow range if we are willing to accept approximate answers. Such bracketing by simple known solutions often allows us to quickly establish limits of the field in cases where exact answers are difficult to obtain.

In the previous example, the field components were calculated directly and simply. This was possible because of the simplicity of the problem, for example, there was only one coordinate of integration. In general, the complexity of a three-dimensional integration of the three vector field components becomes extremely awkward even in simple cases. For this reason, we often prefer to find the scalar potential.

Example 6.2

Charge is uniformly distributed in a cylindrical volume of height $2L$ and radius a. Calculate the electric potential on the axis of the cylinder.

In solving this problem, we could set up a triple integral over the charge elements shown in Fig. 6.2. While we will ultimately do the same three integrations let us proceed by solving three simpler partial problems, as shown in Examples 6.2a to 6.2c.

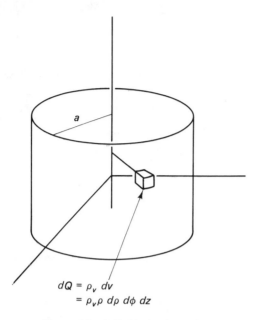

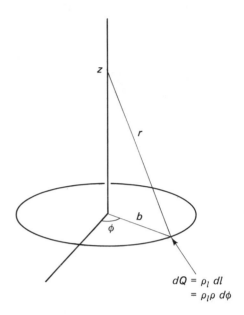

Figure 6.2 Cylindrical volume charge.   **Figure 6.3** Charged ring.

Example 6.2a

Figure 6.3 shows a ring of charge with radius b and with linear charge density ρ_ℓ. Find the potential at points on the axis of the ring.

The magnitude of the ring charge element is $dQ = \rho_\ell d\ell = \rho_\ell b\, d\phi$. Since b and z are constant, so is $r = (b^2 + z^2)^{1/2}$, so these factors can be removed from the integral for V.

$$V_{\text{ring}} = \frac{1}{4\pi\varepsilon_0} \int_0^{2\pi} \frac{\rho_\ell b}{(b^2 + z^2)^{1/2}}\, d\phi = \frac{\rho_\ell b}{2\varepsilon_0} \frac{1}{(b^2 + z^2)^{1/2}}$$

Because this solution is restricted to points on the axis, it is a function of z only. While this can be used to calculate $E_z = -\partial V_{\text{ring}}/\partial z$, it gives no information about an off-axis field component. However, symmetry about the axis dictates that there can be no radial field component at points on the axis, so that the E_z we can calculate is the entire **E** on the axis.

Example 6.2b

What is the potential at points on the axis of the circular plane of charge, with surface charge density ρ_s, shown in Fig. 6.4?

Radius b in the solution of Example 6.2a, must now be regarded as a variable that has values between 0 and a, and which we denote by ρ. The linear ring of charge, with density ρ_ℓ, is actually the thin band of charge lying in the $z = 0$ plane. Since both interpretations must describe the same total charge, $dQ =$ (charge/length) • length must equal $dQ =$ (charge/area) • area. Thus, $\rho_\ell 2\pi\rho = \rho_s 2\pi\rho\, d\rho$, which gives the relation that $\rho_\ell =$

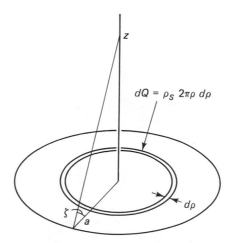

Figure 6.4 Planar charge distribution.

$\rho_s \, d\rho$. This is substituted into the expression for V_{ring} and, since ρ now varies continuously, the result is integrated over all values of ρ:

$$V_{\text{plane}} = \frac{1}{2\varepsilon_0} \int_0^a \frac{\rho_s \rho \, d\rho}{(\rho^2 + z^2)^{1/2}}$$

$$= \frac{\rho_s}{2\varepsilon_0} [(a^2 + z^2)^{1/2} - z]$$

It is informative to check this result in the limit that the plane becomes very large, that is, when $a \to \infty$. In that event, E should equal the field found in Exercise 3.1, for the infinite plane, $\mathbf{E}_\infty = \rho_s/2\varepsilon_0 \mathbf{a}_z$. We must proceed very carefully in moving from the finite-plane result, V_{plane}, to the case of an infinite plane. It would be an error to first take the limit that $a \gg z$, since this gives $V_{\text{plane}} \to \rho_s a/2\varepsilon_0$, from which $E_z = -\partial V_{\text{plane}}/\partial z = 0 \neq \mathbf{E}_\infty$. This discrepancy arises because, when a becomes large, the term $(a^2 + z^2)^{1/2}$ is dominant in the expression for V_{plane} and results in our dropping the term $-z$. However, for large a, $(a^2 + z^2)^{1/2}$ approaches the constant value a, so that the term $-z$ has the largest *rate of change*. Since the field equals the rate of change of potential, this is the most important (i.e., most rapidly changing even though smaller) term. It follows, therefore, that the proper procedure is to *first* find

$$E_z = -\frac{\partial V_{\text{plane}}}{\partial z} = \frac{\rho_s}{2\varepsilon_0} (1 - \sin \zeta)$$

and to *then* take the limit that $a \to \infty$, or $\sin \zeta \to 0$.

Example 6.2c

Find the potential on the axis of the finite cylindrical volume charge shown in Fig. 6.5.

Imagine that the circular area of charge, with surface density ρ_s, found in Example 6.2b, is actually a thin disc section of charge from the cylindrical volume distribution with volume charge density ρ_v. This is shown in Fig. 6.5. Since both interpretations de-

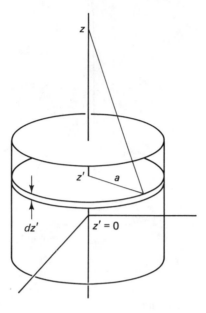

Figure 6.5 Section of cylindrical volume charge.

scribe the same total charge, we have $dQ = \rho_s \pi a^2 = \rho_v \pi a^2 dz'$, from which $\rho_s = \rho_v dz'$. Here the charge location is denoted by z' to distinguish it from the field point z. When substituting ρ_s in the expression for V_{plane}, we must also make the change that $z \rightarrow z - z'$, since we will be moving away from the plane $z' = 0$. Then, integrating over all z' values, we have

$$V = \frac{\rho_v}{2\varepsilon_0} \int_{-L}^{L} \{[(z - z')^2 + a^2]^{1/2} - (z - z')\} dz'$$

The solution and discussion of this equation is considered in a problem at the end of this chapter.

6.2 ELECTRICAL FORCES

Consider a system having electrical charge distributed on various bodies in space. In equilibrium, the sum of the forces on each body, and indeed on each charge, is zero. This means that the electrostatic force on each is balanced by an equal force, for example, of chemical or of mechanical origin.

Example 6.3

The two discs in Fig. 6.6 are separated by 10 cm. Each has a mass of 1 gm and they are resting on a surface with a coefficient of friction $\eta = 0.3$. How much charge can be removed from one disc and placed on the other before the discs move toward each other?

When charge Q has been transferred from one disc to the other, the discs experience an attractive Coulomb force $F_1 = Q^2/4\pi\varepsilon_0 r^2$. The frictional force opposing their motion is $F_2 = \eta mg$, where $g = 9.8$ m/sec^2, so that motion is imminent when these forces are equal:

$$Q = r(4\pi\varepsilon_0 \eta mg)^{1/2} = 0.57 \times 10^{-7} \text{ C}$$

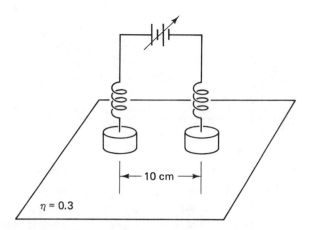

Figure 6.6 Charged discs on a frictional
surface.

In many cases, the nonelectrical forces are ignored because they are great enough to serve as rigid, effectively invariant, constraints. For example, if $Q < 0.57 \times 10^{-7}$ C, the only significant property of the frictional force in Example 6.3 is that it is great enough to eliminate any motion of the charged discs. The electrostatic force causing the motion is $\mathbf{F}_E = Q\mathbf{E} = -Q\nabla V = -\nabla QV = -\nabla W_E$, where W_E is the electric energy of the charge (or body with charge) Q. We see that $\mathbf{F}_E$ is directed so as to reduce the electrical energy.

If the mechanical constraining forces $\mathbf{F}_c$ are also variable, or can be overcome, then motion can take place and the charged bodies will move under the influence of the local field *and* the restraining forces. Variation of $\mathbf{F}_c$ corresponds to a change of mechanical energy of the configuration, that is, $F_c = -\nabla W_m$. The net force on the body or system is then

$$F = F_E + F_c = -\nabla(W_E + W_m) = -\nabla W \tag{6.3}$$

Equilibrium occurs when $F = 0$ or when the total system energy is minimum.

Example 6.4

Two $10 \times 10 \times 1$ cm pieces are cut from different sections of a dielectric material and are found to have $\varepsilon_1 = 2\varepsilon_0$ and $\varepsilon_2 = 4\varepsilon_0$. The material is known to be elastic; it resists a small compression or expansion x with a Hooke's law force (force/volume $= -kx$), where $k = 200$ N/m^4.

The dielectric pieces are placed between parallel plates, as shown in Fig. 6.7, and the capacitor is charged until $Q = 10^{-7}$ C. Determine the equilibrium thickness of the two pieces and find the system capacitance.

This configuration is equivalent to two capacitors in series. For rigid dielectrics, we would have

$$C_1^0 = 2\varepsilon_0 A/t = 17.7 \text{ pF} \quad \text{and} \quad C_2^0 = 4\varepsilon_0 A/t = 35.4 \text{ pF}$$

so that

$$C^0 = \frac{C_1^0 C_2^0}{C_1^0 + C_2^0} = 11.73 \text{ pF}$$

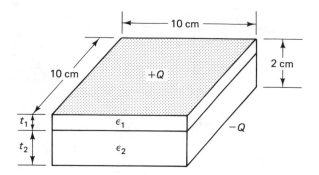

Figure 6.7 Capacitor with compressible dielectrics.

Now, however, we expect that relaxation of the constraint of a fixed dielectric thickness results in a shifted boundary, causing the capacitance to change.

Since $D = \rho_s$ in a parallel-plate capacitor, we have $D = Q/A = 10^{-5}$ C/m^2, which is the same in both media. The energy densities (energies per volume) in the dielectrics are, therefore,

$$U_i = \frac{1}{2} D_i E_i = \frac{D^2}{2\varepsilon_i}$$

so that

$$U_1 = \frac{D^2}{4\varepsilon_0} \quad \text{and} \quad U_2 = \frac{D^2}{8\varepsilon_0}$$

As $U_2 < U_1$, the system energy will be reduced by expanding dielectric 2 and contracting dielectric 1; we therefore expect to find that equilibrium has $t_1 < 1$ cm and $t_2 > 1$ cm. For the greatest reduction of electrostatic energy, we would have $t_1 = 0$ and $t_2 = 2$ cm, but this cannot occur because distortion of the dielectrics, according to Hooke's law, raises the system energy by an amount equal to its mechanical strain energy.

The mechanical energy density is $\mathcal{V}_i = k x_i^2/2$, where x_i is the displacement of the ith medium from its unstrained state. Since $t_1 = 0.02 - t_2$, we calculate that

$$\mathcal{V}_2 = \frac{1}{2} k(t_2 - 0.01)^2$$

$$\mathcal{V}_1 = \frac{1}{2} k(t_1 - 0.01)^2 = \frac{1}{2} k[(0.02 - t_2) - 0.01]^2 = \frac{1}{2} k(0.01 - t_2)^2$$

Now we can write the total system energy W as

$$W = (U_2 + \mathcal{V}_2)t_2 A + (U_1 + \mathcal{V}_1)(0.02 - t_2)A$$

$$\frac{W}{A} = \frac{D^2}{\varepsilon_0}\left(\frac{t_2}{8} + \frac{0.02 - t_2}{4}\right) + 0.01k(t_2 - 0.01)^2$$

Equilibrium occurs when W is a minimum, so we set $\partial W/\partial t_2 = 0$ and find $t_2 = 0.01 + D_2/0.16k\varepsilon_0 = 1.35$ cm.

Returning to the original problem, we calculate $C_1 = 2\varepsilon_0 A/t_1 = 27.24$ pF and $C_2 = 4\varepsilon_0 A/t_2 = 26.23$ pF, so that $C = C_1 C_2/(C_1 + C_2) = 13.36$ pF.

Relaxing the constraint of dielectric rigidity has resulted in a material distortion of $(t_2 - t_1)/t = 35$ percent and in a capacitance increase of nearly 14 percent.

6.3 POISSON'S EQUATION

In all the electrostatic problems considered thus far, the fields and potentials have been determined from their known charge sources. This is not the only situation met in practice and it is necessary to solve problems in which other field parameters are specified. For example, instead of the charge distribution, the potential or its derivatives can have known values in part of the volume or on the surfaces surrounding the region of interest. For these cases, a very powerful technique has been developed in which standard forms of solution are modified in prescribed ways to fit the particular problem under consideration.

Derivation of the basic equations is quite simple, starting from Maxwell's electrostatic equations. As we saw in Chapter 2, $\nabla \times \mathbf{E} = 0$ leads to $\mathbf{E} = -\nabla V$. Substituting this in $\nabla \cdot \mathbf{D} = \rho$ yields

$$\nabla \cdot \varepsilon \mathbf{E} = -\varepsilon \nabla \cdot \nabla V = \rho \quad \text{or} \quad \nabla^2 V = -\rho/\varepsilon \tag{6.4}$$

where ε is taken to be a constant, and the notation $\nabla^2 V$ denotes the divergence of the gradient of the scalar potential function. This is *Poisson's equation;* it relates the second spacial derivatives of V to the volume charge density. Because techniques have been developed for finding solutions of differential equations, Eq (6.4) is a useful vehicle for evaluating V in many cases of interest.

Example 6.5: The Semiconductor *pn* Junction

Doped semiconductors are charge-neutral materials that have free negative (*n*-type) or free positive (*p*-type) charge carriers and oppositely charged atomic cores. When an *n*-type and a *p*-type material are intimately joined, forming a *pn* junction, the free carriers diffuse into the opposite material, where they disappear due to recombinations with the local opposite carriers. However, this diffusion leaves a region, called the *depletion region,* because it is depleted of free carriers, in which the uncompensated charged dopant atoms can manifest their charges. As a result, charge densities appear that are closely equal to the doping levels: $+eN$ C/m^3 in the *n*-type and $-eP$ C/m^3 in the *p*-type, where N and P are the densities of singly charged dopant atoms, and e is the electron charge magnitude. This is shown in Fig. 6.8(a).

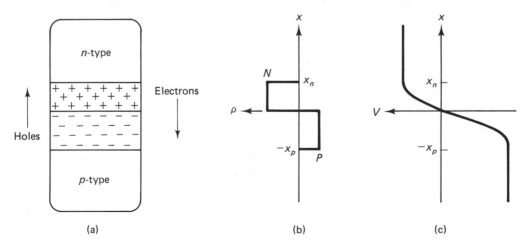

Figure 6.8   The *pn* junction.

The *pn* junction can be simply modeled by assuming a one-dimensional abrupt junction. In this model, the variations occur in only the x direction, and the charge densities are uniform over the depletion regions, which have lengths x_n and x_p on the n and p sides, respectively. Outside the depletion regions, the charge densities abruptly drop to zero. This is shown in Fig. 6.8(b). On each side of the junction, we can write Poisson's equation:

$$\frac{d^2V_n}{dx^2} = -\frac{eN}{\varepsilon} \qquad 0 < x < x_n$$

$$\frac{d^2V_p}{dx^2} = +\frac{eP}{\varepsilon} \qquad -x_p < x < 0$$

These equations are directly integrable, giving the solution form

$$V_i = a_i x^2 + b_i x + c_i$$

Taking the second derivative of this and substituting above, we obtain the values of a_n and a_p. Since E is zero on the outer edge of the depletion regions, we have

$$\frac{dV_n(x_n)}{dx} = 0 = \frac{dV_p(-x_p)}{dx}$$

which yields values of b_n and b_p. Without any loss of generality, we can set $V(0) = 0$, so that $c_n = 0 = c_p$, giving

$$V_n(x) = -\frac{eN}{2\varepsilon}(x^2 - 2x_n x) \qquad 0 < x < x_n$$

$$V_p(x) = +\frac{eP}{2\varepsilon}(x^2 + 2x_p x) \qquad -x_p < x < 0$$

$V(x) = V_n(x) + V_p(x)$ is shown in Fig. 6.8(c).

Note that since D_{normal} is continuous, we must have

$$\frac{d}{dx}V_n(0) = \frac{d}{dx}V_p(0)$$

which yields the result $Nx_n = Px_p$. This is a statement of the equality of the total negative and positive charge per unit area of the junction. From $V(x)$, we can obtain the potential barrier height V_B, the initial potential to be overcome before forward conduction can begin.

$$V_B = V(x_n) - V(-x_p) = \frac{e}{2\varepsilon}(Nx_n^2 + Px_p^2)$$

6.4 LAPLACE'S EQUATION

The theory of linear differential equations tells us that to any special solution of Eq. (6.4), such as we found in Example 6.5, we can add general solutions of the equation

$$\nabla^2 V = 0 \qquad\qquad\qquad\qquad (6.5)$$

as long as these also satisfy the boundary conditions of the problem. Equation (6.5) is *Laplace's equation*. The general solutions of Laplace's equation are particularly valuable as problems are frequently met for which $\rho = 0$ and for which V must be determined from its behavior on the boundaries of the region of interest.

This equation was first presented and used by the Marquis Pierre Simon de Laplace (1749–1827) in his five-volume series *Mechanique Celeste*, which was published between 1799 and 1825. In that work, he summarized, organized, and treated mathematically the extant knowledge of celestial mechanics. The analysis was so advanced and complex that it confounded even the most accomplished analysts of his day. As brilliant as he was technically, so Laplace was also politically astute, managing to rise to a high position in the monarchy before the French revolution, in the revolutionary First French Republic, and during the Napoleonic period following the revolution.

In order to apply Laplace's equation in a variety of cases, we will establish some general forms of its solutions. The Laplacian operator ∇^2 can be expressed in many coordinates; in the rectangular system it is particularly simple:

$$\nabla^2 V = \frac{\partial^2 V}{\partial x^2} + \frac{\partial^2 V}{\partial y^2} + \frac{\partial^2 V}{\partial z^2} = 0 \qquad (6.6)$$

A general solution can be found by reducing this partial differential equation in (x, y, z) to three ordinary differential equations, each of which depends on only one of the coordinate variables. This is known as *separating the variables,* and proceeds by taking a trial form for V, which is the product of three functions, each of which is only dependent upon one of the variables:

$$V(x, y, z) = X(x)Y(y)Z(z) \qquad (6.7)$$

In substituting Eq. (6.7) into Eq. (6.6), it must be noted, for example, that X and Y are constants with respect to z differentiation, so that we have

$$YZ\frac{\partial^2 X}{\partial x^2} + XZ\frac{\partial^2 Y}{\partial y^2} + XY\frac{\partial^2 Z}{\partial z^2} = 0$$

Dividing this by $V = XYZ$ gives

$$\frac{1}{X}\frac{d^2 X}{dx^2} + \frac{1}{Y}\frac{d^2 Y}{dy^2} + \frac{1}{Z}\frac{d^2 Z}{dz^2} = 0 \qquad (6.8)$$

This is a strange result since each of the terms is a function of only a single independent variable, and is therefore independent of the others. Such a sum of independent terms can be zero only if *each* of the terms is individually a constant, and the sum of the constants is zero.

$$\frac{1}{X}\frac{d^2 X}{dx^2} = -k_x^2 \qquad (6.9a)$$

$$\frac{1}{Y}\frac{d^2 Y}{dy^2} = -k_y^2 \qquad (6.9b)$$

$$\frac{1}{Z}\frac{d^2 Z}{dz^2} = -k_z^2 \qquad (6.9c)$$

$$k_x^2 + k_y^2 + k_z^2 = 0 \qquad (6.10)$$

Expressing the constants as $-k^2$ is merely a choice of convenience; since it is not yet known whether the k are real, imaginary, or complex, it does not imply any physical nature to the solution. However, this form is convenient for the most common situation where k is a positive real constant.

It is easily shown that an equation of the form

$$\frac{d^2X}{dx^2} = -k_x^2 X \qquad (6.11a)$$

has the solution

$$X(x) = X_1 \cos k_x X + X_2 \sin k_x X \qquad (6.11b)$$

where X_1 and X_2 are constants. From Eq. (6.10), it is clear that not all the k^2 can be real and positive. We will only consider the case where there is no dependence on one coordinate, such as for situations that are long and invariant in one dimension, say along the z axis. Then $k_z = 0$ and, from Eq. (6.10),

$$k_y^2 = -k_x^2 \qquad (6.12)$$

so that the equation for Y becomes

$$\frac{d^2Y}{dy^2} = -k_y^2 Y = k_x^2 Y \qquad (6.13a)$$

This has the solution

$$Y = Y_1 e^{k_x y} + Y_2 e^{-k_x y} \qquad (6.13b)$$

where the Y_i are constants. Note two features of Eqs. (6.11b) and (6.13b):

1. Both involve the same k (here it is k_x), so that the subscript (x) can be dropped, as there is no need to distinguish it.

2. Equations (6.9a) and (6.9b) are identical, so that the choice of $X(x)$ to have sinusoids, with the result that $Y(y)$ has exponentials, is arbitrary. We might, just as well, have started with $Y(y)$ as being sinusoidal and then derive that $X(x)$ contains the exponential terms. Either set of solutions is correct; the form of the problem to be solved determines which is the most advantageous choice to make.

Example 6.6

Find the potential function $V(x, y)$ in the infinite channel shown in Fig. 6.9. The side walls are at zero potential and the base is held at V_0.

The potential is clearly independent of z, so that $k_z = 0$ and we have the case just discussed. This problem is unbounded along y and has a repeated value along x, between the walls, features that lend themselves to choosing $X(x)$ to have the sinusoidal form and $Y(y)$, the exponential form.

$$X(x) = X_1 \sin kx + X_2 \cos kx$$

$$Y(y) = Y_1 e^{ky} + Y_2 e^{-ky}$$

$$V(x, y) = X(x)Y(y)$$

Once these functional forms of the solution are determined, the given problem determines the values of the coefficients X_j and Y_j, and of k.

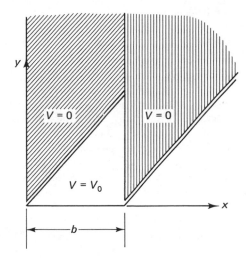

Figure 6.9   Potential channel.

To fit the general solutions to the specific problem at hand, we use the given values of V on the surfaces bounding the volume in which we are interested. The boundary conditions are

$$V = 0 \quad \text{when } x = 0, \qquad \text{all } y > 0$$
$$V = 0 \quad \text{when } x = b, \qquad \text{all } y > 0$$
$$V = 0 \quad \text{when } y \to \infty, \quad 0 < x < b$$
$$V = V_0 \quad \text{when } y = 0, \qquad 0 < x < b$$

As the first two conditions apply for all y, they can only be satisfied by the function $X(x)$. Substitution gives

$$0 = X_1 \sin k(0) + X_2 \cos k(0) = X_2$$
$$0 = X_1 \sin kb + X_2 \cos kb = X_1 \sin kb$$

where the first result ($X_2 = 0$) has been used in the second. For $X_1 \sin kb = 0$, we cannot have $X_1 = 0$ or the entire solution is zero. Therefore, the sine term must vanish. This occurs when $kb = n\pi$, where n can be any integer. This means that there are an infinite number of X solutions given by

$$X_n = X_{1n} \sin k_n x \qquad k_n = n\frac{\pi}{b}$$

A point that is often troublesome is the generality of coefficient values obtained from specific boundary conditions. In the general expression for $X(x)$, we have X_1 and X_2 as constants. The boundary condition at $x = 0$ indicates that, for this problem, on this plane, we must have $X_2 = 0$. Then, since X_2 is a constant, it must have this same zero value at all points.

Example 6.6 (cont.)

From the condition that V vanishes as $y \to \infty$, we must conclude that $Y_1 = 0$ or else we would have $Y(y) \to \infty$. Our solution is then of the form

$$V = \sum_n V_n \sin k_n x \, e^{-k_n y} \qquad k_n = n\frac{\pi}{b} \qquad V_n = X_{1n} Y_2$$

Because the sum of solutions of a linear differential equation is also a solution, and as there is no reason to exclude any, all the possible n solutions have been included.

The final boundary condition is that $V = V_0$ when $y = 0$ and $0 < x < b$. Before applying this condition, we note that values of V outside of this region are of no interest. Since the solution will not be applied outside the region $0 \leq x \leq b$, we can specify it in any convenient way in the external region. Therefore, since $V = 0$ on the vertical boundary planes, we will not affect the solution between the planes if we take the potential along $y = 0$ to be the square wave shown in Fig. 6.10.

It is left as an exercise for the reader knowledgeable in the methods of Fourier series to establish that this wave can be expressed as

$$V_{\text{square wave}} = \frac{4V_0}{\pi} \sum_{n\,\text{odd}} \frac{1}{n} \sin k_n x \qquad k_n = n\frac{\pi}{b} \qquad (6.14)$$

so that the potential that satisfies all the boundary conditions is

$$V = \frac{4V_0}{\pi} \sum_{n\,\text{odd}} \frac{1}{n} e^{-k_n y} \sin k_n x \qquad k_n = n\frac{\pi}{b}$$

6.5 POLAR COORDINATES

As many important problems lend themselves directly to solution in curvilinear coordinates, we should be familiar with the solutions to Laplace's equation in these coordinate systems. In this section, we will discuss the case for boundary conditions in circular cylindrical coordinates. In this system, Laplace's equation is

$$\nabla^2 V = \frac{1}{\rho}\frac{\partial}{\partial\rho}\left(\rho\frac{\partial V}{\partial\rho}\right) + \frac{1}{\rho^2}\frac{\partial^2 V}{\partial\phi^2} + \frac{\partial^2 V}{\partial z^2} = 0 \qquad (6.15)$$

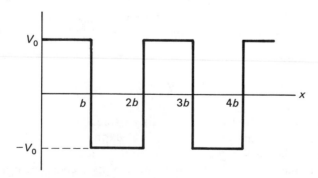

Figure 6.10   Square-wave potential.

We apply the method of separation of variables and, as before, consider the case that there is no z-dependence. Then $\partial^2 V/\partial z^2 = 0$ and we can take $V(\rho\phi) = R(\rho)\Phi(\phi)$. Substituting this expression and multiplying by $\rho^2/R\Phi$ yields, for the two functions,

$$\frac{\partial^2 \Phi}{\partial \phi^2} = -k^2 \Phi \qquad\qquad (6.16a)$$

$$\rho \frac{\partial}{\partial \rho}\left(\rho \frac{\partial V}{\partial \rho}\right) = k^2 R \qquad\qquad (6.16b)$$

It is not difficult to show that these are satisfied by

$$V = \Phi(\phi)R(\rho) = (A \cos k\phi + B \sin k\phi)(M\rho^k + N\rho^{-k}) \qquad (6.17a)$$

The angular coordinate introduces a restriction on values of k. Since V is single-valued, we must have $\Phi(\phi + 2\pi) = \Phi(\phi)$, which demands that the k be an integer:

$$k = n \qquad n = 0, 1, 2, 3, \ldots \qquad\qquad (6.17b)$$

in which case, the most general expression for V must be the sum of all the possible solutions:

$$V = \sum_n (A_n \cos n\phi + B_n \sin n\phi)(M_n\rho^n + N_n\rho^{-n}) \qquad n = 0, 1, 2, \ldots$$

$$(6.17c)$$

Example 6.7

A dielectric rod with circular cross-section radius a is between the parallel plates of a very large capacitor and has its axis normal to the electric field direction. It's dielectric permittivity is ε. Determine the field configuration.

Before the rod is put in place, there exists a uniform electric field in the capacitor. We take this field to be in the positive x direction. While we recognize that the induced charge on the dielectric will alter the $\mathbf{E}$ field in its vicinity, we can assume that the field will still be uniform at great distances from the rod. This is expressed, in terms of the geometry of Fig. 6.11, as

$$V = -E_0 x = -E_0\rho \cos \phi \qquad \rho \gg a$$

(When $\rho \gg a$, this gives $\mathbf{E} = -\nabla V = \mathbf{a}_x E_0$, which is the uniform field along $\mathbf{a}_x$.)

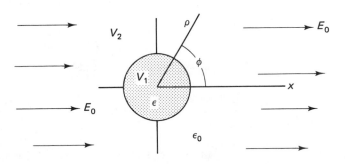

Figure 6.11   Dielectric cylinder in a uniform field.

In a problem such as this, which has distinct dielectric regions, we must assign a different potential function to each region and then relate these functions through the interregional boundary conditions. Thus, on Fig. 6.11, we assign the function V_1 within the cylindrical dielectric and V_2 outside of it. At the surface of the cylinder, the field boundary conditions are expressed in terms of the potentials through $\mathbf{D} = \varepsilon\mathbf{E}$ and $\mathbf{E} = -\nabla V$. In addition, V must be continuous since it represents the work done in moving a test charge to the field point, and this cannot change abruptly over the infinitesimal displacement from one medium to the other.

$$D_{2n} = D_{1n} \qquad \varepsilon_0 \frac{\partial V_2}{\partial \rho} = \varepsilon \frac{\partial V_1}{\partial \rho} \qquad \text{at } \rho = a$$

$$E_{2t} = E_{1t} \qquad \frac{1}{\rho}\frac{\partial V_2}{\partial \phi} = \frac{1}{\rho}\frac{\partial V_1}{\partial \phi} \qquad \text{at } \rho = a$$

$$V_2 = V_1 \qquad\qquad\qquad \text{at } \rho = a$$

We should note that the conditions

$$E_{2t}\,(\rho = a) = E_{1t}\,(\rho = a) \quad \text{and} \quad V_2\,(\rho = a) = V_1\,(\rho = a)$$

are equivalent. This is seen as follows: the fields are derived by differentiating the V. Therefore, adding a constant to one of these V, say V_2, so that $V_1 = V_2$ at some selected point on the cylinder surface, does not alter the solution. However, having done this, we find that the condition $E_{1t}\,(\rho = a) = E_{2t}\,(\rho = a)$ means that the tangential derivatives of the V must also be equal on that surface. Since the rates of change of the V, along the surface, are equal, and the V themselves are equal at one point, they will be equal at every point on $\rho = a$. Therefore, only one of these boundary conditions can be used.

This problem has the axial symmetry discussed with respect to Eq. (6.17), so the potentials in both regions have the form

$$V = \sum_n (A_n \cos n\phi + B_n \sin n\phi)(M_n\rho^n + N_n\rho^{-n})$$

In region 1, we must have all the $N_n = 0$, so that V_1 does not become infinite at the origin ($\rho = 0$). We might also expect to have all the $M_n = 0$ in region 2, so that V_2 does not become infinite when $\rho \to \infty$, except that a uniform field at large distances imposes the condition that V_2 does become infinite, but in a specific way. Since $V\,(\rho \to \infty) = -E_0\rho \cos \phi$, it is linearly proportional to ρ. We should recognize that this is not a truly infinite potential since ρ only increases until the distant capacitor plates are reached; we treat the plates as being at infinity because $\rho_{\text{plates}} \gg a$. Therefore, in V_2, we must retain $M_1 = -E_0$ to satisfy the field condition when $\rho \gg a$. Lumping the remaining coefficients, $A_n' = A_n M_n$, $A_n'' = A_n N_n$, etc., we have the sums over n:

$$V_1 = \sum_n (A_n' \cos n\phi + B_n' \sin n\phi)\rho^n$$

$$V_2 = \sum_n (A_n'' \cos n\phi + B_n'' \sin n\phi)\rho^{-n} - E_0\rho \cos \phi$$

At $\rho = a$, we have $D_{1n} = D_{2n}$ and $V_1 = V_2$, so that

$$\varepsilon \sum_n n(A_n' \cos n\phi + B_n' \sin n\phi)a^{n-1} = \varepsilon_0 \sum_n - n(A_n'' \cos n\phi + B_n'' \sin n\phi)a^{-n-1}$$
$$- \varepsilon_0 E_0 \cos \phi$$

$$\sum_n (A_n' \cos n\phi + B_n' \sin n\phi)a^n = \sum_n (A_n'' \cos n\phi + B_n'' \sin n\phi)a^{-n} - E_0 a \cos \phi$$

Following the method of Fourier series, we multiply these equations by $\sin \ell\phi \, d\phi$ or by $\cos \ell\phi \, d\phi$ and since they hold for all ϕ, we integrate from 0 to 2π. Using the relations

$$\oint \sin n\phi \cos \ell\phi \, d\phi = 0 \quad \text{for all } n, \ell$$

$$\oint \sin n\phi \sin \ell\phi \, d\phi = 0 \quad \text{for } n \neq \ell \tag{6.18}$$

$$\oint \cos n\phi \cos \ell\phi \, d\phi = 0 \quad \text{for } n \neq \ell$$

(where $\oint$ denotes integration over a full period), we find that the boundary conditions are satisfied if

$$A'_n = B'_n = A''_n = B''_n = 0 \qquad \text{for } n \neq 1$$

$$B'_1 = 0 = B''_1$$

$$\frac{\varepsilon}{\varepsilon_0} A'_1 = -\frac{A''_1}{a^2} - E_0$$

$$A'_1 a = \frac{A''_1}{a} - E_0 a$$

The reader can verify these results by substitution. As a result,

$$V_1 = -2E_0 \frac{\varepsilon_0}{\varepsilon + \varepsilon_0} \rho \cos \phi \qquad\qquad \rho \leq a$$

$$V_2 = E_0 a^2 \frac{\varepsilon - \varepsilon_0}{\varepsilon + \varepsilon_0} \frac{\cos \phi}{\rho} - E_0 \rho \cos \phi \qquad \rho \geq a$$

Note that the field is uniform within the dielectric rod:

$$\mathbf{E}_1 = -\nabla V_1 = \mathbf{a}_x 2E_0 \frac{\varepsilon_0}{\varepsilon + \varepsilon_0}$$

6.6 UNIQUENESS THEOREM

We have developed several different means to calculate the electric fields under different conditions and in different geometries. In Chapter 3, we used Gauss' law for symmetrical distributions of electric charge; there and in this chapter, we used direct summation and integration, over the charge sources, with Coulomb's law. The same holds true for magnetic fields. In Chapter 4, we used Ampère's law to find the magnetic fields of some symmetrical current distributions, and in Chapter 7, we will use direct integration of the Biot–Savart relation over current sources. These techniques will also be carried farther in the time-varying cases, for example, to study waves. In this chapter, we have found differential equations for the potential functions and shown that they could be solved directly using standard forms. With all these different methods available, we must now establish an important theorem that, regardless of which method is used to find the potential, and therefore the fields, the same result will be determined, that is, the solution is unique. The answer obtained by one of these methods can be in a different form, and may be hard to recognize as equivalent to another solution, but it must be the same.

We start the proof with a mathematical identity derived from the divergence theorem for a vector $\mathbf{G}$:

$$\oint \mathbf{G} \cdot \mathbf{n} \, dS = \int \nabla \cdot \mathbf{G} \, dv \qquad (6.18a)$$

where the volume v is enclosed by the surface S. If $\mathbf{G}$ has the form $\mathbf{G} = \psi \nabla \psi$, where, for now, ψ is an arbitrary scalar function,

$$\oint \psi \nabla \psi \cdot \mathbf{n} \, dS = \int \nabla \cdot (\psi \nabla \psi) \, dv \qquad (6.18b)$$

Rewriting the left-hand side and expanding the right-hand side gives

$$\oint \psi \frac{\partial \psi}{\partial n} \, dS = \int (\nabla \psi)^2 \, dv + \int \psi \nabla^2 \psi \, dv \qquad (6.18c)$$

Returning to the fields problem, suppose there are two solutions for the potential function, V_i and V_j. They may have been obtained by a different method, but they describe the same source distribution within the volume, and have the same time variation. Both functions must satisfy Poisson's or Laplace's equation, so that if we take

$$\psi = V_j - V_i \qquad (6.19)$$

then $\nabla^2 \psi = \nabla^2 V_j - \nabla^2 V_i = 0$, and the last term in Eq. (6.18c) vanishes. Now consider two different conditions under which we might have solved for V_i and V_j.

Dirichlet Condition.   The problem is one in which the potential is specified on the surface S. It is illustrated by Example 6.6. Both V_i and V_j must be the same on S, so that $\psi = V_j - V_i = 0$ on S and the surface integral, on the left-hand side of Eq. (6.18c) is zero.

Neumann Condition.   The problem is stated such that the normal derivative of the potential is given on the bounding surface S. In the electric case, $\partial V/\partial n = -E_n$, so that the Neumann condition is a specification of the fields normal to the bounding surface S, as illustrated by the dielectric boundary of Example 6.7. In this case, both V_i and V_j must have the specified derivative, so that $\partial \psi/\partial n = \partial V_j/\partial n - \partial V_i/\partial n = 0$, again making the surface integral, on the left-hand side of Eq. (6.18c), equal to zero.

In both cases, Eq. (6.18c) reduces to $\int (\nabla \psi)^2 \, dv = 0$. This integrand is a positive quantity at every point of the volume, so that the integral can only be zero if $\nabla(V_j - V_i) \equiv 0$. This in turn implies that $V_j - V_i = c$, that is, constant. Since the fields are derived from V by derivatives, a constant difference does not appear and we can take $c = 0$, or $V_i = V_j$.

Let us review what this proof has established. It deals with cases in which we seek a solution for the fields and potentials in a volume v, bounded by a surface S, where we have a specified charge distribution (which may be zero) within the volume, with a specified time variation (which, in the static case, may be zero). Then the solution for V is unique if either of two additional conditions is satisfied. These are either the Dirichlet condition, that the solution for V reduces to given values of V on the boundary S, or the Neumann condition, that the solution has $\partial V/\partial n$, which reduces to given values on the boundary S.

Example 6.8

When an electric field is applied to a conductor containing a spherical cavity, the cavity acquires a surface charge density $\rho_s = -\rho_0 \cos \theta$, where θ is measured from the field direction, as shown in Fig. 6.12. Find the electric field and potential within the cavity.

We recall that $E = D = 0$ in the conductor and, in that case, Gauss' law gives the result (see Example 5.7) that for the field outside the metal (i.e., in the cavity), $D_n = \rho_s$. For this geometry, $\mathbf{n} = -\mathbf{a}_r$, so $D_n = -D_r = \rho_s$. From $\mathbf{D} = \varepsilon_0 \mathbf{E} = -\varepsilon_0 \nabla V$, we have $\rho_s = -\varepsilon_0 \partial V/\partial r$ at $\rho = a$.

$$\frac{\partial V(r = a)}{\partial r} = -\frac{\rho_0}{\varepsilon_0} \cos \theta$$

The specification of surface charge density on a metal is equivalent to giving the normal derivative of V on the surface S; within the spherical hole, the problem has a Neumann boundary condition.

The above equation applies only at $r = a$. However, its form suggests that we try

$$V = -\frac{\rho_0}{\varepsilon_0} r \cos \theta$$

Note that integration of the boundary condition is not a proper operation, since it only holds at $r = a$, but it is, nevertheless, useful. This expression for V satisfies the condition for the volume charge density that $\rho/\varepsilon_0 = -\nabla^2 V = 0$, and, on the bounding surface, it gives the correct value of $D_n = \rho_s$.

This potential, therefore, satisfies all the conditions of a Neumann problem and it must be the correct answer in the cavity. Note that $z = r \cos \theta$, so that we can write $V = -z\rho_0/\varepsilon_0$. From this, we see that

$$\mathbf{E} = -\nabla V = \mathbf{a}_z \frac{\rho_0}{\varepsilon_0}$$

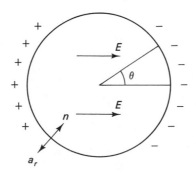

Figure 6.12 Charge distribution on a spherical hole.

The field in the volume is uniform in the z direction. This result was established by a dubious procedure, but we have argued that it satisfies all the boundary conditions and therefore is correct.

While Example 6.8 involves a finite volume, there is nothing in the uniqueness proof that restricts us from letting the dimension of S become infinitely large so that the boundary conditions describe the potential or the fields at very large distances, and the solutions will be for all space.

We should also note that the first integral, on the left-hand side of Eq. (6.18c), vanishes if $\psi = 0$ on part of S and $\partial\psi/\partial n = 0$ on the other parts, so that a mixed Dirichlet–Neumann problem also has a unique potential.

6.7 METHOD OF IMAGES

Advantage can be taken of the uniqueness theorem in some special and important cases. The method of images is one that uses the fact that some difficult problems can be taken to be portions of other problems that have greater symmetry and therefore have simpler solutions. In Example 6.9, we use the fact that the potential is zero at all points of the plane which is the perpendicular bisector of the line connecting equal and opposite charges.

Example 6.9

Write an expression for the potential at all points due to a point charge $+Q$ that is above an infinite grounded conducting plane.

Figure 6.13 shows the charge $+Q$ located at $z = h$ and a charge $-Q$ located at $z = -h$. The plane $z = 0$ then has $V = 0$. For points above the plane, the region $z \leq 0$ could be replaced by the grounded conducting plane, since the semiinfinite volume $z > 0$ still satisfies the same Dirichlet boundary condition at $z = 0$ (i.e., that $V = 0$ on the plane), and at $r = \infty$ (again, $V = 0$ when we are infinitely far from the source charges). Therefore, *in the region $z > 0$*, the potential we seek is given by that of the two charges. By using circular cylindrical coordinates, with ρ and ϕ in the plane normal to z,

$$V = \frac{Q}{4\pi\varepsilon_0} \left\{ \frac{1}{[(z - h)^2 + \rho^2]^{1/2}} - \frac{1}{[(z + h)^2 + \rho^2]^{1/2}} \right\}$$

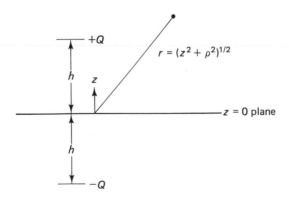

Figure 6.13 Method of images.

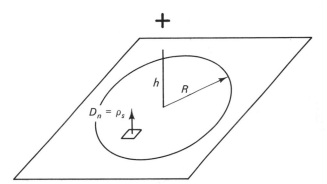

Figure 6.14 Induced charge density on the conducting plane.

Clearly, the boundary conditions are satisfied, since $V(z = 0) = 0$ and $V(\infty) = 0$. When the field point approaches the charge Q, the last term approaches the constant $-1/2h$, and the first term gives the potential due to the point charge.

The charge $-Q$ at $z = -h$ is called the *image* of the given charge. It does not really exist, but merely describes the effect, in the region $z > 0$, of all the induced charge on the plane $z = 0$. The true area density of this planar induced charge can be found from

$$\rho_s = D_n = -\varepsilon_0 \frac{\partial V(z = 0)}{\partial z} = \frac{Qh/2\pi}{(h^2 + \rho^2)^{3/2}}$$

From this, the charge within a planar radius R of the free charge Q, as shown on Fig. 6.14, is

$$Q_{\text{induced}} = -(Qh/2\pi) \int_0^R \frac{2\pi\rho \, d\rho}{(\rho^2 + h^2)^{3/2}} = Q\left[1 - \frac{h}{(h^2 + R^2)^{1/2}}\right]$$

so that, for example, $Q_{\text{induced}} = -Q/2$ within $R = \sqrt{3}h$ and $Q_{\text{induced}} = -Q$ when $R = \infty$.

Note that these solutions do not apply when $z < 0$, since $E = 0$ in and below the plane; we have taken $V = 0$ in the conductor.

EXERCISES

Extensions of the Theory

6.1. As shown in Fig. 6.15(a) (next page), the potentials from a point charge Q at two points separated by $\mathbf{a}_z\delta$ are $V(r)$ and $V(r) + \nabla V \cdot \mathbf{a}_z\delta$. From this, it can be seen in Fig. 6.15(b), that the potential at a point due to an electric dipole $\mathbf{p} = \mathbf{a}_z q\delta$ may be written as the sum of the potential from the positive charge, $V(r)$, and from the negative charge, $-[V(r) + \nabla V \cdot \mathbf{a}_z\delta]$. Use this to derive an expression for $V(\mathbf{p})$, and show that E of a dipole $\sim 1/r^3$.

6.2. **(a)** Newton's law for uniform circular motion states that $F_r = mv^2/r$. If we suppose that a hydrogen atom has a stationary nucleus and a rotating electron, then F_r is the Coulomb force. Use Newton's law to find a relation between the tangential velocity v and r.

(b) If the orbital angular momentum is quantized so that $mvr = nh/2\pi$, where $h = $ Planck's constant $= 6.626 \times 10^{-27}$ erg/cm, and n is an integer, derive the stable radii and their energies.

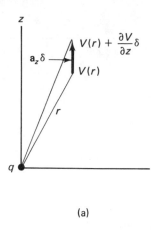
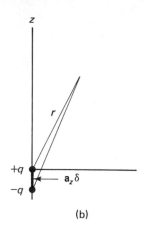

(a)   (b)

Figure 6.15   For Exercise 6.1.

6.3. The method of images can be extended to dielectric boundaries. Consider a charge Q placed in medium ε_1 at a distance z_0 above the plane surface of dielectric ε_2. The interface is at $z = 0$ and both media are semiinfinite. Show that the boundary conditions on $E_{\text{tangential}}$ and D_{normal} are satisfied for the following conditions:

(a) The field in medium 1 is given by the original charge Q and a hypothetical point charge $q_2 = Q(\varepsilon_1 - \varepsilon_2)/(\varepsilon_1 + \varepsilon_2)$ located at $z = -z_0$.

(b) The field in medium 2 is given by removing the original charge and replacing it, at the same location, by a hypothetical charge $q_1 = 2Q\varepsilon_1/(\varepsilon_1 + \varepsilon_2)$.

6.4. We have considered the image problem of a point charge above a conducting plane. Give the solution for the corresponding problem of a line charge above the conducting plane. Are all the boundary conditions satisfied?

6.5. Since a conductor is a medium that has a great amount of free charge, it is sometimes envisioned as a dielectric that can polarize so easily that $\varepsilon \to \infty$. Discuss the results of Exercise 6.3 if medium 2 is such a superdielectric.

6.6. A power transmission system has a conducting line of radius a whose axis is a distance $2a$ above a conducting earth. To find the capacitance of the line with the earth, an equivalent configuration can be analyzed in which the original line and the earth are replaced by two line-charge mirror images located at a distance $\sqrt{3}a$ from the earth plane. Show that these charges make the surface of the original cylindrical conductor an equipotential. (See Exercise 6.4.)

6.7. A very thin spherical shell has a uniform charge density ρ_s. A small piece of the shell is removed without disturbing the remainder of the charge. Show that the field strength at the center of the hole is $\rho_s/2\varepsilon_0$. *Hint:* Use superposition.

6.8. For Example 6.4,

(a) Derive a general expression for the capacitance as a function of Q.

(b) Since $Q = CV$, write this as $C = C(V)$.

(c) At what value of V is C least sensitive to voltage fluctuation?

Problems

6.9. Using Coulomb's law, find the fields in Exercise 3.5.

*6.10. A uniform surface charge density ρ_s exists on the hemispherical surface: $r = r_0$ and $\theta \leq \pi/2$. What is the electric potential at the center?

6.11. A uniform volume charge density ρ_0 exists in the hemispherical volume $r \leq r_0$ and $\theta \leq \pi/2$. What is the electric potential along the polar axis (z)?

*6.12. Solve Problem 3.16 for the quantitative flux densities.

6.13. The square $-a \leq x \leq a$, $-a \leq y \geq a$, $z = 0$ contains the charge density $\rho_s = \rho_0|x|\,|y|/a^2$. Set up the integrals to find $\mathbf{E}$ and V at points on the x axis and on the y axis.

*6.14. The square area $0 < x < 1$, $0 < y < 1$, $z = 0$ has a uniform surface charge density ρ_s and the area $-1 < x < 0$, $0 < y < 1$, $z = 0$ has a uniform surface-charge density $-\rho_s$.
(a) Find the potential at points on the z axis. *Hint:* Take $V = -\int \mathbf{E} \cdot \mathbf{a}_z\,dz$.
(b) Find the potential at points on the y axis. *Hint:* Take $V = -\int \mathbf{E} \cdot \mathbf{a}_y\,dy$.

6.15. Start with the expression for $\mathbf{E}$ of a finite line charge, from Example 6.1, and show that it approaches that of a point charge when $z \to \infty$.

*6.16. For the distribution of Problem 6.13, find $\mathbf{E}$ and V on the z axis.

6.17. For Example 6.2c, complete the integral for the cylindrical volume-charge distribution. Show that when $z \to \infty$, the field approaches that of a point charge: $Q = 2L\pi a^2 \rho_v$. Note the order in which the limit is taken and the differentiation is done.

*6.18. An air-filled parallel-plate capacitor has plates of area A and separation x, and is permanently connected to a battery of voltage V. Write an expression for the total energy of this capacitor, and show that the pressure (force/area) on the plates is given by $\varepsilon_0 V^2/2x^2$.

6.19. A parallel-plate capacitor is partly immersed in an insulating dielectric liquid, as shown in Fig. 6.16. When the plates are connected to a 600-volt battery,

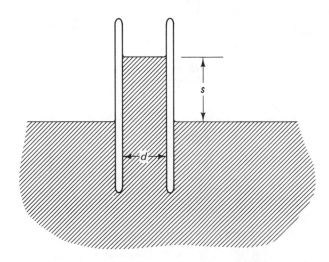

Figure 6.16   For Exercise 6.19.

the level of the liquid rises. Given that $d = 2$ mm for the plates and $\varepsilon_r = 10$ for the liquid, how much does the liquid rise between the plates? The liquid density is 1 gm/cm³. Neglect surface tension.

6.20. Using the solution to Example 6.4, determine the lateral shrinkage or expansion of the dielectrics. Assume that the metal plates are oversized, so that field uniformity is maintained throughout.

6.21. The dielectrics of Example 6.4 are each cut with dimensions $10 \times 5 \times 2$ cm and placed in a capacitor, as shown in Fig. 6.17. Their vertical size is constrained to be constant.
  (a) Calculate the undistorted reference capacitance C_0.
  (b) Determine the equilibrium widths of medium 1 and medium 2 if the total width remains constant.
  (c) What is the new capacitance?

6.22. An air-filled parallel-plate capacitor is charged by a battery V that is then disconnected. A dielectric slab with relative permittivity ε_r and weight G is partly lifted into the capacitor, as shown in Fig. 6.18. Determine the length of the slab, z, which is between the plates at equilibrium. *Hint:* The capacitor charge Q is constant.

6.23. Solve Problem 6.22 if the battery is left connected when the dielectric is inserted. *Hint:* Now voltage V is constant.

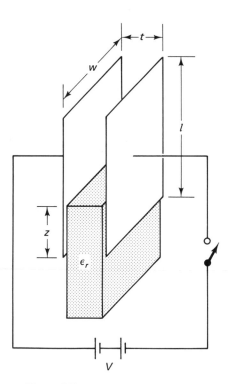

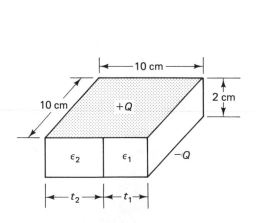

Figure 6.17 For Exercise 6.21.   **Figure 6.18** For Exercises 6.22 and 23.

6.24. Solve Example 6.4 assuming that the permeability also changes when the dielectric is distorted, that is, $\varepsilon_r = \varepsilon_{r_0}(1 - \alpha x/t)$, where $\alpha = 0.1$.

*6.25. A vacuum-tube diode can be modeled as having two parallel plane electrodes: the cathode at $z = 0$, which emits electrons and is at ground potential, and the anode at $z = z_0$, which collects electrons and is at the potential $+V_0$. Determine the potential distribution in the interelectrode region if the electron density is $\rho = \rho_0(2 - z/z_0)$.

6.26. A vacuum-tube triode can be modeled as having a cylindrical cathode with radius ρ_c at potential $-V_0$, which emits electrons; an anode with radius $\rho_a > \rho_c$ at potential $+V_0$; and an open-mesh grid with radius ρ_g, where $\rho_a > \rho_g > \rho_c$, at potential V_g, where $-V_0 < V_g < +V_0$. The two regions have different uniform charge densities ρ_1 and ρ_2. Determine the potentials in both regions.

6.27. Two planar surfaces of WO_2, at $x = -1$ cm and at $x = +1$ cm, are both held at potential V_0. When heated, WO_2 emits electrons. The electron density in the region between the surfaces is $\rho = \alpha x^2$ C/m^3. Find $V(x)$ between the plates.

6.28. Find the solution to Problem 6.27 if $\rho = \alpha(1 - x^2)$.

6.29. The potentials of the plates in Problem 6.27 are changed so that $V(-1$ cm$) = V_0$ and $V(+1$ cm$) = -V_0$. Determine $V(x)$ if, between the plates, it is found that $\rho = \alpha x$.

*6.30. Solve Example 6.6 if the base potential is no longer the constant V_0, but is maintained at the potential $V = V_0 \sin(2\pi x/b)$.

*6.31. The channel of Example 6.6 is modified by changing its extension along the y axis. Solve the problem if a cap, at potential V_0, is placed at $y = a$.

6.32. Solve Problem 6.31 if the cap is at zero potential.

6.33. For Example 6.6, find the charge density on the planes $x = 0$ and $x = b$. Find the total charge/length on the plane $y = 0$.

*6.34. Solve Laplace's equation in the space between two coaxial cylindrical conductors at potentials 0 and V_0 and corresponding radii $b > a$. Compare the potential with that found in Example 3.3.

*6.35. A boss on a metal surface has the form of a long semicircular ridge of radius $\rho = a$, as shown in Fig. 6.19. The plate is at zero potential and another plate at potential V_0 is held parallel to it at a distance $d \gg a$. Solve for the potential at points above the plane. What is the ratio of the electric field at the boss to

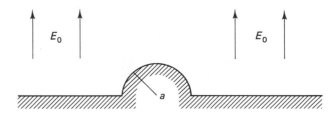

Figure 6.19 For Exercise 6.35.

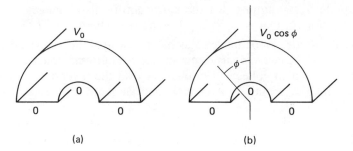

Figure 6.20 For Exercise 6.38.

that on the plate at a point far from the boss? *Hint:* solve Laplace's equation in cylindrical coordinates.

6.36. Solve Problem 6.35 if the plate with the boss is at potential V_0 and the distant plate is at zero potential.

*6.37. The semiinfinite plane $\phi = 0$ is held at a potential of V_0 and the semiinfinite plane $\phi = \pi/6$ is at $V = 0$. Find V at points $0 \leq \phi \leq \pi/6$. *Hint:* Problems in which V only depends on a single variable can be solved by direct integration of $\nabla^2 V = 0$.

*6.38. Find the potentials in the long semicylindrical cavities shown in Figs. 6.20(a) and (b).

6.39. The conducting planes $\phi = 0$ and $\phi = \pi/6$ are held at zero potential and the cylindrical surface $\rho = \rho_0$ is at $V = V_0$. Find V at points $0 \leq \phi \leq \pi/6$ and $\rho \geq \rho_0$.

6.40. The conducting planes $\phi = 0$ and $\phi = \pi/6$ and the cylindrical surface $\rho = \rho_0$ are held at zero potential. The surface $\rho = 2\rho_0$ is at $V = V_0$. For $0 \leq \phi \leq \pi/6$, find V at points for which
(a) $2\rho_0 \geq \rho \geq \rho_0$
(b) $\rho \leq \rho_0$

6.41. The plane $x = 0$ is at $V = 0$, and the plane $y = 0$ is at $V = V_0$. Find the potential for points for which $x \geq 0$ and $y \geq 0$. *Hint:* This problem can be solved in cylindrical coordinates.

6.42. Find the potential at points in the volume bounded by two coaxial half cylinders with $\rho = a$, as shown in Fig. 6.21. The half cylinder with $0 < \phi < \pi$ has $V = V_0$; the half cylinder with $\pi < \phi < 2\pi$ has $V = -V_0$.

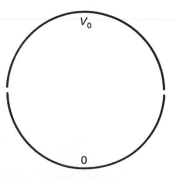

Figure 6.21 For Exercise 6.42.

6.43. Interpret the final result of Example 6.7 and show that for $\rho > a$, the cylinder appears to be a dipole at the origin. What is the dipole moment?

6.44. A charge Q is located at point $(x, y, 0)$. The planes $x = 0$ and $y = 0$ are perfectly conducting. Using the method of images, find the equivalent charge distribution that will produce the same electric field in the region where charge Q resides.

7

TECHNIQUES IN MAGNETOSTATICS

Chapters 4 and 5 discussed magnetostatic fields. Consideration was given to their symmetry properties, their energies, the forces they exert, and their interactions with matter. Ampère's law was used to calculate the fields from current distributions that have a significant amount of symmetry. As mentioned in Chapter 6 with respect to electrostatial fields, although the major conceptual features of magnetic fields have been considered, many situations arise for which additional techniques are needed to solve for the fields. Some of these are considered in this chapter.

7.1 BIOT–SAVART LAW

It was pointed out in Chapter 4 that the magnetic field generated by an infinitesimal current element of length $d\ell$ carrying current I in the direction of the unit vector $\mathbf{i}_0$ is

$$d\mathbf{H} = \frac{I\,d\ell}{4\pi r^2}\mathbf{i}_0 \times \mathbf{r}_0 \tag{7.1}$$

where $\mathbf{r}_0$ and r are, respectively, the unit vector and distance from the current element to the point at which the field is measured. The total field strength $\mathbf{H}$ at the field point is obtained by integrating Eq. (7.1) to add the contributions of the entire circuit of I.

It is interesting to briefly discuss the symmetry properties of Eq. (7.1) before applying it to calculate fields. Observe that the infinitesimal magnetic field source

$i_0 I\, d\ell$ is fundamentally different from the electrostatic field source Q in that it has direction as well as magnitude. (This is inherent in any quantity that represents a flow, or motion, since velocity and displacement are vectors.) The field produced, therefore, displays a more complex symmetry. Since the source is infinitesimal, we expect the field to still reflect spherical symmetry about that source, hence, the magnitude dependence upon $1/4\pi r^2$ as in the electrostatic case. Now, however, there is also an axial, or cylindrical, symmetry about the direction of flow. Figure 7.1 shows a sphere about the source point and a cylinder about the current axis; their intersections are coaxial circular contours. This suggests that any of the family of such circles derived from the different size spheres and cylinders have the full symmetry of the source and, therefore, represent field lines; they are given by $i_0 \times r_0$. In addition, the elemental contribution to the magnetic field should be proportional to the strength of the source: $I\, d\ell$.

The direction of **H** could just as well be taken as opposite to that shown in Fig. 7.1 (i.e., $-i_0 \times r_0$) and still meet all the symmetry requirements. For this reason, **H** is called an *axial vector,* indicating that its physical nature only defines it as a rotation about the axis, but does not specify its direction of rotation about that axis. This is different from the electric field that is directed from positive to negative, which is called a *polar vector.* The direction of the axial field vector is taken by convention to be that of the right-hand rule, as given in Eq. (7.1).

We will frequently use a shorter notation for the current element: instead of $i_0 I\, d\ell$ we will denote it by $\mathbf{I}\, d\ell$. If volume currents are flowing, we recognize that Eq. (7.1) still applies if we substitute, for $\mathbf{I}\, d\ell$, the quantity $\mathbf{J}\, dv$. In either case, the total field can be found by integrating over all the current loops or over the total current distribution.

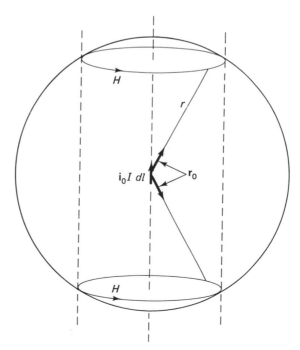

Figure 7.1   Symmetry of magnetic-field lines.

7.2 CALCULATIONS FROM CURRENT

The following example illustrates the use of Eq. (7.1).

Example 7.1

Calculate the field on the axis of a solenoidal winding of length $2L$ and radius a, carrying current I through n turns per meter of length.

Example 7.1a

We first solve a simpler problem and then extend it to the desired one. In particular, we will determine the field from a single turn of the solenoid.

Figure 7.2 shows the current loop with two diametrically opposite elements, $\mathbf{I}\,d\ell = \mathbf{a}_\phi Ia\,d\phi$, and the field contributions from these elements, $d\mathbf{H} = \mathbf{I}\,d\ell \times \mathbf{r}_0/4\pi r^2$. Because both elements are at the same distance from the field point, and all the angles are equal,

$$|dH_1| = |dH_2| = \frac{Ia\,d\phi}{4\pi r^2}$$

Both contributions, dH_1 and dH_2, lie in the plane of the z axis and the line connecting the current elements. Their off-axis components are equal and opposite, so that they cancel. This is true for the transverse fields of all current elements considered in pairs, so there will be zero transverse field. We need only compute the longitudinal axial field

$$H_z = \int \cos \gamma \, dH = \frac{Ia}{4\pi} \int_0^{2\pi} \frac{a\,d\phi}{r^3} = \frac{Ia^2}{2r^3}$$

where we have taken $\cos \gamma = a/r$ and have removed $r^3 = (a^2 + z^2)^{3/2}$ from the integral because it is independent of ϕ.

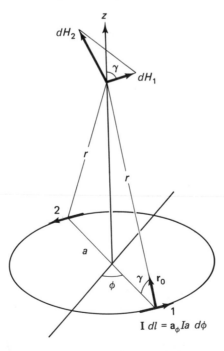

$$\mathbf{I}\,dl = \mathbf{a}_\phi Ia\,d\phi$$

Figure 7.2 Field of a current loop.

Recalling that the magnetic moment of a current loop is $m = I \times \text{area} = I\pi a^2$, the result of Example 7.1a can be expressed as $H_z = m/2\pi r^3$. In Exercise 6.1, it was shown that the field from an electrostatic dipole (charge $+q$ at $z = \delta/2$, charge $-q$ at $z = -\delta/2$) is $D_z = p/2\pi r^3$, where $p = q\delta$ is the electric dipole moment. Since both fields vary as $1/r^3$, the current loop is called a *magnetic dipole*.

An important physical difference between electric field sources and magnetic field sources lies in the fact that there is no magnetic counterpart to the electric *monopole*, that is, the point charge for which $D \sim Q/r^2$. Although an infinitesimal current element appears to have this $1/r^2$ variation, according to Eq. (7.1), such an element can only produce its field when it is part of a larger circuit. As we have seen in the previous example, the fact that the current loop must be closed leads to a higher-order inverse radial dependence. Later in this chapter, we will discuss the *magnetic pole*, which is more closely analogous to the electric point charge. However, we will see that this is merely a convenient mathematical way of representing certain magnetic phenomena, particularly some effects of magnet material.

Example 7.1b

Figure 7.3 shows the loop of Example 7.1a embedded in the solenoid we now wish to consider. The thin loop is taken to have a longitudinal extent dz', and the current flowing in that width is $dI = nI\,dz'$. With this, the field produced by the element dz' is given by the loop field of Example 7.1a, modified to $dH_z = a^2\,dI/2r^3$. We also note that we now have $r = [a^2 + (z - z')^2]^{1/2}$. With these changes, and since $\sin\theta = a/r$,

$$H_z = \frac{1}{2}nIa^2 \int_{-L}^{L} \frac{dz'}{r^3} = \frac{1}{2}nIa \int_{-L}^{L} \frac{\sin\theta\,dz'}{r^2}$$

Introducing the angle θ allows us to make a greatly simplifying change of the integration variable. Figure 7.4 shows the change of θ as the point of integration moves from

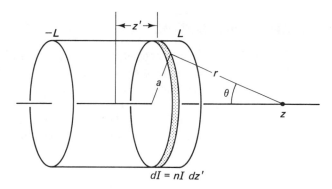

Figure 7.3   Solenoidal winding.

$dI = nI\ dz'$

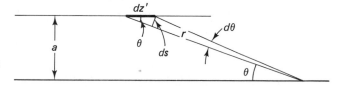

Figure 7.4   Relationship of integration variables.

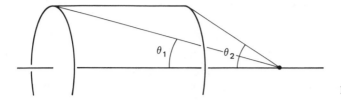

Figure 7.5 Defining angles.

one element along the solenoid, a distance dz' to the next. From the figure, it is seen that the element $ds = r\,d\theta = dz'\sin\theta$, and, since $\sin\theta = a/r$, we find that $dz'/r^2 = d\theta/a$. The expression for H_z becomes

$$H_z = \frac{nI}{2}\int_{\theta_1}^{\theta_2} \sin\theta\,d\theta = \frac{nI}{2}(\cos\theta_1 - \cos\theta_2)$$

where θ_1 and θ_2 are defined in Fig. 7.5. While this result only holds on the axis, it has some notable features:

- In the center of the solenoid, we have $-\cos\theta_2 = \cos\theta_1$, giving

$$H_{z\,(\text{center})} = nI\,\cos\theta_{\text{center}}$$

If the coil is very long, then $\theta_{\text{center}} \to 0$ and $\cos\theta_{\text{center}} \to 1$, and we have the result for an infinite solenoid: $H = nI$.

- In the end plane of the solenoid, we have $\theta_2 = \pi/2$, so $\cos\theta_2 = 0$ and

$$H_{z\,(\text{end})} = \frac{1}{2}nI\,\cos\theta_{\text{end}}$$

For a long solenoid, the angles are small so $\cos\theta_{\text{end}} \approx \cos\theta_{\text{center}}$, and the field in the end plane is approximately half that in the center.

7.3 VECTOR POTENTIAL

The electric scalar potential and the electrostatic field are related by $\mathbf{E} = -\nabla V$. Since the curl of any gradient is zero, this leads to $\nabla \times \mathbf{E} = -\nabla \times \nabla V = 0$, which is one of Maxwell's static equations. In analogy with that case, a potential can also be defined in the magnetostatic case. Since the divergence of any curl is zero, Maxwell's equation $\nabla \cdot \mathbf{B} = 0$ is satisfied if $\mathbf{B}$ is the curl of another vector $\mathbf{A}$. To obtain $\mathbf{A}$, we note, from Eq. (7.1), that the flux density is

$$\mathbf{B} = \int d\mathbf{B} = \frac{\mu}{4\pi}\int \frac{I\,d\boldsymbol{\ell} \times \mathbf{r}_0}{r^2} \tag{7.2}$$

While a direct and rigorous derivation is too involved to warrant presentation here, it can be shown that Eq. (7.2) can be written as

$$\mathbf{B} = \nabla \times \mathbf{A} \qquad \mathbf{A} = \frac{\mu}{4\pi}\int \frac{\mathbf{I}}{r}\,d\boldsymbol{\ell} \tag{7.3a, b}$$

$\mathbf{A}$ is called the *magnetic vector potential*.

An alternative approach to deriving and understanding Eqs. (7.3) is from Maxwell's other magnetostatic equation $\nabla \times \mathbf{H} = \mathbf{J}$. From Eq. (7.3a), this is

$$\mu \mathbf{J} = \nabla \times \mu \mathbf{H} = \nabla \times \mathbf{B} = \nabla \times (\nabla \times \mathbf{A}) = \nabla(\nabla \cdot \mathbf{A}) - (\nabla \cdot \nabla)\mathbf{A}$$

$$(7.4a)$$

We pointed out in Chapter 2 that the curl and divergence are independent properties of a vector. We know that $\nabla \times \mathbf{A} = \mathbf{B}$, but we do not yet have a condition on $\nabla \cdot \mathbf{A}$. We are, therefore, free to choose $\nabla \cdot \mathbf{A} = 0$. We will see later that this is consistent with the time-varying condition. As a result, we find, from Eq. (7.4a), that $\mathbf{A}$ satisfies the vector form of Poisson's equation:

$$\nabla^2 \mathbf{A} = -\mu \mathbf{J} \qquad (7.4b)$$

Equation (7.4b) indicates that each component of $\mathbf{A}$ satisfies the scalar Poisson equation with the corresponding scalar component of $\mathbf{J}$ as its source, just as the scalar electric potential V satisfies the Poisson equation with the scalar charge density ρ as its source. In that case, the equation and its solution are

$$\nabla^2 V = -\frac{\rho}{\varepsilon} \qquad V = \frac{1}{4\pi\varepsilon} \int \frac{\rho}{r} \, dv$$

Here we have the three scalar component equations:

$$A_x = \frac{\mu}{4\pi} \int \frac{J_x}{r} \, dv \qquad (7.5a)$$

$$A_y = \frac{\mu}{4\pi} \int \frac{J_y}{r} \, dv \qquad (7.5b)$$

$$A_z = \frac{\mu}{4\pi} \int \frac{J_z}{r} \, dv \qquad (7.5c)$$

Since the solution of Eq. (7.4b) must be the vector sum of the individual component solutions, it follows that

$$\mathbf{A} = \frac{\mu}{4\pi} \int \frac{\mathbf{J}}{r} \, dv \qquad (7.5d)$$

Figure 7.6 shows an infinitesimal volume of current with density $\mathbf{J}$. The current crossing its surface is $d\mathbf{I} = \mathbf{I}\,d\ell = \mathbf{J}\,dS$, so that

$$\mathbf{J}\,dv = \mathbf{J}\,dS\,d\ell = \mathbf{I}\,d\ell \qquad (7.6)$$

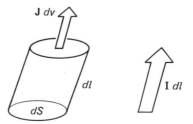

Figure 7.6 Equivalent current elements.

Substituting in Eq. (7.5d) results in Eq. (7.3b). Because of the identical forms of the V and $\mathbf{A}$ equations, all the techniques employed in the previous chapter to solve Laplace's and Poisson's equation for V, for example, using the electric-field boundary conditions, can also be used with $\mathbf{A}$, using the magnetic-field boundary conditions.

Equations (7.3b) and (7.5) give $\mathbf{A}$ in terms of the currents and, as in the electrostatic case, a considerable simplification is often obtained in magnetic problems by finding the potential $\mathbf{A}$ and then using it to determine $\mathbf{B}$. This is so because each element of $\mathbf{A}$ is parallel to the corresponding element $\mathbf{I}$ or $\mathbf{J}$. Before illustrating such a solution, let us verify that the general form, Eqs. (7.3), does lead to the field of a small current element, according to Eq. (7.1) or Eq. (7.2).

Example 7.2

Figure 7.7(a) shows a small current element, $\mathbf{I}\,d\ell = I\,d\ell\mathbf{a}_z$, at the origin. Find $\mathbf{A}$ and demonstrate that it leads to Eq. (7.2).

For the sufficiently small current element, its length can be taken to be infinitesimal, so that $d\ell = \ell$ and the integral in Eq. (7.3b) need not be retained.

$$\mathbf{A} = \frac{\mu I\ell}{4\pi r} = \mathbf{a}_z\frac{\mu I\ell}{4\,\pi r} \tag{7.7}$$

This gives $\mathbf{A}$ in terms of the cartesian coordinate unit vector $\mathbf{a}_z$ and the spherical coordinate r. To take the curl of $\mathbf{A}$, we must express it in a single consistent coordinate system. Therefore, from Fig. 7.7(b), we write $\mathbf{a}_z$ in spherical coordinates:

$$\mathbf{a}_z = \mathbf{a}_r \cos\theta - \mathbf{a}_\theta \sin\theta \tag{7.8}$$

so that

$$\mathbf{A} = \frac{\mu I\ell}{4\pi r}(\mathbf{a}_r \cos\theta - \mathbf{a}_\theta \sin\theta)$$

$$= \mathbf{a}_r A_r(r,\theta) + \mathbf{a}_\theta A_\theta(r,\theta)$$

and

$$\mathbf{B} = \nabla \times \mathbf{A} = \mathbf{a}_\phi\frac{\mu I\ell}{4\pi r^2} \sin\theta$$

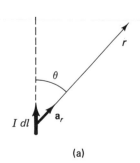

$I\,dl$

(a)

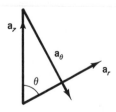

(b)

Figure 7.7 (a) Current element (b) Unit vector transformation.

Note from Fig. 7.7(a) that $\mathbf{I} \times \mathbf{a}_r = I\mathbf{a}_z \times \mathbf{a}_r = I \sin \theta \, \mathbf{a}_\phi$. Substituting into the expression for $\mathbf{B}$, gives

$$\mathbf{B} = \frac{\mu}{4\pi r^2} \ell \mathbf{I} \times \mathbf{a}_r$$

which is the desired expression.

Example 7.3

Calculate the vector potential in the central plane of a straight current-carrying wire segment in air.

If I is along the z axis, then all contributions to $\mathbf{A}$ are also parallel to $\mathbf{a}_z$. The vector integration, therefore, reduces to a scalar integration of the current elements: $I \, d\ell = I \, dz$. Referring to Fig. 7.8, we have

$$\mathbf{A} = \mathbf{a}_z \frac{\mu_0}{4\pi} I \int_{-L}^{L} \frac{dz}{r} = \mathbf{a}_z \frac{\mu_0 I}{4\pi} 2 \int_0^L \frac{dz}{(z^2 + \rho^2)^{1/2}}$$

$$= \mathbf{a}_z \frac{\mu_0 I}{2\pi} \ln \frac{L + (L^2 + \rho^2)^{1/2}}{\rho} = \mathbf{a}_z A_z(\rho)$$

since ρ is a constant under this integration. Taking the curl of this expression yields, after some algebraic manipulation,

$$\mathbf{B} = \nabla \times \mathbf{A} = \mathbf{a}_\phi \frac{\mu_0 I \, \cos \theta_L}{2\pi\rho}$$

For an infinite wire, $\theta_L \approx 0$ and $\cos \theta_L \approx 1$, giving the result found in Chapter 5.

Example 7.4

Calculate $\mathbf{A}$ at distant points in the xz plane, from the small square current loop shown in Fig. 7.9(a). The sides are of length L and the current is I.

First, note that at points in the xz plane, the vector potential from side AB is equal and opposite to that from side CD, so they sum to zero. This is so because the vector direction of $\mathbf{A}$ is parallel to the $\mathbf{I}$, which are antiparallel in the two sides, and because the magnitudes are equal since these sides are symmetrically located with respect to the xz plane.

For substitution into the expression for $\mathbf{A}$ found in Example 7.2, we note, for sides AD and BC, that $\theta = \pi/2$ in Eq. (7.8). From Fig. 7.9(b) for large R,

$$r_{BC} \approx R + \frac{L}{2} \sin \phi \qquad r_{DA} \approx R - \frac{L}{2} \sin \phi$$

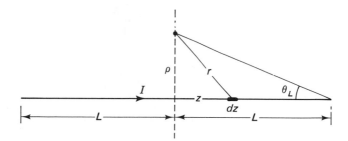

Figure 7.8   Straight current segment.

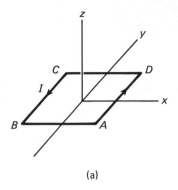

(a)

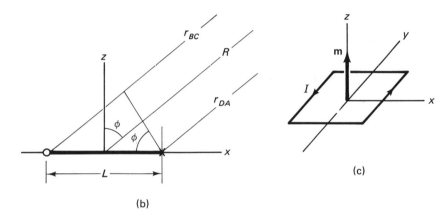

(b)

(c)

Figure 7.9   Magnetic dipole.

The currents and, therefore, the **A** of sides DA and BC are oppositely directed, so the contributions from the two elements are to be subtracted:

$$\mathbf{A} = \frac{\mu IL}{4\pi} \mathbf{a}_y \left[\frac{1}{R - \dfrac{L}{2}\sin\phi} - \frac{1}{R + \dfrac{L}{2}\sin\phi} \right]$$

$$= \frac{\mu IL}{4\pi} \mathbf{a}_y \left[\left(R - \frac{L}{2}\sin\phi \right)^{-1} - \left(R + \frac{L}{2}\sin\phi \right)^{-1} \right]$$

$$= \frac{\mu IL}{4\pi R} \mathbf{a}_y \left[\left(1 + \frac{L}{2R}\sin\phi \right) - \left(1 - \frac{L}{2R}\sin\phi \right) \right]$$

$$+ \text{ terms that are higher powers of } \frac{L}{2R}\sin\phi$$

$$\mathbf{A} = \frac{\mu IL^2}{4\pi R^2} \sin\phi \, \mathbf{a}_y$$

where we have used the binomial expansion with $L/2R \ll 1$.

This expression for **A** can be put in a standard form in terms of a vector $\mathbf{m} = I\mathscr{A} = IL^2\mathbf{n}$, where $|\mathscr{A}|$ is the loop area and $\mathbf{n} = \mathbf{a}_z$ is normal to $\mathscr{A}$, given by the right-hand thumb when the fingers curl with I, as shown in Fig. 7.9(c). With this change, we have submerged the dependence of our result on the specific shape of the current loop. If the field point is sufficiently far so that details of the loop shape are unimportant, we can generalize the answer to any small loop that is defined in terms of **m**. We recognize that the direction of **A** is actually about the loop axis; it was only for ease of calculation that we chose a square loop and found **A** in the xz plane. We then have

$$\mathbf{A} = \frac{\mu}{4\pi r^2}\mathbf{m} \times \mathbf{r}_0$$

The field produced by this magnetic dipole can be calculated by taking $\mathbf{B} = \nabla \times \mathbf{A}$. With $\mathbf{m} = m\mathbf{a}_z$, and using Eq. (7.8) to express $\mathbf{a}_z$ in spherical coordinates,

$$\mathbf{H} = \mathbf{B}/\mu = \frac{m}{4\pi r^3}(2\cos\theta\,\mathbf{a}_r + \sin\theta\,\mathbf{a}_\theta)$$

When $\theta = 0$, this reduces to the solution found in Example 7.1a. In fact, these expressions for **A** and **H** are extensions of that example for field points off the axis, in the case that L (or a in Example 7.1a) $\ll R$.

7.4 MAGNETIC SCALAR POTENTIAL

We have studied the electric scalar potential $\mathbf{E} = -\nabla V$, and the magnetic vector potential $\mathbf{B} = \nabla \times A$. Both V and **A** satisfy Poisson's equation or, in the absence of sources, Laplace's equation, so that, formally, they involve similar methods of solution. In practice, however, the vector nature of **A** makes it much more difficult to solve many real magnetics problems than with the scalar V in the corresponding electric cases. Therefore, in an effort to attain computational ease in the magnetic case, a magnetic scalar potential has also been employed in a number of important cases. To see when this can be done, let us begin with the definition of this potential ψ:

$$H = -\nabla\psi \tag{7.9}$$

Since the magnetic scalar potential is employed almost entirely in the solution of static or quasistatic problems, we will consider only the time-invariant case. In that event, Maxwell's equations separate into two electric and two magnetic statements. The latter are now

$$\mathbf{J} = \nabla \times \mathbf{H} = -\nabla \times \nabla\psi = 0 \tag{7.10a}$$

$$\nabla \cdot \mathbf{B} = \mu\nabla \cdot \mathbf{H} = -\mu\nabla^2\psi = 0 \tag{7.10b}$$

where, in Eq. (7.10a), we have noted that the curl of any gradient is identically zero. This equation states that the definition given by Eq. (7.9) can only be applied when $\mathbf{J} = 0$, that is, away from the real current field sources.

Example 7.5

Determine the magnetic scalar potential from a long line of current along the z axis.

In this case, the answer is already known, from Eq. (4.7), to be $\mathbf{H} = \mathbf{a}_\phi I/2\pi\rho$. Then $\mathbf{H} = -\nabla\psi = -\mathbf{a}_\phi\, \partial\psi/\rho\, \partial\phi$, from which we find $\psi = -I\phi/2\pi$ (neglecting the integration constant).

A problem can arise when employing ψ; we illustrate it by applying Ampère's law in integral form:

$$I = \oint \mathbf{H} \cdot d\boldsymbol{\ell} = -\oint \nabla\psi \cdot d\boldsymbol{\ell} = -\oint d\psi_\ell = \frac{I}{2\pi}(\phi_{\text{final}} - \phi_{\text{initial}})$$

If the path of integration is a circle about the current line, then $\phi_{\text{final}} - \phi_{\text{initial}} = 2\pi$ and $\oint \mathbf{H} \cdot d\boldsymbol{\ell} = I$, as expected. In evaluating the integral, we note that we could have $\phi_{\text{final}} - \phi_{\text{initial}} = n2\pi$, where $n = 1, 2, 3, \ldots$. If we are to maintain the physical meaning of Ampère's law and the single-valuedness of physical quantities, we must prevent complete circling of the current. We can do this by placing a hypothetical infinitesimally thin radial cut in the (ρ, ϕ) plane, as seen in Fig. 7.10, which cannot be crossed. Integrating around the path, we now have only the single value $\phi_{\text{final}} - \phi_{\text{initial}} = 2\pi$.

Example 7.5 illustrates a fuller meaning of the restriction to not apply the scalar potential in the region of the sources. Not only must we not be *at* $\mathbf{J}$ (note that ψ, in Example 7.5, is undefined when $\rho = 0$), but we must not allow our solution region to completely enclose the current sources.

In Eq. (7.10b), μ was constant, with the result that ψ satisfies Laplace's equation. Of course, its solution must be subject to appropriate boundary conditions between magnetic media with differing μ. Since the normal component of $\mathbf{B}$ is continuous at an interface between medium 1 and medium 2,

$$\mu_1 \frac{\partial\psi_1}{\partial n} = \mu_2 \frac{\partial\psi_2}{\partial n} \tag{7.11a}$$

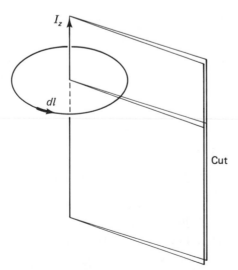

Figure 7.10 Spacial cut to maintain single-valued magnetic scalar potential.

where n is the coordinate normal to the boundary. And, since the tangential component of **H** is continuous,

$$\frac{\partial \psi_1}{\partial \tau} = \frac{\partial \psi_2}{\partial \tau} \rightarrow \psi_1 = \psi_2 \qquad (7.11\text{b, c})$$

where τ is a coordinate tangent to the boundary. To obtain Eq. (7.11c) from Eq. (7.11b), note that any constant value can be added to ψ without affecting the fields, since they are derived by taking derivatives of ψ. Therefore, if such a constant is added to, say, ψ_1, so that $\psi_1 = \psi_2$ at a point of the boundary, then the condition that the two ψ have equal derivatives *along* the interface means that the ψ themselves will remain equal at every point of the boundary.

Example 7.6

Determine the shielding attenuation factor $H_{\text{outside}}/H_{\text{inside}}$ for a long hollow magnetic cylinder in a uniform transverse applied field.

From our previous work on electrostatic potentials, we know that the axially symmetric solution of Laplace's equation in cylindrical coordinates has the general form

$$V = \sum_n (A_n \cos n\phi + B_n \sin n\phi)(M_n \rho^n + N_n \rho^{-n})$$

Applying this to Fig. 7.11(a), we recognize that the coefficients of ρ^n would be zero in region 1 if ψ_1 were to remain finite as $\rho \rightarrow \infty$. As in Example 6.7, the exception to this is the term $H_0 \rho \cos \phi$, which describes the uniform applied field. Similarly, the coefficients of ρ^{-n} must be zero in region 3 as ψ_3 is finite when $\rho \rightarrow 0$.

The symmetry of Fig. 7.11(b) about the plane $\phi = 0$ demands that $H_\phi(\phi) = -H_\phi(-\phi)$, that is, that $H_\phi(\phi) \sim \partial \psi / \partial \phi$ is an odd function of ϕ. This implies that $\psi(\phi)$

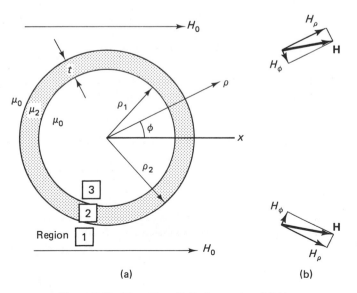

Figure 7.11   Magnetic cylinder in an external field.

is an even function, so that it can have no terms varying as $\sin \phi$. We then find that $H_\rho(\phi) = \partial \psi/\partial \rho$ is an even function of ϕ, also as demanded by the symmetry shown in Fig. 7.10(b). Using these results in the general solution yields

$$\psi_1 = -H_0 \rho \cos \phi + \sum b_n \rho^{-n} \cos n\phi$$

$$\psi_2 = \sum (c_n \rho^n + d_n \rho^{-n}) \cos n\phi$$

$$\psi_3 = \sum a_n \rho^n \cos n\phi$$

The boundary conditions of Eqs. (7.11), at $\rho = \rho_2$ and at $\rho = \rho_1$, taking $\mu = \mu_2/\mu_0$, give

$$-H_0 \rho_2 \cos \phi + \sum b_n \rho_2^{-n} \cos n\phi = \sum (c_n \rho_2^n + d_n \rho_2^{-n}) \cos n\phi$$

$$-H_0 \cos \phi + \sum -nb_n \rho_2^{-n-1} \cos n\phi = \mu \sum (nc_n \rho_2^{n-1} - nd_n \rho_2^{-n-1}) \cos n\phi$$

$$\sum a_n \rho_1^n \cos n\phi = \sum (c_n \rho_1^n + d_n \rho_1^{-n}) \cos n\phi$$

$$\sum na_n \rho_1^{n-1} \cos n\phi = \mu \sum (nc_n \rho_1^{n-1} - nd_n \rho_1^{-n-1}) \cos n\phi$$

As in Example 6.7, these equations can be multiplied by $\cos m\phi \, d\phi$ and integrated from 0 to 2π. With Eqs. (6.18), this yields the result that, for each equation, the coefficients of each $\cos n\phi$ term are individually equal. These yield, after some manipulation,

$$-H_0 \rho_2^2 - b_1 = \mu c_1 \rho_2^2 - \mu d_1$$

$$c_n \rho_1^{2n} = d_n \frac{\mu - 1}{\mu + 1} = b_n \frac{\mu - 1}{2\mu} \qquad n \neq 1$$

$$d_n \rho_1^{-2n} = c_n \frac{\mu - 1}{\mu + 1} = a_n \frac{\mu - 1}{2\mu} \qquad \text{for all } n$$

Noting that these equations must hold for all values of μ, including $\mu = 1$ (when $\mu_2 = \mu_0$, so the magnetic shield vanishes), we find that

$$a_n = b_n = c_n = d_n = 0 \qquad \text{for all } n \neq 1$$

$$a_1 = -4\mu_0 H_0 \frac{\rho_2^2}{\rho_2^2(\mu + 1)^2 - \rho_1^2(\mu - 1)^2}$$

If we are only interested in the field in region 3, there is no need to explicitly display the coefficients b_1, c_1, and d_1. Within the shielded volume,

$$\mathbf{H}_{\text{inside}} = -\nabla \psi_3 = -\nabla a_1 \rho \cos \phi = -\nabla a_1 x = -a_1 \mathbf{a}_x$$

showing that the internal field is uniform. The field reduction factor is

$$\frac{H_{\text{outside}}}{H_{\text{inside}}} = \frac{H_0}{H_{\text{inside}}} = -\frac{H_0}{a_1}$$

In practice, when a high-quality magnetic shield is made, nickel–iron alloys are used, which have μ ranging from several hundreds to hundreds of thousands. As a result, it is safe to take $\mu \gg 1$ in the expression for $\mathbf{a}_1$, giving

$$\frac{H_{\text{outside}}}{H_{\text{inside}}} = \mu \frac{\rho_2^2 - \rho_1^2}{4\rho_2^2}$$

Let $\rho_2 = \rho_1 + t$, where t is the cylinder wall thickness. If it is thin-walled ($t \ll \rho$), we have $\rho_2^2 - \rho_1^2 \approx 2\rho t$, where we need not distinguish the small difference between ρ_1 and ρ_2. Our final result is then

$$\frac{H_{\text{outside}}}{H_{\text{inside}}} = \frac{\mu t}{2\rho}$$

Using representative values of $\mu = 40,000$, $t = 0.062$ inch, and $\rho = 3$ inch (t and ρ can be in any units, as long as they are both in the same units, since their ratio will be invariant), yields $H_{\text{outside}}/H_{\text{inside}} = 413$, showing the considerable level of magnetic field reduction that can be obtained.

This is the technique of shielding systems that are sensitive to ambient magnetic fields. Modern electron-beam devices, such as photomultipliers and cathode-ray tubes, are used in many applications where great precision is required, for example, medical imaging, and reading two-dimensional patterns for image intensification and processing. Because these devices are particularly affected by small low-frequency alternating fields, they frequently need to be shielded. It is usual practice to shield the display tubes of high-quality oscillographs.

In electronic circuits, one can frequently avoid noise and interference, or pack elements more densely in a small package, by enclosing certain key components in magnetic shields. Since the most troublesome source of magnetic fields in many circuits is the input power transformer, it is not uncommon to find this component with an enclosing shield. A concern over dc and low-frequency magnetic shielding leads naturally to concern for higher-frequency effects. The problem of radio-frequency interference (RFI) was alluded to in Problem 5.29; it is a vital pervasive problem in our modern society. Unfortunately, we cannot deal with it here.

7.5 MAGNETIC POLES

In the previous section, boundary problems in the presence of magnetic material were treated by reference only to the conditions on H and on B. Through the parameter $\mu/\mu_0 = 1 + \chi = 1 + M/H$, the material property M was implicitly included. In many applications involving ferromagnetic material such as iron, cobalt, nickel, and their alloys, M is not uniform, and varies widely, so that μ may not be a useful parameter unless its use is accompanied by an understanding of the response of ferromagnetic material. One aspect of this understanding involves the concept of magnetic poles, which we derive through use of the magnetic scalar potential.

We begin by taking

$$\nabla \cdot \mathbf{B} = \mu_0 \nabla \cdot (\mathbf{H} + \mathbf{M}) = 0 \qquad (7.12a)$$

Since $H = -\nabla\psi$, this becomes

$$\nabla^2\psi = -\rho_m \qquad \text{where } \rho_m = -\nabla \cdot \mathbf{M} \qquad (7.12b)$$

Thus, in cases where the magnetization is not uniform, we find that ψ satisfies Poisson's equation with an effective density of *magnetic charge* given by $\rho_m = -\nabla \cdot \mathbf{M}$.

To distinguish the magnetic charge from electric charge, we refer to ρ_m as a density of magnetic *poles*. In a volume dv, where $\mathbf{M}$ is changing, the magnetic-pole strength is $dQ_m = \rho_m\, dv$, and ψ could be found from Q_m by the same methods used to find the electrostatic V from *its* charge sources $dQ = \rho\, dv$.

Historically, the development of magnetic theory paralleled that of electrostatics in the assumption that isolated magnetic charge exists, making it natural to consider ρ_m and a magnetic scalar potential. With the realization that magnetic poles always occur in pairs, that is, that the lowest irreducible magnetic field source is the dipole, and that an equivalent representation of the dipole is a current loop, the asymmetry of electric and magnetic fields was recognized. Today, there are established theories of the fundamental particles of nature that predict what properties a magnetic monopole should have if it exists, and considerable effort has been expended to search for it using studies of high-energy nuclear colliding machines and cosmic rays. Such a discovery would alter the fundamental understanding of the symmetry and forces that constitute the universe.

As no such particle has been found, Eq. (7.12b) must be regarded as a mathematical fiction. However, it describes material behavior in a way that allows simple and direct visualizing and calculation of otherwise complicated properties. Since Eq. (7.12b) is derived from Maxwell's equations, it can be used in a consistent manner to obtain correct results. Particularly when studying ferromagnetic material, where M is very large and properties are nonlinear, the assumed existence of poles is common and useful. This is illustrated in the following sections.

Of particular importance is the change of $\mathbf{M}$ that occurs at the boundary of an otherwise uniformly magnetized body. Figure 7.12(a) shows such a situation, with $\mathbf{M} = M_0 \mathbf{a}_z$ within the body and $\mathbf{M} = 0$ outside of it. For simplicity, $\mathbf{M}$ is taken to be normal to the surface. The change of M and of $\rho_m = -\nabla \cdot \mathbf{M} = -dM/dz$ are

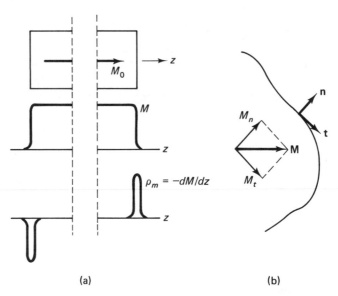

(a)   (b)

Figure 7.12   Surface magnetic poles.

shown schematically, on a much magnified scale, at the surface. It is seen that there is an equivalent surface pole density on the area $A = \int dx\,dy$. At the right edge, this is given by

$$\rho_{m,s} = \frac{Q_m}{A} = \frac{1}{A}\int \rho_m \, dv = \int \rho_m \, dz = -\int_{M_0}^{0} dM = M_0 \qquad (7.13)$$

Note that since $\rho_m = -\nabla \cdot \mathbf{M} = -\partial M/\partial z$, we find ρ_m to be positive at the right edge of the sample, where M decreases. Similarly, it is negative at the left edge of the sample, where M increases. Rather than speak of positive and negative poles, we further distinguish them from the electrical case by referring to *north* and *south* poles. (These names developed historically from the definition that the north pole of a bar magnet is attracted toward the geomagnetic North Pole of the earth.) Figure 7.12(b) shows a case where $\mathbf{M}$ is *not* parallel to the surface normal $\mathbf{n}$. In this case, the tangential component of $\mathbf{M}$ does not change in the direction of $\mathbf{t}$, so that $\rho_{m,s} = \mathbf{M} \cdot \mathbf{n}$.

From Eqs. (7.12), we see that $\nabla \cdot \mathbf{H} = -\nabla \cdot \mathbf{M} = \rho_m$, so that $\mathbf{H}$ diverges from magnetic poles, that is, surface pole distributions produce magnetic fields, in exact analogy to the electric fields generated by surface charge on a polarized dielectric. Externally to the body, this is the magnetic field for which magnetized material is known. Internally, this field opposes the magnetization that produced it; as it tends to reduce that magnetization, it is called the *demagnetizing field*.

Example 7.7

A thin disc of magnetic material is placed normal to a large uniform applied field. Examine the effect of pole formation on the internal fields.

If the applied field is large, then the body will be uniformly magnetized parallel to the field, that is, $\mathbf{M} = M_0 \mathbf{a}_z$, as shown in Fig. 7.13. From the previous discussion, the magnetic surface pole density is $\rho_{m,s} = M_0$, so there are two equal and opposite pole distributions on the parallel body surfaces. Recall from the electrostatic case that two parallel planes of charge produce a field equal to the surface charge density ρ_s [see Exercise 3.3 and the discussion leading to Eq. (3.24a)]. Here the same principle holds, with $\rho_{s,\text{magnetic}} = M$, so that the demagnetizing field must be

$$\mathbf{H}_{\text{demagnetizing}} = -M_0 \mathbf{a}_z \qquad (7.14a)$$

In the z direction, the internal flux density is

$$B = \mu_0(H + M) = \mu_0[(H_{\text{applied}} - M_0) + M_0] = \mu_0 H_{\text{applied}} \qquad (7.14b)$$

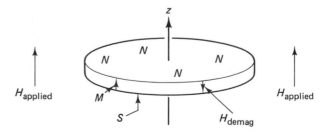

Figure 7.13 Thin magnetic disc.

where H has been expressed as the sum of the applied and demagnetizing fields. The final result, $B = \mu_0 H_{applied}$, is the same as that outside the body.

From an alternative approach, we note that **B** must be the same in and out of the body, since the boundary condition of **B** states that its normal component (which, in this case, is the entire **B**) is continuous.

7.6 DEMAGNETIZING FACTOR

In the cases just discussed, **H** and **M** were uniform throughout the body. In fact, for arbitrarily shaped magnetic bodies, the density of poles and their demagnetizing fields will not be uniform, for example, near a surface irregularity such as that shown in Fig. 7.14(a) As a result, **B** and **M** will not be uniform. In ferromagnetic material, the demagnetizing field is often very large and, since it opposes the applied field and the internal magnetization, it causes reversal of some of the magnetic moments. Instead of the uniformly magnetized body of Fig. 7.12, typical equilibrium distributions of magnetization in regular magnetic bodies are shown in Figs. 7.14(b) and (c). These are views of the surface magnetic structure looking directly onto a magnetic material. The small end regions and vertical domains are called *reverse* or *closure domains*. While **M** is not uniform throughout the magnetic

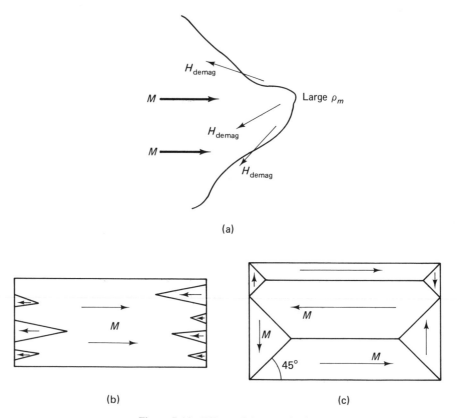

(a)

(b)   (c)

Figure 7.14   Effects of demagnetization.

body, it is uniform within each magnetic domain. These configurations serve to lesser the pole densities and demagnetizing fields by reducing the divergence of M and by distributing the poles over greater distances. Note that the domains are not necessarily equal, often depending upon microscopic structural details of the material. The boundaries between domains in which the magnetization has differing orientations are called *domain walls*; they are shown as lines in Fig. 7.14.

Example 7.8

Show that there are no poles produced by the domain structure of Fig. 7.14(c), so that the demagnetizing fields are zero.

We start with the relation

$$-\rho_m = \nabla \cdot \mathbf{M} = \frac{\partial M_n}{\partial n} + \frac{\partial M_t}{\partial t} \tag{7.15}$$

where the divergence is written in terms of derivatives along the orthogonal axes n and t, which denote directions normal and tangential to the domain walls, respectively. If we move vertically between the horizontally oriented domains, that is, along $\mathbf{n}$, then, even though M_t reverses, M_n does not, so that $\rho_m = 0$. If we move perpendicular to the 45° walls, we have $\partial M_n/\partial n = 0$ because M_n is the same in both domains. In both cases, $\partial M_t/\partial t = 0$.

The occurrence of demagnetizing fields in ferromagnetics is often a critical aspect of their magnetic behavior. As indicated by the examples just given, the reduction of $H_{\text{demagnetizing}}$ determines the structure of magnetic domains, particularly at the edges of the body. While it is therefore important to calculate these fields, in general, this is an almost impossible task because of the inhomogeneity of the pole and field distributions. In one particular class of geometries, however, the demagnetizing field is simply found. In an ellipsoidal body that is magnetized along one of its axes, the pole distribution is such that the induced field is also uniform and is given by

$$\mathbf{H}_{\text{demagnetizing}} = -N\mathbf{M} \tag{7.16a}$$

where N is the demagnetizing factor along *that* axis. For a general ellipsoid, shown in Fig. 7.15, a theorem relates the factors along each of the axes:

$$N_i + N_j + N_k = 1 \tag{7.16b}$$

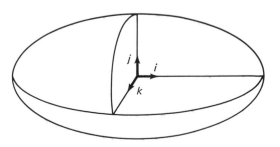

Figure 7.15   General ellipsoid.

Example 7.9

Calculate the three demagnetizing factors for a large disc.

For this example to have meaning, we must assume the commonly used approximation that a flat disc is an ellipsoid with one very short axis (along the disc normal) relative to the other two. We call this the i axis. In Example 7.7, Eq. (7.14a), we found that along this axis, $H_{\text{demagnetizing}} = -M_0$, which means that $N_i = 1$. From Eq. (7.16b), we must then have $N_j = 0 = N_k$.

This result can be understood physically by recognizing that when a large thin disc is magnetized in its plane there are only a few poles on the thin edges, and these are sufficiently far from the bulk of the disc so that their fields are negligible, making $H_{\text{demagnetizing}} \approx 0$.

Example 7.10

Calculate the three demagnetizing factors for a thin round wire.

As before, we represent the wire by an ellipsoid that has one very large i axis, the length of the wire, and that is short along the other two axes. Using the reasoning of the last paragraph of Example 7.9, we see that, for magnetization along the length of the wire, there are very few very distant poles, so $N_i = 0$. The other two orthogonal axes are perpendicular to the wire axis. For a round wire, they are equal, so that Eq. (7.16b) gives $2N_j = 1$, or $N_j = \frac{1}{2} = N_k$.

From these examples, we see that the ellipsoidal approximation can be used to approximate the demagnetization factor of nonellipsoidal bodies. The factor depends upon the relative dimensions of the body; the upper and lower curves in Fig. 7.16 show the demagnetizing factors for the cases that two of the ellipsoid axes are equal, being either greater or smaller than the third axis. N is a function of the ratio of the axis length along which the sample is magnetized to the perpendicular axis length. Other cases, that is, where all the axes are different, fall in the shaded region.

Demagnetization factors are derived here from the divergence of **M**, using the concept of the magnetic scalar potential. It is worth noting that in the case of dielectrics, the scalar electric potential results in an identical depolarizing factor from the divergence of **P**. From a practical point of view, $|\mathbf{P}|$ is generally small enough that depolarization effects are not significant compared to other factors that enter into the

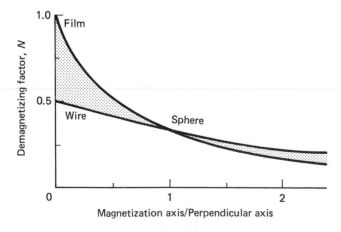

Figure 7.16 Demagnetizing factors of ellipsoids of revolution.

polarization behavior. However, there is a class of materials, called *ferroelectrics,* which have large permanent electric moments, similar to the magnetic moments of ferromagnetics, in which depolarization effects are relevant.

7.7 MAGNETIC CIRCUITS

Magnetic materials are generally divided into *soft* materials, which are used in transformers, relays, motors, etc., and *hard* materials, which are generally permanent magnets. While that nomenclature originally derived from the mechanical properties of the two classes, these mechanical properties were by-products of material selection and alteration so as to make their resistance to magnetic change similarly *soft* (i.e., easy) or *hard* (i.e., difficult). Because magnetic structures can involve several types of materials, for example, with different values of μ and different magnetizations, or can contain both hard and soft materials, and because the material parameter μ is a function of flux density, or may not be uniquely defined, the analysis of these magnetic structures can be very complex. In many cases, the magnetic structure can be treated as an analog of a corresponding electric circuit. Like most analogies, it is inexact and limited, but allows a useful understanding of the behavior and design of magnetic systems.

To illustrate development of the analog, let us refer to the configuration of Fig. 7.17, which shows what might be a two-coil electromagnet or a transformer with an air gap, consisting of materials of different permeabilities μ_i, different lengths ℓ_i, and different cross-sectional areas A_i. It will be shown shortly that the use of air gaps in magnetic structures lends significant advantage and versatility. We apply Ampère's law, $\oint \mathbf{H} \cdot d\ell = I$, around the dotted path shown. Since the path elements are discrete, the contour integral becomes a summation over the sections.

$$\sum H_i \ell_i = \sum N_i I_i = \sum \mathcal{V}_i \tag{7.17}$$

As this magnetic circuit is driven by the current linking it, we have taken our first electrical analog by defining the magnetomotive force (mmf) $\mathcal{V}_i = N_i I_i$ as corresponding to the electrical voltage, or electromotive force (emf).

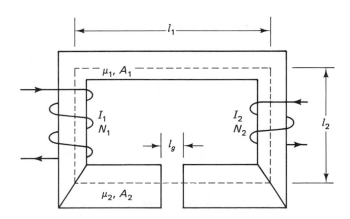

Figure 7.17   Simple magnetic structure.

The left-hand side of Eq. (7.17) can be expressed in a more useful form by multiplying and dividing each term by $\mu_i A_i$ and combining terms:

$$H_i \ell_i = \mu_i H_i A_i \frac{\ell_i}{\mu_i A_i} = B_i A_i \mathcal{R}_i = \Phi_i \mathcal{R}_i \tag{7.18}$$

where Φ_i is the total flux in medium i, and $\mathcal{R}_i$ is called the *reluctance* of that medium. Equation (7.17) then has the form

$$\sum \mathcal{V}_i = \sum \Phi_i \mathcal{R}_i \qquad \text{(loop equation)} \tag{7.19}$$

We complete the electrical analog if we take the flux Φ_i to correspond to the electrical current and $\mathcal{R}_i$ to the electrical resistance. Note that the magnetic reluctance is

$$\mathcal{R} = \ell / \mu A \tag{7.20}$$

and the electrical resistance is $R = \ell / \sigma A$, so that μ plays the same role in flux conduction as σ does in current conduction.

Equation (7.19) is the magnetic counterpart of the electrical-circuit loop equation $\sum V_i - \sum R_i I_i = 0$. To complete the circuit analysis, we establish the counterpart of the electrical node equation $\sum I_i = 0$. This follows from Maxwell's equation on the conservation of flux, $\oint \mathbf{B} \cdot \mathbf{n} \, dS = 0$, which states that all flux entering a volume (a circuit node) must leave it, that is,

$$\sum \Phi_i = 0 \qquad \text{(node equation)} \tag{7.21}$$

Example 7.11

Solve and discuss the magnetic circuit of the system shown in Fig. 7.17.

Note from the figure that the current mmf's are additive since the fields they generate are in the same direction around the loop. Since there is a single flux path, all the $\Phi_i = \Phi$, and we need only apply Eq. (7.18) to the loop shown in Fig. 7.18.

$$\Phi = \frac{\mathcal{V}_1 + \mathcal{V}_2}{\mathcal{R}_1 + \mathcal{R}_2 + \mathcal{R}_g}$$

If the system is constructed with high-quality material, then μ_1 and $\mu_2 \gg \mu_0$, and, even though ℓ_1 and $\ell_2 \gg \ell_g$, we normally have $\mu_i / \ell_i \gg \mu_0 / \ell_0$, so that $\mathcal{R}_1$ and $\mathcal{R}_2 \ll \mathcal{R}_g$. Keeping only this leading term in the denominator gives

$$\Phi_\infty = \frac{\mathcal{V}_1 + \mathcal{V}_2}{\mathcal{R}_g}$$

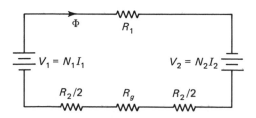

Figure 7.18 Equivalent magnetic circuit of Fig. 7.17.

from which

$$B_\infty = \mu_0 \frac{N_1 I_1 + N_2 I_2}{\ell_g} = \mu_0 H_\infty$$

Note that B_∞ and H_∞ are the fields that would be found in the air gap if the core materials had $\mu = \infty$. Alternatively, they are the fields if all the turns were wrapped at the gap, or if there were an infinitely long air-core solenoid with ampere–turns/meter given by $NI/\ell_g = (N_1 I_1 + N_2 I_2)/\ell_g$, where NI is the effective number of ampere–turns. While μ is not infinitely large, actual circuits commonly have $\mathcal{R}_1$ and $\mathcal{R}_2 \ll \mathcal{R}_g$, so that the air gap dominates, with the high-permeability material serving only to conduct all the flux to that gap.

7.8 MATERIAL PROPERTIES

If ℓ_g is very small or μ_1 and μ_2 are not so very large, $\mathcal{R}_g$ can be sufficiently small so that we cannot ignore $\mathcal{R}_1$ and $\mathcal{R}_2$ and we must account for the actual properties of the core materials. Because these properties are nonlinear and dependent upon magnetic history and material preparation and treatment, it is inadvisable to attempt an exact analysis. Instead, we seek a graphical solution of the magnetic circuit. This is based on Eq. (7.17), stating that the input mmf $(= \Sigma \mathcal{V}_i)$ must equal the circuit mmf's $(= \Sigma H_i \ell_i)$.

Example 7.11 (cont.)

For simplicity in the following discussion, we take a single constituent core with uniform cross-sectional area. Then $A_1 = A_2$ and the total magnetic core length is $\ell_m = 2(\ell_1 + \ell_2) - \ell_g$. In addition, as before, we define $N_1 I_1 + N_2 I_2 = NI$. Since there is a single circuit loop, the flux Φ is constant throughout and the loop equation is obtained by equating the input and circuit mmf's:

$$NI = H_m \ell_m + H_g \ell_g$$

from which

$$H_m = \left(\frac{NI}{\ell_g} - H_g\right)\frac{\ell_g}{\ell_m} = (H_\infty - H_g)\eta \tag{7.22}$$

$H_\infty = NI/\ell_g$ is defined in the first part of Example 7.12, and $\eta = \ell_g/\ell_m$ is often referred to as the *geometric demagnetizing*, or *reducing, factor*.

This expression relates H_m, the field in the core, to H_g, the gap field, with parameters being the factor η and H_∞, the *infinite permeability* field. Note that the right-hand side of this relation is purely dependent on the geometric construction and the excitation; it does not contain any reference to the magnetic material being used. The particular values of H_m and H_g to be used in satisfying Eq. (7.22) must depend on the properties of the magnetic material, which is brought into the analysis through the graphical construction in Fig. 7.19.

The figure is begun by drawing the "Air-gap line," that is, the relation that holds in air, $B = \mu_0 H$, extending from points $(B, H) = (0, 0)$ to (B_∞, H_∞), as shown. For a given B, a point on this line represents the corresponding value of H in the gap, that is, H_g, and the distance from the right edge, that is, from H_∞, gives $(H_\infty - H_g)$, also shown on the figure. According to Eq. (7.21) multiplication of this difference by the reducing factor η gives H_m for each value of B, so that H_m falls on a second line in Fig. 7.19, shown as H_{core}.

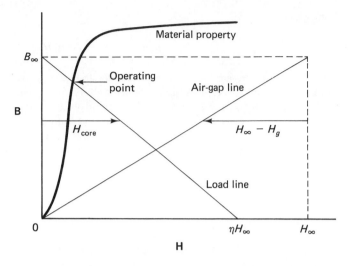

Figure 7.19   Solution of the nonlinear magnetic circuit.

In analogy with electrical usage, we call this plot of H_{core} for each value of B the load line of the magnetic circuit; it gives the relation between B and H in the core material, as determined by the physical structure of the magnetic circuit. The material-property curve also shows the inherent relation between H_m and B_m of the core material, so that the intersection of the load line and this *normal magnetizing* curve defines the operating point.

Note that at this point, the flux density is reduced from B_∞ and we no longer have H_∞ in the gap. The gap field is the intersection of the "Air-gap-line" and the horizontal line through the operating point.

The more vertical the material $B–H$ curve, the closer the operating point approaches B_∞. Values of $\mu = B/H$ taken from this curve are shown in Fig. 7.20 for a soft cast steel. Use of μ from Fig. 7.19 is often preferable to employing the normal magnetizing curve, from which it is derived. An initial guess of B_m can be made, and by comparing the corresponding H_m values from the load line [or from Eq. (7.22)] and from Fig. 7.20, the operating point can be found in a few iterations.

The particular form of the $B–H$ curve used in Figs. 7.19 and 7.20 may seem strange in view of the discussion in Chapter 4, indicating the varied hysteresis properties of magnetic material. Rather than deal with such complexity, which would make problem-solving a practical impossibility, the normal magnetizing curve is used. This is obtained by connecting the tips of successively larger hysteresis loops, as shown in Fig. 7.21. Clearly, the ratio $\mu = B/H$ is different at each point of each hysteresis curve, so the ratio taken from this curve must be regarded as approximate; in the fully cyclic alternating case, this μ can be close to the average μ of the hysteresis loop whose maxima are at the same values. These features point up the crude nature of magnetic-circuit calculations and the danger of overrefining them, although the insight obtained by such calculations is invaluable in understanding the relevant features of magnetic systems.

One advantage of using an air gap can be seen from Fig. 7.19. By adjusting η, through varying ℓ_g, one can select the point of intersection of the load line and the

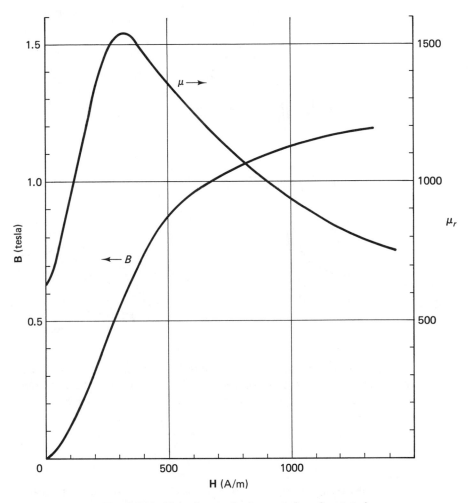

Figure 7.20 Normal magnetization curve for soft cast steel.

material characteristic so as to obtain the maximum value of $\mu = B/H$, or a particular lower value, or an operating point near the saturation shoulder of the curve. All of these have advantages in particular applications of magnetic devices.

7.9 MAGNETIC FORCE AND ENERGY

Changing a magnetic field causes a variation in the energy density, given by $dU_m = H\,dB$. If $B = \mu H$, then the energy per unit volume is $U_m = BH/2$, and to find the total energy, one must integrate U_m over the field volume. In the case of a magnetic circuit, this integral becomes a sum over the various circuit component volumes:

$$W = \frac{1}{2} \sum B_i H_i v_i = \frac{1}{2} \sum (B_i A_i)(H_i \ell_i) = \frac{1}{2} \sum \Phi_i \mathcal{V}_i \qquad (7.23)$$

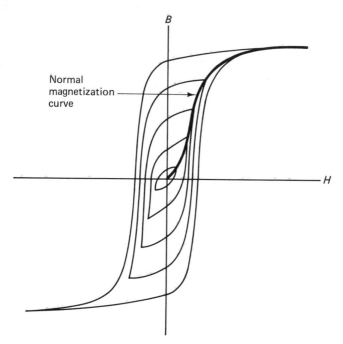

Normal magnetization curve

Figure 7.21 Derivation of the normal magnetization curve.

Therefore, when the parameters change in a magnetic circuit, there is an energy change, and, as in the electrostatic case [see Eq. (6.3)], the constraining force F_x, which opposes change of the circuit parameter x is

$$F_x = -\frac{dW}{dx} \tag{7.24}$$

Example 7.12

The electromechanical relay shown in Fig. 7.22 has 100 turns in its control windings. Soft cast iron, whose characteristics are shown in Fig. 7.20, is used to make the core and armature; their combined length is 20 cm and they have a cross-sectional area of 2 cm². The pivot end of the armature contains a constant air gap with an effective length of 1 mm.

The end of the armature firmly contacts the core when the relay is fully excited. The relay is designed so that the spring separates the armature and core when the current falls below 9 A.

What is the required spring force when the relay is closed?

The force exerted by the spring must just equal the force with which the armature is held in place when $I = 9$ A. Equation (7.24) can be used to find this force if it is known how the energy W changes with ℓ, the air-gap dimension. For this, the core and gap fields must be found. Note that even though the 1-mm gap is at the hinge end, and is not shown in the figure, it is, in any event, a gap in the magnetic circuit (the only one when the relay is closed), so that $H_\infty = NI/\ell_g = 10^2 \cdot 9/10^{-3} = 9 \times 10^5$ A/m, and $B_\infty = \mu_0 H_\infty = 1.131$ T. Finally, $\eta = \ell_g/\ell_m = 0.1/20 = 5 \times 10^{-3}$, so that $\eta H_\infty = 4500$ A/m. The circuit load line is shown schematically in Fig. 7.23; when that line is drawn in Fig. 7.20, the intersection with the material characteristic gives $B = 0.935$ T and $H_m = 610$ A/m.

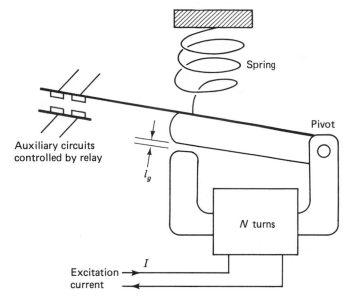

Figure 7.22   Electromechanical relay.

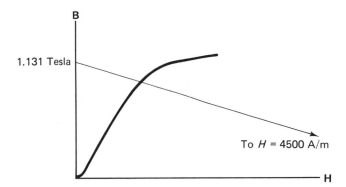

Figure 7.23   Graphical solution of the
magnetic circuit.

Since B is the same in the gap as in the core, the system energy can be written in a
form that emphasizes the gap length:

$$W = \sum \Phi_i \mathcal{V}_i = B^2 A \ell_g / \mu_0 + BH_m v_{\text{mag. core}}$$

The force is found by substituting this expression into Eq. (7.24), with $x = \ell_g$. In doing
this, note that the relative change of B and H_m are much smaller than that of ℓ_g, so they can
be considered constant for an infinitesimal change of ℓ_g. As a result, only the first term
varies on the right-hand side of the expression for W, giving

$$F = \frac{\delta W}{\delta \ell_g} = B^2 A / \mu_0 = 139 \text{ N}$$

7.10 PERMANENT-MAGNET CIRCUITS

In the previous examples, an applied mmf was used to generate the flux. In permanent magnets, the material generates its own field from the poles at interfaces or at air gaps that are part of the circuit. As a result, these fields are demagnetizing fields, and, as indicated by Eqs. (7.14a) and (7.16a), H in the magnetic material is oppositely directed from M and B. Analysis of such situations therefore requires use of the second quadrant, or negative-H portion, of the material $B–H$ curve, called the *demagnetization curve*. Representative curves are shown in Fig. 7.24 for two commonly used permanent-magnet materials: alnico (whose name indicates its constituents) and an alloy of samarium and cobalt. In a zero applied field, the materials exhibit their remnant value B_r, and as H decreases, (that is, increases negatively), B decreases until it reaches zero when H equals the coercive field, H_c.

Around the periphery of Fig. 7.24 are values of $B/\mu_0 H = \mu_{r(\text{effective})}$, the effective relative permeability. To understand the significance of this parameter, consider the circuit in Fig. 7.25, showing a permanent-magnet material with an air gap.

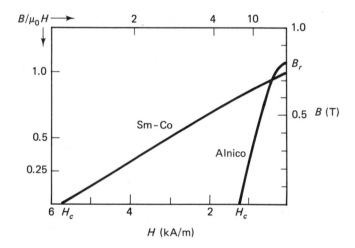

Figure 7.24   Demagnetization curves for permanent magnetic materials.

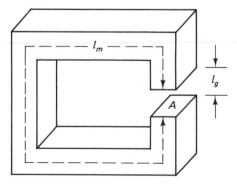

Figure 7.25   Permanent magnet.

In the absence of a driving mmf, the sum of the mmf's in the core and gap is zero, so that

$$H_g \ell_g = -H_m \ell_m \tag{7.25}$$

Conservation of flux demands that

$$B_g A_g = B_m A_m \tag{7.26}$$

Dividing these equations, noting that $B_g / H_g = \mu_0$ and taking the core and gap areas to be equal, gives

$$\mu_{r(\text{effective})} = \frac{B_m}{\mu_0 H_m} = -\frac{\ell_m}{\ell_g} = -\frac{1}{\eta} \tag{7.27}$$

where η is the geometric demagnetizing factor introduced earlier. Equation (7.27) represents the geometric load line, which is to be drawn through the origin of Fig. 7.24 and the correct point on the periphery. Its intersection with the material characteristic demagnetization curve gives the operating point.

Example 7.13

Reconsider the relay shown in Fig. 7.22 and studied in Example 7.12. When the relay is fully open, the gap at the end of the armature is 3 mm and the exciting current is disconnected. The remnant magnetization of this cast steel is 0.5 T, and, because it is a soft material, its demagnetization curve is a straight line with $H_c = 200$ A/m, that is, $B = 0.5 + 2.5 \times 10^{-3} H$. Find the necessary spring force under these conditions.

The geometric factor is $1/\eta = \ell_m / \ell_g = 20/0.4 = 50$. The intersection of the load line, $B = -50 H$, with the demagnetization line described above, occurs at $H = 0.0095$ A/m and $B = 0.45$ T. Proceeding as with Eqs. (7.23) and (7.24),

$$W = \sum \Phi_i \mathcal{V}_i = B^2 A \ell_g / \mu_0 + B H_m v_{\text{core}}$$

and

$$F = \frac{\delta W}{\delta \ell_g} = B^2 A / \mu_0 = 32 \text{ N}$$

By combining the results of Examples 7.12 and 7.13, the spring constant is found to be

$$K = \frac{\Delta F}{\Delta x} = \frac{139 - 32}{4 - 1} \approx 36 \frac{\text{N}}{\text{mm}}$$

We should note one feature that has been omitted from Eq. (7.26). Flux flowing about the curved path of the magnetic circuits we have considered, has a tendency to leave the magnetic material and flow in the air; the two paths, air and iron, represent parallel reluctances. However, because of the high ratio of μ / μ_0, the core reluctance is much smaller than that of the air, so that most flux remains in the magnetic material, just as electric current flows in the parallel path of lowest resistance in proportion to the resistance ratio. However, in crossing the air gap in Example 7.13, the medium of flux conduction is air and in and near the gap there is a tendency for the flux lines to spread beyond the area A. This flux spreading is called *fringing*. A quantitative evaluation of the extent and effect of fringing can be done in any specific case, although it may be quite laborious. A quick estimate of fringing can be made by increasing the diameter or lateral dimensions of A by the size of ℓ_g, since the greater the gap, the more extensive is the fringe field.

An important quality factor for a permanent-magnet material is its *energy product,* obtained by multiplying Eqs. (7.25) and (7.26) (and ignoring the negative sign):

$$B_m H_m = B_g H_g \frac{v_{\text{gap}}}{v_{\text{core}}} = \frac{\text{gap energy}}{\text{magnetic volume}} \qquad (7.28)$$

The energy product gives the available energy per unit volume of magnetic material. Usable forces depend upon the gap energy, so that a high BH product implies the ability to obtain useful devices with relatively little magnetic material, thereby reducing weight, bulk, cost, etc.

Although we are using SI units in this text, the technical field of magnetics is one in which gaussian units are used at least as often. The conversion can be straightforward since

$$10^4 \text{ gauss} = 1 \text{ tesla} \quad \text{and} \quad 1 \text{ oersted} = 10^3/4\pi \text{ A/m} = 79.578 \text{ A/m}$$

In these units, permanent-magnet materials can be obtained that have energy products of greater than 30×10^6 G–Oe (0.24×10^6 T–A/m). The area of improved hard magnetic materials is one of great research effort and recent years have seen vast improvements, with the consequent ability to make permanent-magnet devices that were inconceivable before, for example, medical in-body pumps and actuators. The advent of small, efficient, inexpensive, permanent-magnet motors, coupled with modern microelectronics has opened new areas of control and power usage that will surely have profound effects on the way we use technology. The BH product of permanent-magnet materials has been increasing exponentially since the beginning of the twentieth century, and as it is still well below the theoretical limit, there is reason to believe that we will continue to see exciting changes in this area.

7.11 CORE LOSSES

Although the approximations of magnetic circuit theory take account of some material nonlinearity, they ignore the double-valued hysteresis behavior of ferromagnetic material, that is, the fact that magnetic change is not fully reversible. When a magnetic sample is carried between two magnetic states and then back, the forward and reverse paths are not the same, and some area is enclosed by the path in the B–H plane, as shown in Fig. 7.26(a). In moving from point C to point D along path 1 in the figure, the change of magnetic energy density, given by Eq. (5.17a), is

$$dU_1 = \int_C^D H_1 \, dB_1 = \text{Area to the left of curve 1} \qquad (7.29a)$$

In returning to its initial point along path 2, the energy density changes by

$$dU_2 = \int_D^C H_2 \, dB_2 = -\int_C^D H_2 \, dB_2 = -(\text{Area to left of curve 2}) \qquad (7.29b)$$

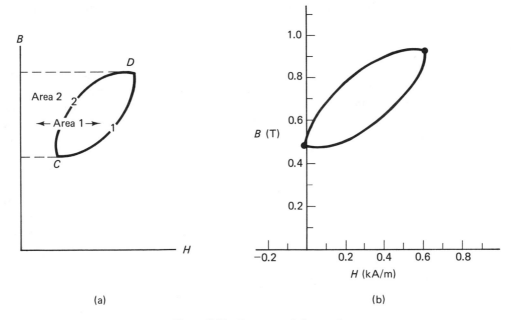

Figure 7.26 Ferromagnetic hysteresis.

When the material is taken about the complete cycle, its change of energy per unit volume is

$$dU = dU_1 + dU_2 = \text{Area enclosed by } B\text{--}H \text{ loop of the cycle} \quad (7.29c)$$

This energy is absorbed by the magnetic material in the course of changing its magnetization, and it is then dissipated, generally, by conversion to heat, although in some situations, there can be considerable acoustic energy as well, and some radiation.

Example 7.14

In Examples 7.12 and 7.13, we examined different stages in the operation of a magnetic relay. In practice, such a relay will cycle many times, continually moving between its closed state, $(B_m, H_m) = (0.935 \text{ T}, 610 \text{ A/m})$, and its open state, $(B_m, H_m) = (0.476 \text{ T}, -0.01 \text{ A/m})$, as shown in Fig. 7.26(b). While we have calculated the end points of this loop, the intermediate points are unknown, so that the area and the energy per cycle cannot be estimated. In any event, this energy, $W = v \oint B \, dH$, must be supplied each cycle by the current source.

Core, or hysteresis, loss is a serious commercial problem for power industries throughout the world. Each time a transformer is cycled, which happens once each $1/60$ second in the Americas and once each $1/50$ second in most of the rest of the world, a part of the energy is lost, that is, converted into useless form. Considering that electrical power may be transformed six or more times between generation and ultimate use, the total power lost is considerable. In addition, magnetic material is cycled in the almost unimaginable number of large and small motors in commercial, industrial, and home use, in relays of all sorts, and in fluorescent lamp ballasts.

Large motors and transformers can be made very efficient (in some cases 95 to 98 percent efficient), so that the loss is quite small compared to the total amount of power it handles. However, smaller equipment is generally less efficient and these occur in vastly greater numbers. All together, hundreds of billions of watt–hours are lost each year in the United States alone, costing billions of dollars. And this figure does not include the additional ohmic heating due to the extra current that must flow through conductors to supply the core-loss energy, or the manufacturing and maintenance costs of cooling components that become hot as a result. Particularly in an age of concern over the limits of natural energy availability, this is a matter of great importance.

Engineers, physicists, metallurgists, etc. have spent considerable effort in examining the magnetization-reversal mechanism in ferromagnetics, hoping that a better understanding will lead to improved and more efficient materials and techniques. And considerable improvement has been made, through new materials and more critical use of materials and design of transformers and motors. It is worth noting, however, that even so, the mechanism of only a fraction of the hysteresis loss is explainable in terms of generally accepted theories of the behavior of ferromagnetics. The remaining fraction, ranging from 30 to 80 percent, depending on the material and mode of use, represents an area of exciting active research, not only for the advancement of science and the understanding of nature, but for the direct betterment of our increasingly energy-consuming civilization.

EXERCISES

Extensions of the Theory

7.1. Diamagnetism is the property of certain materials that display a negative magnetic susceptibility χ. In fact, all material has an inherent negative susceptibility and we only find $\chi > 0$ when other mechanisms, such as reorientation of an inherent permanent magnetic moment, occur that dominate. The inherent diamagnetism can be understood through the Faraday–Lenz law. In this exercise, assume that interatomic forces restrain the orbits from rotating in the field.

Consider an atomic orbital magnetic moment resulting from having an electron rotating about a circular atomic orbit that has a radius of 0.6 A, with an angular frequency of 10^{14} Hz. A field, which is applied perpendicular to the plane of the orbit, increases uniformly to 2 T over a period of 10 sec.

(a) Calculate the change of energy of the electron from the induced electromotive force. If this appears in the form of kinetic energy, determine the new rotational frequency and the new magnetic moment. Show that this has the effect of adding a magnetic moment that is opposite to the applied field.

(b) If the atom is in an ideal gas that exists at standard temperature and pressure, what is the susceptibility of the gas (be careful of units)? Show that $\mu_r < 1$.

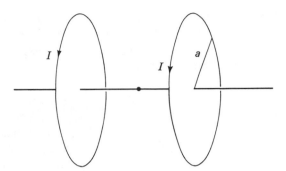

Figure 7.27   For Exercise 7.2.

7.2.   Two identical circular coaxial coils are shown in Fig. 7.27. The currents I are parallel. By adding the fields of the individual coils,
   (a) find H on the axis midway between the coils.
   (b) show that $dH_z/dz = 0$ at that midpoint (see the note).
   (c) show that $d^2H/dz^2 = 0$ at the midpoint, if the coils are separated by a distance equal to their radius.
   (d) show that $d^3H/dz^3 = 0$ at the midpoint (see the note).
   Note: For parts (a) and (d), symmetry arguments can be used to show that all odd-powered terms in the expansion $H(z)$ must vanish.
      Helmholtz coils are the coil pair situated as in part (c). They provide a particularly simple and convenient way to obtain a uniform field over a relatively large volume. Since the field at a point on the axis can be written as a Taylor series,

$$H(z) = H(z_0) + (z - z_0)\frac{dH(z_0)}{dz} + \frac{(z - z_0)^2}{2!}\frac{d^2H(z_0)}{dz^2}$$
$$+ \frac{(z - z_0)^3}{3!}\frac{d^3H(z_0)}{dz^3} + \frac{(z - z_0)^4}{4!}\frac{d^4H(z_0)}{dz^4} + \cdots +$$

it follows from the results of this exercise that the field at the center of a Helmholtz coil varies very slowly, that is, as the fourth power of the small displacement.

*7.3.   **(a)** Extend Example 7.3 by calculating the vector magnetic potential $\mathbf{A}$ of a finite linear conductor at a point not in its midplane.
   (b) Calculate $\mathbf{B}$ in the plane $z = 0$ and find the region where the difference from that of an infinitesimal current element is less than 10 percent.
   (c) Determine the region where the difference from an infinite current line is less than 10 percent in that same equatorial plane.

Problems

7.4.   If the current is reversed in one of the loops of Exercise 7.2, show that the field at the midpoint has a uniform gradient when the Helmholtz configuration is used.

7.5.   Calculate the magnetic field at the center of a square current loop of side a and carrying current I.

*7.6. Calculate the magnetic field at the center of a circular current loop of radius a and carrying current I. Compare your result with that from Problem 7.5.

7.7 **(a)** In example 7.4, we found an expression for the field from a magnetic dipole at points that are very far from the loop.

$$\mathbf{B} = \frac{\mu m}{4\pi r^2}(\mathbf{a}_r 2 \cos\theta + \mathbf{a}_\theta \sin\theta)$$

For a circular current loop with radius a and current I, this applies when $r \gg a$. If we stretch the application of that result to the limit that it applies when $r > a$, calculate the total flux crossing the plane area $\theta = \pi/2$ and $r > a$.

   (b) To the extent that the result of part (a) is meaningful, flux closure implies that this same total flux must pass through the area of the loop itself. Making the assumption that B is uniform in the planar loop area, calculate this value of B. Compare your answer with that from Problem 7.6.

*7.8. A thin disc of radius a has a uniform surface charge density ρ_s. The disc rotates about its normal axis with angular speed ω.

   (a) Recalling that $K = dI/ds$, find $K(\rho)$, the surface current density about the axis.

   (b) Calculate the magnetic field at points on the axis of rotation.

7.9. A long cylindrical conductor of radius b carries a uniform current density $\mathbf{J}$ along its length, except for an off-centered cylindrical hole of radius a, as shown in Fig. 7.28. Using the principle of superposition, show that the magnetic field is uniform in the hole.

7.10. The flat-line transmission system shown in Fig. 7.29 (next page) carries a surface current density $K = K_0 t/(|x| + t/2)$ A/m. Find the magnetic field strength at points (x_0, y) for which $-t/2 \le x_0 \le t/2$ and $y = 0$.

7.11. A square current loop carrying current I is located in the xy plane with its corners located at $(-L/2, -L/2)$, $(-L/2, L/2)$, $(L/2, L/2)$, $(L/2, -L/2)$. Using

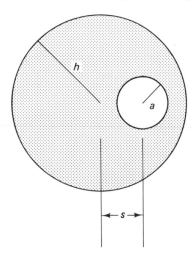

Figure 7.28 For Exercise 7.9.

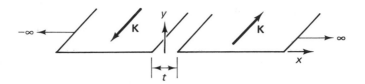

Figure 7.29   For Exercise 7.10.

the result of Example 7.3, find the magnetic field along the x axis at points $x = 0, L/8, L/4, 3L/8$, and $0.45L$. Compare this variation with the results of Problems 7.5. to 7.7.

*7.12. What is the magnetic vector potential on the axis of a circular current loop? How is this answer related to the field value on the axis?

7.13. Find the boundary conditions for **A** from those of **B** and **H**.

7.14. Substitute $\mathbf{B} = \nabla \times \mathbf{A}$ in Maxwell's equations and show that, in the absence of any electrostatic fields, $\mathbf{E} = -\partial \mathbf{A}/\partial t$.

7.15. Show that the magnetic flux through the area of a closed curve is equal to the line integral of **A** about that curve.

7.16. Show that only one of the following can be a magnetostatic vector potential. For that one, derive expressions for the magnetic field and the current density.
   (a) $\mathbf{A} = A(x)\mathbf{a}_x$
   (b) $\mathbf{A} = A(x)\mathbf{a}_y$

*7.17. A long cylindrical spaceship 20 m in diameter has a steel surface that is 10 cm thick. Though not selected for its magnetic properties, this steel has $\mu = 10\mu_0$. In addition, the ship's nuclear drive is encased in a long cylindrical shell that is 3 m in diameter, 30 cm thick, and has $\mu = 1000\mu_0$. The ship passes through a region where the magnetic field is 100 Oe (= 8000 A/m). What must be the field sensitivity of the ship's nuclear drive if it is to be unaffected by the field?

*7.18. Solve Laplace's equation for the magnetic scalar potential produced by the surface current density $K\mathbf{a}_z$ flowing on the plane $x = 0$. Find H. *Hint:* Assume $\psi = \psi(y)$ only.

7.19. A cylindrical surface current density $K\mathbf{a}_\phi$ flows on the surface $\rho = a$.
   (a) Solve Laplace's equation for ψ and find an expression for H. *Hint:* Assume $\psi = \psi(z)$ only.
   (b) Solve Laplace's equation for ψ_1 in the region $\rho < a$, and for ψ_2 in the region $\rho > a$.
   (c) Using the boundary condition for H at $\rho = a$, relate the constants of your two solutions.
   (d) Why was it necessary to break this problem into parts (a) and (b) instead of finding a single potential ψ in both regions?
   (e) Use the uniqueness theorem from Chapter 6 to justify the assumption that $\psi = \psi(z)$ only in the two regions.

*7.20. A long solenoid has n turns per meter and carries a current I. A long iron slug of cross-sectional area A is partly inserted in the solenoid. The iron has $\mu = 100\mu_0$. Find the force on the slug and its direction.

7.21. The poles of an electromagnet are designed to provide a region of space in which the magnetic field is given by $H_z = H_0 - \xi z^2$. A magnetic sphere with radius a and with relative permeability μ_r is hung at the origin from a spring whose force is given by $F_z = F_0 - kz$. What is the equilibrium position of the sphere? Neglect its weight.

8

NUMERICAL METHODS

8.1 INTRODUCTION

Thus far in this text, we have been dealing with closed-form solutions to electric- and magnetic-field problems, that is, solutions expressed through exact mathematical formulas. This is the historical approach, in keeping with the development by great physicists and mathematicians of the past. The advantage is a clear exposition and understanding of the behavior of the fields. Analytic functions are satisfying because they are familiar. For example, if the potential $\Phi = e^{-\alpha y} \sin \beta x$, one can picture the changes of Φ as x and y vary. This familiarity has grown out of experience with the behavior of the exponential and sinusoidal functions, and comprehension may be difficult in cases where the potential is not expressible as a simple function. To help, we construct tables of values of $\Phi(x, y)$ at as many points as needed. Graphical presentations of these data, done either by hand or on a computer, through such means as contour plots for potential and arrows and colors for field density, make the solutions comprehendible and attractive.

These arguments were recognized from the earliest days of electromagnetic theory. In his *Treatise on Electricity and Magnetism,* published in 1873, James Clerk Maxwell states:

> In certain classes of cases, such as those relating to spheres, there are known mathematical methods by which we may proceed. In other cases we cannot afford to despise the humbler method of actually drawing tentative figures on paper. . . .

This latter method I think may be of some use, even in cases in which the exact solution has been obtained, for I find that an eye-knowledge of the forms of the equipotential surfaces often leads to a right selection of mathematical method of solution.

Maxwell's text includes instructions for manually constructing such figures from the field expressions, and he illustrates his work with several. Figure 8.1 is one of his drawings, showing the lines of force between two charged plates.

On the other hand, it is often not possible to obtain field expressions from which to construct diagrams. We have been able to solve only those problems where

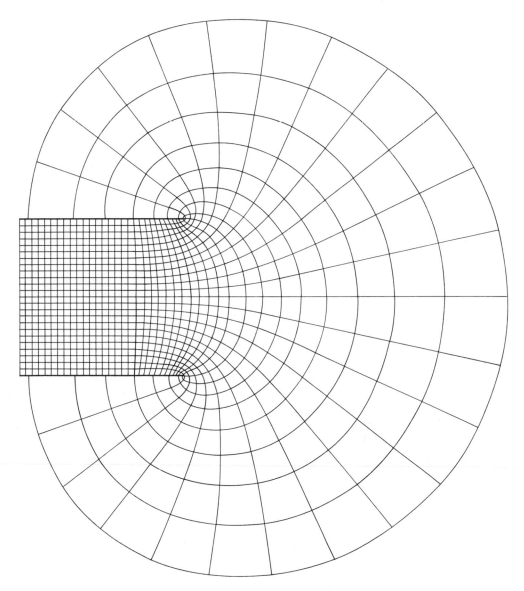

Figure 8.1   Equipotentials and lines of force between two plates.

physical response is linear and where there is a high degree of symmetry. Problems of practical interest generally require approximations in order to make the mathematics tractable, that is, amenable to solution. Techniques have been developed to analyze some problems that do not lend themselves to solution in closed mathematical form. For example, in dealing with boundary-value problems in Chapter 6, we have had solutions in the form of infinite series, which demand numerical substitution to clarify their content. And there are approximation techniques with which difficult configurations are treated as departures from ideal solvable problems, and the differences are then determined. Even in these cases, however, the goal is to obtain concise formulas for the fields, and the numerical methods merely serve to elucidate those formulas.

It requires a different approach, perhaps a different state of mind, to surrender the expectation of a continuous closed-form analytical solution and to seek numerical values directly. However, not only does this allow elucidation of meaning where analytic solutions are complex, but its methods of solution make no distinction between such cases and those where no analytic expression can be found, greatly widening the range of problems that can be solved.

8.2 LIMITS AND METHODS

In using numerical methods, values of the variable of interest are sought directly at points of space. In our case, this will be either the electric scalar potential Φ or the magnitude of the vector magnetic potential $|\mathbf{A}|$. Since the infinite continuum of points cannot be calculated, a scale must be established to determine those points at which the potential will be determined; the finer the scale, the more values are needed and the more time-consuming the calculations. Furthermore, the precision of knowing Φ or A at each point is an important consideration, depending on how rapidly it varies and on the accuracy demanded. By using modern digital computers, the fields and potentials can be found to as many significant figures as desired, within computer limits, but this involves increasingly detailed computations. We see, therefore, that there are limits to what can be done computationally. A great number of very precise values requires large tabulations, or large memory storage in computers, and requires lengthy time periods to calculate.

The painstaking detail of the manual calculations of Fig. 8.1 is obvious and illustrates why, in the past, these practical difficulties greatly limited the availability of numerical solutions. However, recent advances in computational power, in terms of both computer hardware and the software that drives it, have made numerical techniques practical and realizable. Such high degrees of precision and accuracy are now possible that we can consider these solutions to be exact. The use of this approach is rapidly growing; it has already reached a state of development in which the modern engineer and scientist should understand how to use it to advantage.

Basically, one is faced with the conflict of spatial resolution and field precision versus computer-memory capability and computation time. Considerable savings are obtained by limiting consideration to two-dimensional problems and we will generally observe this limit. Fortunately, there are major problems of technical interest involving only two dimensions, and many three-dimensional cases can be represented by two-dimensional sections.

In practice, the two-dimensional spatial region of interest is divided into small subdivisions, commonly either by a square grid or by a triangular mesh, as shown in Fig. 8.2. The field quantity of interest is then found at the nodes of this screen. Depending on the availability and cost of memory and computer time, the screen scale can be reduced to obtain greater resolution (the bold and light lines in Fig. 8.2 show two levels of resolution) and the computational precision can be increased, in principle without limit.

Solving for electromagnetic fields in numerical form at various field points amounts to acquiring a tabulation of numbers that frequently serve more to confuse than clarify. Plotting equipotential contours is an elegant way of interpreting the solution. Graphical presentation of the equipotentials shows whether the solution is reasonable with respect to problem symmetries and limitations. It presents the direction of the flux flow and a description of its magnitude. In magnetics, flux is along the equipotentials, and in electrostatics, flux is normal to equipotentials. These conclusions follow by considering only a two-dimensional variation of the potentials. In this case many problems of general interest have $\mathbf{A}$ in the direction of $\mathbf{a}_z$, so that

$$\mathbf{B} = \nabla \times \mathbf{A} = \nabla \times \mathbf{a}_z A = \mathbf{a}_x \frac{\partial A}{\partial y} - \mathbf{a}_y \frac{\partial A}{\partial x} \qquad (8.1a)$$

$$\mathbf{E} = -\nabla \Phi = -\mathbf{a}_x \frac{\partial \Phi}{\partial x} - \mathbf{a}_y \frac{\partial \Phi}{\partial y} \qquad (8.1b)$$

If the x axis is positioned along the local direction of an equipotential (constant A or Φ), no change in the potential will occur along x, so that $\partial\{\cdot\}/\partial x = 0$. It is readily seen, then, that $\mathbf{B}$ is parallel to $\mathbf{a}_x$, that is, along the equipotential, and $\mathbf{E}$ is parallel to $\mathbf{a}_y$, that is, normal to the equipotential. Moreover, in interpreting plots of equipotentials, we can use the fact that the flux is stronger where the potential contours are crowded. Since the field equals the rate of change of the potential, in such locations, the potential changes by the same amount over a shorter distance, so that the derivative is higher. In contrast, the flux is weaker where the equipotentials spread out.

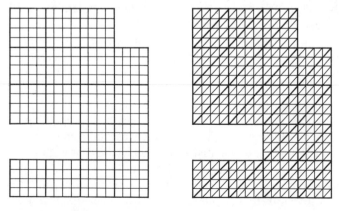

Figure 8.2 Screens for numerical calculation.

Most computing systems supporting graphics have facilities, through libraries, for plotting equipotentials, and consulting the manuals will reveal the exact commands for linking the library with the programs we will present here, and for calling the plotting routine. Commonly, what is required as input are the potentials and coordinates of various points inside a rectangular frame within which the plots are sought. With this, a short program is presented at the end of this chapter that plots the equipotentials.

In this chapter, techniques are developed for computer solution of problems in electrostatics and magnetostatics. In later chapters, these will be extended to dynamic cases. While there are a host of numerical methods available for solving equations, with and without digital computers, only a few that are most prominent are presented. Algorithms are given for carrying out the specific operations of each. These are in a generalized format known as *pseudocode,* although they are also shown imbedded in Pascal programs to solve specific problems. Note also that in this chapter we are departing from our custom of using V to denote the electric potential; instead we use Φ. In addition, we will use r to denote position in the two-dimensional plane, rather than our customary notation of ρ for the polar radius.

8.3 DIFFERENTIAL AND INTEGRAL EQUATIONS

In the preceding chapters, the electromagnetic equations were expressed in integral and in differential form; both can be used to solve field problems analytically and numerically. Some problems are better solved with one approach and some with the other, the primary difference being in the manner of treating and including independent and dependent sources and boundary conditions.

Given current and charge sources, integral equations express their effects at often distant spatial points; the integration represents superposition of the fields of elemental point-like sources. From Chapter 3,

$$\Phi = \frac{1}{4\pi\varepsilon} \int \frac{\rho_v}{r} \, dv \qquad (8.2a)$$

where ρ_v is the volume-charge density. This expression implicitly contains the far-field boundary conditions, that is, Φ tends to zero as r goes to infinity, so that these are said to be *natural conditions* of integral problem formulations. Problems arise, however, when inhomogeneities or polarizable media are present, because secondary sources, that is, electrostatic charges induced on material interfaces and the magnetization of permeable parts, generate additional fields. Employing integral methods in situations that have several different regions, therefore, requires preliminary determination of the secondary sources. This poses a number of difficulties, which we cannot discuss in this brief treatment.

In two dimensions, a point charge actually represents a line of charge normal to the plane of the configuration, with line-charge density ρ_ℓ. In this case, the result has been established in Eqs. (3.19) and (3.20), that the field and potential, in polar coordinates are $\mathbf{E} = \mathbf{a}_r \rho_\ell / 2\pi\varepsilon r$ and $\Phi = (\rho_\ell / 2\pi\varepsilon) \ln r$ (neglecting the zero potential reference). For a line of charge with finite cross-sectional area dS, the charge/

length is $\rho_v\,dS$. The integral formulation sums the contribution of the individual elements, so that Eq. (8.2a) must be replaced by

$$\Phi = \frac{1}{2\pi\varepsilon}\int \rho_v \ln r\,dS \tag{8.2b}$$

While integral equations, like Eqs. (8.2a) and (8.2b), express action at a distance, the differential counterpart, the Poisson equation

$$-\varepsilon\nabla^2\Phi = \rho_v \tag{8.3}$$

expresses a local effect, that is, how the potential Φ at a point changes in the vicinity of that point. In a rather simplistic view, Eq. (8.3) can be used to find how the value at one point affects its neighbor, and how its effect sequentially ripples out to a distant point of interest. Since the effect at each field point arises from all the other field points, in this view, the individual field points cannot be treated in isolation. All the field points must be solved together, so that the number of variables to be determined can be very large. The equations for such multivariable problems are generally presented in terms of an array of coefficients referred to as a *matrix*.

Equations (8.2) and (8.3) are for the electrostatic field. For magnetic fields, the corresponding equations are, from Eqs. (7.4) and (7.5),

$$\mathbf{A} = \frac{\mu}{4\pi}\int \frac{\mathbf{J}}{r}\,dv \tag{8.4a}$$

and

$$-\frac{1}{\mu}\nabla^2\mathbf{A} = \mathbf{J} \tag{8.4b}$$

where $\mathbf{A}$ is the magnetic vector potential, μ is the permeability, and $\mathbf{J}$ is the current density.

In two-dimensional problems, $\mathbf{J}$ generally has only one component, that is, current flow is normal to the plane of the configuration diagram. Since $\mathbf{A}$ is parallel to $\mathbf{J}$, we can write $\mathbf{J} = \mathbf{a}_z J$ and $\mathbf{A} = \mathbf{a}_z A$. Reasoning as before with the electric potential, for the transition from three to two dimensions, Eqs. (8.4) reduce to the scalar equations:

$$A = \frac{\mu}{2\pi}\int J \ln r\,dS \tag{8.5a}$$

$$-\frac{1}{\mu}\nabla^2 A = J \tag{8.5b}$$

Formally, these are identical to Eqs. (8.2b) and (8.3), so that the procedure for computing A from Eqs. (8.5) is the same as for Φ; the same computational methods can be employed, except for changing ρ to J and the term $1/\varepsilon$ to μ.

8.4 MATRIX ALGEBRA

It was mentioned earlier that because of the large number of variables to be solved simultaneously, it is necessary to utilize the mathematical properties of matrices to formulate numerical solution methods. A brief discussion of matrix properties is presented in this section.

By general definition, a *matrix* is an array of numbers. A matrix $[A]$ of size $m \times n$ is an array of m rows and n columns. The element A_{ij} of the matrix is the term that resides at row i, column j. An example of a matrix is

$$[A] = \begin{bmatrix} 2 & 3 \\ 1 & 1.5 \end{bmatrix}$$

which is a 2×2 matrix. A_{11} is 2. Matrix algebra defines the rules for summation and multiplication of matrices. If $[P] = [A] + [B]$, then

$$P_{ij} = A_{ij} + B_{ij} \qquad (8.6)$$

where corresponding terms have been added. As an example,

$$\begin{bmatrix} 1 & 2 \\ 0.5 & 1 \end{bmatrix} + \begin{bmatrix} 5 & 6 \\ 2 & 3 \end{bmatrix} = \begin{bmatrix} 6 & 8 \\ 2.5 & 4 \end{bmatrix}$$

This requires that both A and B be of the same size.

On the other hand, in multiplication, if $[P] = [A][B]$, where $[A]$ is $m \times n$, and $[B]$ is $n \times q$, then $[P]$ is $m \times q$ and given by

$$P_{ij} = \sum_{k=1}^{n} A_{ik} B_{kj} \qquad (8.7)$$

For multiplication, it is required that $[B]$ have the same number of rows as $[A]$ has columns. For example,

$$\begin{bmatrix} 1 & 2 & 1.5 \\ 0.5 & 1 & 3 \\ 1.5 & 2 & 1 \end{bmatrix} \begin{bmatrix} 5 & 6 \\ 2 & 3 \\ 1 & 2 \end{bmatrix} = \begin{bmatrix} 10.5 & 15 \\ 7.5 & 12 \\ 12.5 & 17 \end{bmatrix}$$

An easy way to remember how to multiply is to note that P_{ij} is created through element-by-element multiplication of row i from A with column j from B. For the previous matrix, P_{12} is the multiple of row 1 of A (1 2 1.5) with column 2 of B (6 3 2), or $1 \times 6 + 2 \times 3 + 1.5 \times 2 = 15$. A frequently useful matrix is the identity matrix $[I]$, with ones for all the main diagonal elements (the main diagonal is from upper left to lower right) and zeros off the diagonal. For a given size, this matrix acts like unity in multiplication.

$$\begin{bmatrix} 1 & 2 \\ 0.5 & 1 \end{bmatrix} \begin{bmatrix} 1 & 0 \\ 0 & 1 \end{bmatrix} = \begin{bmatrix} 1 & 2 \\ 0.5 & 1 \end{bmatrix}$$

Matrix multiplication gives us a convenient way of depicting equations and is of great practical importance in solving equations. For example, the equation pair

$$2x + y = 4$$
$$x + 2y = 5 \tag{8.8a}$$

is given in matrix form by

$$\begin{bmatrix} 2 & 1 \\ 1 & 2 \end{bmatrix} \begin{bmatrix} x \\ y \end{bmatrix} = \begin{bmatrix} 4 \\ 5 \end{bmatrix}$$

or

$$[A][X] = [b] \tag{8.8b}$$

where $[A]$ is the matrix of coefficients. An extremely convenient way to solve this equation for x and y is to use the inverse matrix, defined by

$$[A]^{-1}[A] = [A][A]^{-1} = [I] \tag{8.9}$$

If A^{-1} is known, then the unknown vector $[X]$ is given by

$$[X] = [I][X] = [A]^{-1}[A][X] = [A]^{-1}[b] \tag{8.10a}$$

The inverse of

$$\begin{bmatrix} 2 & 1 \\ 1 & 2 \end{bmatrix} \quad \text{is} \quad \frac{1}{3} \begin{bmatrix} 2 & -1 \\ -1 & 2 \end{bmatrix}$$

since

$$\frac{1}{3} \begin{bmatrix} 2 & 1 \\ 1 & 2 \end{bmatrix} \begin{bmatrix} 2 & -1 \\ -1 & 2 \end{bmatrix} = \begin{bmatrix} 1 & 0 \\ 0 & 1 \end{bmatrix}$$

The solution of Eq. (8.8) is, therefore,

$$\begin{bmatrix} x \\ y \end{bmatrix} = \frac{1}{3} \begin{bmatrix} 2 & -1 \\ -1 & 2 \end{bmatrix} \begin{bmatrix} 4 \\ 5 \end{bmatrix} = \begin{bmatrix} 1 \\ 2 \end{bmatrix} \tag{8.10b}$$

This is a *direct method of solution*. Iterative schemes are indirect. These keep improving the solution until they are stopped. An example of an iterative scheme is Gauss' method, where arbitrary initial values are assigned to the unknowns, and these are cyclically improved. For the previous example, if we take $x = 0$ and $y = 0$ as a starting guess, then for iteration 1, using row 1 of Eq. (8.8a), $x = (4 - y)/2 = (4 - 0)/2 = 2$. Now using row 2, $y = (5 - x)/2 = (5 - 2)/2 = 1.5$. For the second iteration, $x = (4 - 1.5)/2 = 1.25$ and $y = (5 - 1.25)/2 = 1.875$. After several such iterations, the correct solution is reached to whatever accuracy is desired.

For later reference, it is convenient to define the transpose of a matrix, which is the matrix obtained by interchanging rows and columns. Examples are

$$[A] = \begin{bmatrix} 1 & 2 & 1.5 \\ 0.5 & 1 & 3 \\ 1.5 & 2 & 1 \end{bmatrix} \qquad [A]^t = \begin{bmatrix} 1 & 0.5 & 1.5 \\ 2 & 1 & 2 \\ 1.5 & 3 & 1 \end{bmatrix}$$

$$[B] = \begin{bmatrix} 5 & 6 \\ 2 & 3 \\ 1 & 2 \end{bmatrix} \qquad [B]^t = \begin{bmatrix} 5 & 2 & 1 \\ 6 & 3 & 2 \end{bmatrix}$$

$$[C] = \begin{bmatrix} 4 \\ 2 \\ 3 \end{bmatrix} \qquad [C]^t = \begin{bmatrix} 4 & 2 & 3 \end{bmatrix}$$

A single 3×1 matrix, such as $[C]$, is frequently used to denote a vector. The magnitude of the vector is found from $|C|^2 = [C]^t[C] = 29$.

For our purposes, mainframe computing systems usually have libraries that allow the computation of $[X]$ in relationships like Eqs. (8.8), from the definition of $[A]$, $[b]$, and the number of unknowns n. Direct and iterative methods are available. Therefore, once we assemble the matrix equation defined by $[A]$, $[b]$, and n, we can call one of the methods of solution to obtain an answer.

8.5 NUMERICAL SOLUTION OF INTEGRAL EQUATIONS

In its most elementary form, the solution of electromagnetic problems by integral methods involves finding the fields when all sources are known. It is, perhaps, simplest to explain the procedure through an example, and for this we consider the two-component system shown in Fig. 8.3(a), where each conductor has a square cross section. Unlike the situation where the elements have circular cross sections, this problem cannot be solved analytically. If these are conductors carrying current, then a study of the magnetic field is required if we are interested in line inductance. If they represent charged dielectrics, then a study of the electric field is required if we are interested in capacitance. We have seen that both cases reduce to solving the same equations. For the moment, we take the current conductor–magnetic field case.

The governing equation in our two-dimensional magnetostatics comes from the potential of a long current filament, given by Eq. (8.5a):

$$A = \frac{\mu}{2\pi} \int J \ln r \, dS$$

The wires are *discretized* by dividing them into finite filaments, each so small that it can be considered point-like. Then the integral becomes a sum of the contributions from each filament:

$$A = \frac{\mu}{2\pi} \sum J_i \ln r_i \, dS_i \tag{8.11}$$

where J_i is the current per meter of area in filament i, which has area dS_i, and which is at a distance r_i from the point at which we are computing the potential A.

Observe that difficulties will arise if the field point is on or close to a filamental wire i. First, when r approaches zero, the logarithmic function is undefined. Furthermore, by the discretizing process, we intend that the contribution of each subdivision arises entirely from its center point. This is a good approximation if $r_i \gg \sqrt{dS_i}$, so that if r is small, the wire subdivisions will have to be made even smaller. While this problem can be overcome using special integration processes, it can be ignored by restricting the solution to field points exterior to sources (wires).

Example 8.1

In the two-conductor transmission line shown in Fig. 8.3(a), each conductor has a square cross section and carries a uniform current density. Determine the vector-potential distribution preparatory to finding the inductance.

Figure 8.3(b) shows each wire subdivided into 16 filaments. For this discretization of the conductors, Algorithm 8.1 (at the end of the chapter) gives the potential at coordinate point (x, y) exterior to both conductors. The input data for a problem of this nature merely consist of the contents of Table 8.1, which gives the coordinates (the middle) of each filamental wire and the current it contains.

Grid construction is shown in Fig. 8.3(c). It is only necessary to place this grid over one-fourth of the field region because symmetry about the planes $x = 0$ and $y = 0$ assures that symmetrical solutions will be found in the other quadrants. In this way, only one-fourth of the total calculations are required to arrive at a complete solution. Note that this grid has nothing to do with the scale of discretizing the conductors in Fig. 8.3(b). The potentials at the grid points are calculated by adding the potential contributions from all the discrete wire elements. This is done by calling upon Algorithm 8.1 for each grid point. The potentials are stored in a separate file. Once the computation is complete, we can call for a plotting routine, giving as input the coordinate points of the grid and the potentials. The plot corresponding to the two-conductor system is shown in Fig. 8.4. Only a part of the field region is shown, but a larger region or a refined mesh could easily be explored.

TABLE 8.1. FILAMENT LOCATIONS

Filament	x	y	J
1	2.25	0.75	1
2	2.75	0.75	1
.	.	.	.
.	.	.	.
.	.	.	.
5	3.5	0.5	1
.	.	.	.
.	.	.	.
.	.	.	.
17	−3.75	0.75	−1
18	−3.25	0.75	−1
.	.	.	.
.	.	.	.
31	−2.75	−0.75	−1
32	−2.25	−0.75	−1

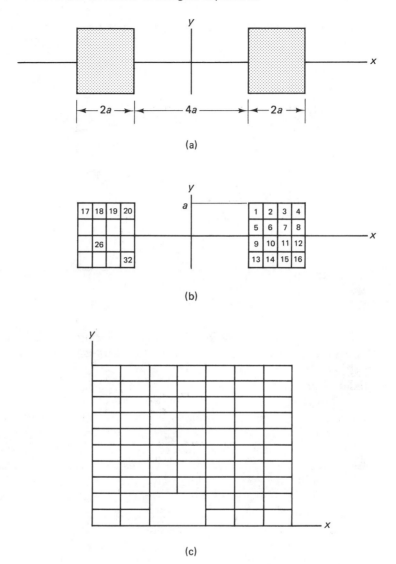

Figure 8.3   (a) Square-conductor system. (b) Numbering of discrete wire elements. (c) Grid for potential solutions.

Calculation of the inductance calls for determination of the vector flux between current elements. Note, however, that the field lines (coincident with the contours of A) penetrate the conductors so that different flux cuts different portions of the current. This means that a weighted flux–current calculation is needed. This is a refinement of induction calculations that we will not explore. The flux can also be expressed as an integral of the vector potential, which results in some simplification. (See Problem 7.15 and Example 8.4 later in the chapter.)

An appendix at the end of this book presents a realization of Algorithm 8.1 imbedded in a Pascal program to solve this problem.

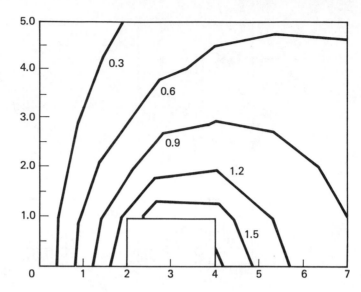

Figure 8.4   Equipotential lines for Example 8.1.

Equation (8.2) through Eq. (8.5) indicate that the algorithm and program could just as well be expressed in terms of the equivalent problem of the potential from oppositely charged dielectrics rods. The electric formulation of this problem suggests a subtle difference between the electric and magnetic problems, as it is related to a limitation of integral solutions. We have assumed a uniform source (current) distribution, whereas in a physical situation, the interaction between sources would result in nonuniformity, unless constraints are present. In the electric case, the charges would congregate toward the middle of the two dielectrics, attracted there by coulomb forces of the opposite charges. In the magnetic case, the current density would increase toward the outside of the two conductors, under the influence of their Lorentz repulsion. This emphasizes the integral method limitation of knowing the source distribution before performing the calculation. Once that is known, as with our assumption of uniformity, the solution proceeds regardless of whether the sources are charge or current.

Problems of this last type are more exactly tackled by first solving for the source distribution and then for the potentials from the now known distribution. To illustrate how this is done, and to demonstrate the integral method in a three-dimensional problem, consider the following example.

Example 8.2

Determine the self-capacitance of the isolated conducting bar shown in Fig. 8.5(a), where t is comparable to the bar height. The bar is 1 meter long.

Referring to Problem 5.11, recall that self-capacitance is the capacitance of a body to a surrounding surface that is infinitely far away. The conductor is taken to be at a potential of 100 V above the distant surface.

The conductor is discretized into 20 rectangles in Fig. 8.5(b). Each volume tS is represented by charge acting at its center and we note that there are only 10 unknown charges,

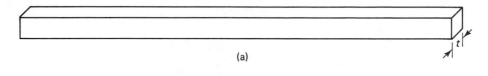

(a)

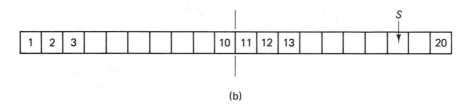

(b)

Figure 8.5 (a) Charged rod. (b) Discretized rod.

since the body is symmetrical in the two halves. As this is a three-dimensional problem, the applicable equation is Eq. (8.2a), so that the potential at section i, due to the charge in section j, is

$$\Phi_{ij} = \frac{1}{4\pi\varepsilon_0} \int \frac{\rho}{r}\, dv = \frac{1}{4\pi\varepsilon}\frac{q_j}{r_{ij}} \tag{8.12a}$$

It has been assumed that the distance r_{ij} is practically constant in the integration over the charge in rectangle j, so that $\int \rho\, dv/r = q_j/r_{ij}$. This assumption cannot be made for the potential at the center of rectangle i, due to its own charge, so it is necessary to proceed differently in this case.

$$\Phi_{ii} = \frac{1}{4\pi\varepsilon_0} \int \frac{\rho}{r}\, dv = \frac{1}{4\pi\varepsilon_0} \int \frac{\rho}{r} t\, dS$$

$$= \frac{1}{4\pi\varepsilon_0}\frac{q_i}{S_i} 3.54\sqrt{S_i} = \frac{3.54 q_i}{4\pi\varepsilon_0 \sqrt{S_i}} \tag{8.12b}$$

where we have used the relationships $dv = t\, dS$, $\rho t = q_i/S_i$, and $\int dS/r = 3.54\sqrt{S}$.

Consider the charged area to be a circle of radius r_0 instead of a rectangle; then, as shown in Fig. 8.6, the area dS can be considered to be a ring of radius r and thickness dr, so that, for the *field point* at the center of the circle,

$$\int dS/r = \int_0^{r_0} 2\pi r\, dr/r = 2\pi r_0 = 2\pi\sqrt{S/\pi} = 3.54\sqrt{S}$$

Superposition of the contributions from each charge gives the known voltage at each section i:

$$\Phi_i = 100 = A_{i1}q_1 + A_{i2}q_2 + \cdots + A_{i10}q_{10} \tag{8.13a}$$

where we account for the symmetrical charge, on the other half of the plate, in computing the coefficients A_{ij}. For example, A_{ii} has a contribution from Φ_{ii}, given by Eq. (8.12b) (with a numerical value of S), and this is summed with a contribution of the type in Eq. (8.12a), coming from the charge q_i on the other half of the conductor. On the other hand, coefficients A_{ij} ($i \neq j$) are made up of two coefficients of the type in Eq. (8.12a), one from the charge q_j and one from its *reflection* charge on the other half of the rod.

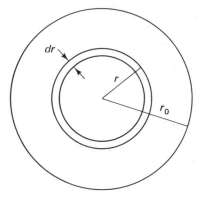

Figure 8.6   Integration for cylindrical self-potential.

Once Eq. (8.13a) is assembled into the matrix form

$$[A][q] = 100[I] \tag{8.13b}$$

where $[I]$ is a column of 1's, it can be solved for the unknown charges $[q] = [q_i]$. The capacitance is computed from $C = Q/V = 2(q_1 + q_2 + \cdots + q_{10})/100$. (The factor 2 accounts for the charges on the second half of the conductor.) Table 8.2 gives the charge variation calculated along the conductor, in normalized units. These values give $C = 29.160\varepsilon_0$ F.

TABLE 8.2. CHARGE DISTRIBUTION

Element numbers	$q/4\pi\varepsilon_0$
1, 20	16.766
2, 19	12.505
3, 18	11.695
4, 17	11.235
5, 16	10.950
6, 15	10.759
7, 14	10.628
8, 13	10.540
9, 12	10.486
10, 11	10.459

In these calculations, there is no need to have all the conductor regions of equal length. For improved accuracy, a finer subdivision could be used only close to the ends, where the variations in charge are high, that is, where the assumption of uniform density on each rectangular subdivision is less valid.

8.6 METHOD OF FINITE DIFFERENCES

The finite-difference method is one of the older differential schemes and is rather easy to understand. A simple rectangular mesh is drawn within the solution region, and the potentials are solved at the nodes of the mesh, such that they satisfy Laplace's equation and the known boundary conditions on the borders and interfaces of that region. Differential methods of solving Laplace's equation suffer from a difficulty that is not met in integral approaches. Particularly with problems that are in infinite space, the physical boundaries are very remote, implying the need of a vast mesh with a large number of nodes. However, practical computational limits restrict the number of points that can be used, so we must approximate the distant condition by moving these boundaries to a finite distance from the charges and surfaces of the more immediate problem. Bringing the very distant potentials to a smaller distance distorts the field distribution, introducing error into the solution. One method to correct for this limitation, without introducing too many distant field points, will be shown in Example 8.4.

The potential at the grid points of the rectangular mesh is computed by discretizing the governing differential equation, that is, by converting Poisson's equation [Eq. (8.3)] to an equation of differences rather than differentials. To do this, consider the region of a typical point 0 on a rectangular grid of dimensions h and k, as shown in Fig. 8.7. Points A and B are midway between the lattice points 3, 0, and 1, so that

$$\frac{\partial \Phi(A)}{\partial x} = \frac{\Delta \Phi(A)}{\Delta x} = \frac{\Phi_1 - \Phi_0}{h} \tag{8.14a}$$

$$\frac{\partial \Phi(B)}{\partial x} = \frac{\Delta \Phi(B)}{\Delta x} = \frac{\Phi_0 - \Phi_3}{h} \tag{8.14b}$$

It then follows that at point 0,

$$\frac{\partial^2 \Phi(0)}{\partial x^2} = \frac{1}{h}\left[\frac{\partial \Phi(A)}{\partial x} - \frac{\partial \Phi(B)}{\partial x}\right] = \frac{1}{h^2}[(\Phi_1 - \Phi_0) + (\Phi_3 - \Phi_0)] \tag{8.15}$$

With a similar expression for the second derivative with respect to y, the discretization of Eq. (8.3) at node 0 is

$$\varepsilon\frac{\Phi_1 - \Phi_0}{h^2} + \varepsilon\frac{\Phi_2 - \Phi_0}{k^2} + \varepsilon\frac{\Phi_3 - \Phi_0}{h^2} + \varepsilon\frac{\Phi_4 - \Phi_0}{k^2} = -\rho_0 \tag{8.16}$$

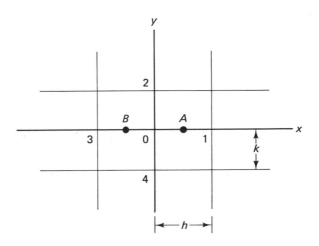

Figure 8.7   Rectangular grid for discretizing Poisson's equation.

Note that for the special case of a square grid, $h = k$, this becomes

$$\Phi_0 = \frac{1}{4}\left(\Phi_1 + \Phi_2 + \Phi_3 + \Phi_4 + \frac{h^2\rho_0}{\varepsilon}\right) \tag{8.17}$$

For every interior point of the mesh in a uniform medium, a condition corresponding to Eqs. (8.16) and (8.17) can be easily written. At boundary points, these must be modified. No equation is necessary at nodes where the potential is already specified (Dirichlet boundary). For nodes on the boundary where the normal derivative is known (Neumann boundary), the appropriate condition can be obtained by taking a hypothetical exterior point. If mesh line 2–0–4 of Fig. 8.7 were such a boundary, with $\partial V/\partial x = p$, then

$$\frac{\partial V}{\partial x} = \frac{\Phi_1 - \Phi_3}{2h} = p \tag{8.18a}$$

so that, for the hypothetical point 3,

$$\Phi_3 = \Phi_1 - 2ph \tag{8.18b}$$

must be substituted into Eqs. (8.16) and (8.17).

Equation (8.16) indicates how to treat cases where the mesh dimensions are not uniform, that is, by having h and k vary with position. Two other cases of nonuniformity must be mentioned: where the value of ε varies, and where ρ_0 changes. To include these in Eq. (8.16), we must use the mean value of ε along each mesh line and the mean value of ρ (the *area* mean) about the node 0. Thus, for the situation shown in Fig. 8.8,

$$\frac{\varepsilon_1(\Phi_1 - \Phi_0)}{h^2} + \frac{(\varepsilon_1 + \varepsilon_0)(\Phi_2 - \Phi_0)}{2k^2} + \frac{\varepsilon_0(\Phi_3 - \Phi_0)}{h^2} + \frac{(\varepsilon_1 + \varepsilon_0)(\Phi_4 - \Phi_0)}{2k^2} =$$
$$-\frac{(\rho_a + \rho_b + \rho_c + \rho_d)}{4} \tag{8.19}$$

Further discussion can best be appreciated through an example.

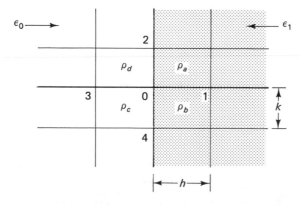

Figure 8.8 Configuration with a nonuniform medium and charge distribution.

Example 8.3

Figure 8.9(a) shows a cross section of a long parallel-plate capacitor. The shaded area has a dielectric with relative permittivity $\varepsilon_r = 5$. What is the capacitance per meter of length?

The capacitor plates are arbitrarily assigned the potentials ± 100 V. Figure 8.9(b) shows the square grid of field points in the first quadrant of the xy plane. From symmetry considerations, we see that a solution in this region solves the complete problem.

Three boundaries satisfy Dirichlet conditions, that is, Φ is known on these surfaces: The x axis, being midway between the plates, is at zero potential. This corresponds to the lower boundary of the mesh (row 7). The distant boundaries, which have zero potential, are represented by setting the potentials along row 1 and column 7 equal to zero.

Since the capacitor plate is at $+100$ V, this value is assigned to row 5, columns 1, 2, and 3. Since the potential distribution is the same on either side of the central vertical plane of symmetry, the potential does not change in the direction *normal* to column 1 of the grid. This plane provides a Neumann boundary condition, that is, $\partial\Phi/\partial n$ is known, specifically $\partial\Phi/\partial x = 0$, so that $\Phi_3 = \Phi_1$, where point 3 is the hypothetical point to the left of the boundary in Fig. 8.9(b).

Finally, note that $\rho = 0$ throughout and, by dividing Eqs. (8.16) through (8.19) by ε_0, then $\varepsilon \rightarrow \varepsilon_r = 5$ in the shaded region and $\varepsilon_0 \rightarrow 1$ elsewhere.

With this, we have an equation for each of the 27 unknown node potentials, so that the system of finite difference equations can be solved.

The system of equations for the undetermined mesh potentials can be solved in a number of ways. While special programs can be written, perhaps the simplest approach, for relatively small problems, is to use the readily available iterative scheme found in most spreadsheet programs. In these programs, the nodes are denoted by their RC (for Row, Column) positions.

Example 8.3 (cont.)

We set up a 7×7 array and enter the appropriate fixed values and Eqs. (8.17) and (8.19), with $\rho_0 = 0$. Figure 8.10 shows the equation entered in site R2C2, which was then filled through to sites R6C6. Along column 1, symmetry makes $\partial\Phi/\partial x = 0$, so that account must be taken of Eq. (8.18) with $p = 0$. Because of local uniformity and the known values of Φ, the only node at which account must be taken of the value of ε is R6C3, and here we enter (with $\varepsilon = 5$), for Φ:

$$= (R[-1]C*(\varepsilon + 1)/2 + RC[-1]*\varepsilon + R[+1]C*(\varepsilon + 1)/2 + RC[+1])/(2*(\varepsilon + 1))$$

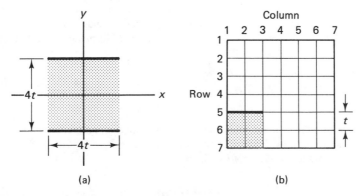

      (a)   (b)

Figure 8.9   (a) Parallel-plate capacitor. (b) Grid for solving.

R2C2	= (R[-1]C+RC[-1]+R[+1]C+RC[+1])/4					
dielCap						
	1	2	3	4	5	6	7
1	0	0	0	0	0	0	0
2	21.15	20.27	17.53	13.02	8.244	3.92	0
3	44.04	42.4	36.84	26.29	16.03	7.45	0
4	70.23	68.44	61.14	39.28	22.15	9.86	0
5	100	100	100	47.53	23.43	9.82	0
6	49.69	49.39	47.86	27.4	14.21	6.01	0
7	0	0	0	0	0	0	0

Figure 8.10   Spreadsheet solution of capacitor problem.

The last divisor is the coefficient of Φ_0, found when solving Eq. (8.19), equal to $2\varepsilon + 2$ at R6C3.

When the equations and values have been entered, the iterative procedure is called. This causes the computer to sequentially carry out the calculations, starting with whatever potential values are present. Because the boundary conditions are fixed, all other values vary as the calculations proceed in continuing sequence. Eventually, it is expected that the potentials approach their final values, and when no node potential changes by more than some predetermined amount (adjustable but otherwise preset by the program), the iterations cease. The tabulated values then satisfy the boundary conditions and Laplace's difference equation to the specified accuracy.

Figure 8.10 shows the potentials found by allowing the worksheet to operate on the equations previously shown. For this small number of nodes, the process requires only a few iterations, taking well under a minute. A plot of the potential surfaces can be constructed from these data, either manually or by using standard graphics programs. A manual construction of some field lines is shown in Fig. 13.6 of Chapter 13. (In that chapter, this problem is used to illustrate certain properties of stripline transmission systems.) From that figure, it can be seen that the approximation of bringing the infinite points to a closer location distorts the field lines and potential surfaces.

Since the original problem was to calculate the capacitance, we have $C = Q/V = Q/200$, where Q is the charge on either capacitor plate. Q can be found by using Gauss' law: $\oint \mathbf{D} \cdot \mathbf{n} \, dS = Q$, where the integral is evaluated on a surface surrounding the capacitor plate. Because the field changes rapidly near the ends of the plates, errors can be introduced and the surface is best taken at some distance. In this case, choose the surface from R2C1 to R2C6 to R6C6 to R6C1.

$$\mathbf{D} \cdot \mathbf{n} = D_n = \varepsilon E_n = \varepsilon \frac{\Delta \Phi}{\Delta n} = \varepsilon \frac{\Phi_2 - \Phi_1}{t} = \varepsilon \frac{\Phi_2}{t}$$

where the distance between two points is $\Delta n = t$, and, because we have taken a path that is adjacent to the Dirichlet boundaries, we have $\Phi_1 = 0$. Since $dS \to t\ell$, where ℓ is a length normal to the plane of the figure,

$$Q = \oint \mathbf{D} \cdot \mathbf{n} \, dS \to 2 \sum \varepsilon \frac{\Phi}{t} t\ell = 2 \sum \varepsilon \Phi \ell$$

The factor 2 is needed because the integration path in two dimensions should completely enclose the capacitor plate, but the sum is over only half. Therefore, the capacitance per unit length is

$$\frac{C}{\ell} = \frac{Q/\ell}{200} = \frac{1}{100} \sum \varepsilon \Phi$$

indicating a summation of the values of $\varepsilon \Phi$ at all nodes of the contour.

Note that nodes R2C1 and R6C1 do not associate with a full area $t\ell$. Because they are at the boundary, they correspond to only $t/2$, so their contributions to the sum must be halved. All nodes have $\varepsilon = \varepsilon_0$, except R6C1 and R6C2, which have $\varepsilon = 5\varepsilon_0$, and element R6C3, which is on the air–dielectric, boundary so its contribution has $\varepsilon_{\text{effective}} = (5 + 1)\varepsilon_0/2$.

Carrying out the summation from Fig. 8.10, we find that $C/\ell = 6.63\varepsilon_0$ F/m. For comparison, note that the neglect of fringing, that is, using the analytic parallel-plate capacitance would give $C_{pp}/\ell = \varepsilon w/d = 5\varepsilon_0(4t)/4t = 5\varepsilon_0$, so that the fringing field is close to one-third of C_{pp}.

8.7 A MAGNETICS EXAMPLE

For magnetics problems, Eq. (8.16) must be adapted by making the substitutions $\Phi \to A$, $\varepsilon \to 1/\mu$, and $\rho \to J$. The corresponding inhomogeneous geometry is shown in Fig. 8.11. Note that the term at the media interface becomes $1/(\varepsilon_1 + \varepsilon_0) \to \mu_1\mu_0/(\mu_1 + \mu_0)$. After some manipulation, it is found that

$$\frac{A_1 - A_0}{h^2\mu} + \frac{(1 + \mu)(A_2 - A_0)}{2k^2\mu} + \frac{A_3 - A_0}{h^2}$$

$$+ \frac{(1 + \mu)(A_4 - A_0)}{2k^2\mu} = \mu_0 \frac{J_a + J_b + J_c + J_d}{4} \qquad (8.20a)$$

where $\mu = \mu_1/\mu_0$. By rearranging terms for ease of calculation,

$$A_0\left(\frac{1 + \mu}{\mu}\right)\left(\frac{1}{h^2} + \frac{1}{k^2}\right) = \left(\frac{A_1}{\mu} + A_3\right)\frac{1}{h^2} + (A_2 + A_4)\left(\frac{1 + \mu}{2\mu k^2}\right) + \mu_0 \frac{\sum J_i}{4}$$

$$(8.20b)$$

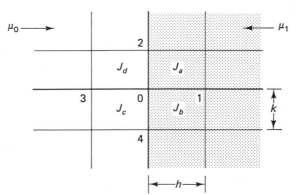

Figure 8.11 Mesh for the inhomogeneous magnetic case.

Example 8.4

Determine the inductance per unit length of the long wrapped magnetic core shown in Fig. 8.12(a). The upper and lower regions contain N windings, carrying current I; the core has permittivity μ.

Figure 8.12(b) shows a grid that can be used to solve this problem. In this figure, the dotted region contains the windings, and the striped region denotes magnetic material. Only the fourth quadrant is shown, as this is the minimum region to represent a complete solution. Note that the grid is not uniform: columns 8 and 9 and rows 6 and 7 are spaced twice as far as their predecessors. This shows one scheme for expanding the region of space of the solution, so as to approximate more closely the distant boundaries, without greatly increasing the number of computation points. When setting Eq. (8.16) or Eq. (8.20) with this scheme, account must be taken of the variation of h and k values.

In setting up the problem, it is clear that row 7 and column 9, representing very distant regions, have $A = 0$. Since it is the symmetry plane, row 1 must also have $A = 0$, and, also from symmetry, we see that $\partial A/\partial x = 0$ along column 1 so that Eq. (8.18b) must be used. For computational simplicity, take $\mu_0 J = 1$ in the current-carrying region, an assumption that will require some later unit corrections. Equation (8.20a) must be adapted at node R2C4, as the region of high μ is only in one quadrant. Figure 8.13 is the realization of

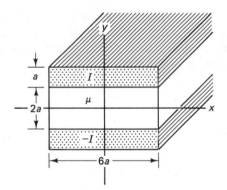

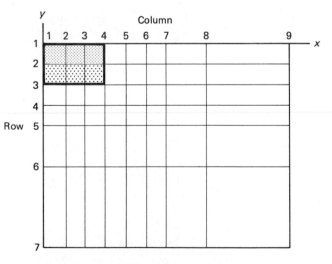

Figure 8.12 (a) Wrapped core. (b) Fourth quadrant grid of solution points.

	2
1	0
2	= (RC[+1]*101/200+R[+1]C+RC[−1]*101/200+0.5)/2.02
3	= (RC[+1]+R[−1]C+R[+1]C+RC[−1]+0.5)/4
4	= (RC[+1]+R[−1]C+R[+1]C+RC[−1])/4
5	= (RC[+1]+R[−1]C+R[+1]C/4+RC[−1])/3.25
6	= (RC[+1]+R[−1]C/4+R[+1]C/16+RC[−1])/2.3125
7	0

Figure 8.13   Computation of column 2 with $\mu = 100$.

Eqs. (8.16) and (8.20b) in column 2 of the mesh. These equations use $\mu = 100$. After iteration, the potential values are shown in Fig. 8.14.

The magnetic field lines, which coincide with the constant-potential lines, are found by interpolating between the values of Fig. 8.14. They are shown in Fig. 8.15.

The flux through a length ℓ of core can be found from

$$\Phi = \int \mathbf{B} \cdot \mathbf{n} \, dS = \int B_x \ell \, dy = \ell \int \frac{\partial A}{\partial y} \, dy = \ell \int dA = 2\ell \, \Delta A \qquad (8.21)$$

where ΔA is the change of A from a point in the windings to the center line ($A = 0$); the factor of 2 is necessary because there is an equal ΔA in the upper half of the system. Examination of Fig. 8.15 discloses that different flux passes through different turns or, from Fig. 8.14, that ΔA depends on the winding point selected. The maximum of A through the windings is at R2C1, with $\Delta A = 3.6$; the minimum is at R3C4, with $\Delta A = 2.0$. Recall that this value is based on having $\mu_0 J = 1$, whereas in actuality $\mu_0 J = \mu_0 NI/6a^2$. Scaling by this factor gives $\Phi = 2\alpha\mu_0 NI\ell/6a^2$, and, for the coefficient of induction,

$$L = N \frac{d\Phi}{dI} = \alpha \frac{\mu_0 N^2 \ell}{3a^2} \qquad \text{where } 2 \le \alpha \le 3.6$$

R3C1	= (2*RC[+1]+R[−1]C+R[+1]C+0.5)/4								
	MagProb								
	1	2	3	4	5	6	7	8	9
1	0	0	0	0	0	0	0	0	0
2	3.58	3.42	2.92	1.86	0.94	0.58	0.42	0.22	0
3	3.27	3.14	2.73	2.02	1.34	0.94	0.73	0.4	0
4	2.73	2.63	2.34	1.89	1.45	1.12	0.91	0.53	0
5	2.39	2.31	2.09	1.78	1.44	1.17	0.98	0.59	0
6	1.67	1.63	1.52	1.37	1.2	1.04	0.92	0.61	0
7	0	0	0	0	0	0	0	0	0

Figure 8.14   Potential values for the magnetic problem.

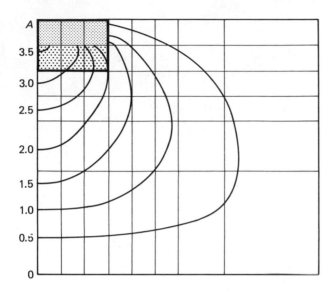

Figure 8.15   Field/equipotential lines for the magnetic problem.

8.8 FINITE ELEMENTS

The finite-element method is a general technique for the solution of differential equations, and is presently the most advanced of the methods for the solution of electromagnetic-field problems. In its precise mathematical form, the method involves complex concepts that give it generality and power. Here we adopt an early simple approach, since it affords greater understanding. We deal initially with Poisson's equation for electrostatic fields subject to either Dirichlet or Neumann boundary conditions.

Following customary and efficient practices, the solution region is divided into small triangles, as shown in Fig. 8.16(a), *within* each of which the potential is assumed to vary linearly. By this is meant that for a general triangle with vertices 1, 2, and 3, shown in Fig. 8.16(b), the potential is given by

$$\Phi = a + bx + cy \qquad x \text{ and } y \text{ within the triangle} \tag{8.22}$$

Boundary

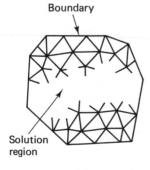

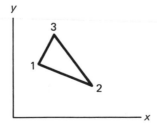

Solution
region

(a)   (b)   **Figure 8.16**   Finite-element triangles.

The smaller the triangles, the more accurate is the assumption of linear variation. (This is easily seen in one dimension, where a curve can be approximated by using short straight lines; the shorter the line lengths, the more closely is the curve approached.) The only limit on accuracy in this regard is the time and data-storage capacity required to calculate for a large number of triangles.

Equation 8.22 for the potential in a triangle can be expressed in terms of the potentials Φ_1, Φ_2, and Φ_3 at the three vertices of the triangle, located at (x_1, y_1), (x_2, y_2), and (x_3, y_3), respectively. At these vertices, Eq. (8.22) gives

$$\Phi_1 = a + bx_1 + cy_1 \tag{8.23a}$$

$$\Phi_2 = a + bx_2 + cy_2 \tag{8.23b}$$

$$\Phi_3 = a + bx_3 + cy_3 \tag{8.23c}$$

Solving these expressions for a, b, and c, we find

$$a = \frac{1}{2A}[\Phi_1(x_2y_3 - x_3y_2) + \Phi_2(x_3y_1 - x_1y_3) + \Phi_3(x_1y_2 - x_2y_1)] \tag{8.24a}$$

$$b = \frac{1}{2A}[\Phi_1(y_2 - y_3) + \Phi_2(y_3 - y_1) + \Phi_3(y_1 - y_2)]$$

$$= b_1\Phi_1 + b_2\Phi\phi_2 + b_3\Phi_3 \tag{8.24b}$$

$$c = \frac{1}{2A}[\Phi_1(x_3 - x_2) + \Phi_2(x_1 - x_3) + \Phi_3(x_2 - x_1)]$$

$$= c_1\Phi_1 + c_2\Phi_2 + c_3\Phi_3 \tag{8.24c}$$

$$A = \frac{1}{2}\begin{vmatrix} 1 & x_1 & y_1 \\ 1 & x_2 & y_2 \\ 1 & x_3 & y_3 \end{vmatrix} \tag{8.24d}$$

where A can be shown to be the area of the triangle.

The new variables b_1, b_2, b_3, c_1, c_2, and c_3 have been defined in Eqs. (8.24) for computational convenience.

$$b_i = (y_{i1} - y_{i2})/2A$$

$$c_i = (x_{i2} - x_{i1})/2A$$

where, for

$$i = 1: i1 = 2 \quad \text{and} \quad i2 = 3$$

$$i = 2: i1 = 3 \quad \text{and} \quad i2 = 1$$

$$i = 3: i1 = 1 \quad \text{and} \quad i2 = 2$$

These mappings are available under the modulo-3 function of computers.

Substituting the values of a, b, and c in Eq. (8.22) gives Φ within the triangle in terms of its vertex values. For a triangle of known vertex coordinates, Algorithm 2, at the end of this chapter, gives a procedure that finds the coefficients b_1, b_2, b_3, c_1, c_2, c_3, and A.

8.9 THE ENERGY FUNCTIONAL

Our goal is to have some appropriate function of the system parameters, called a *functional,* and to optimize it subject to system constraints. That set of vertex potentials is sought that minimizes the system energy subject to the boundary conditions and to Poisson's equation. When found, this set of potentials is the equilibrium solution from which the field solutions can be obtained.

The energy of an electrostatic system can be written in a number of ways. We have seen it in terms of the energy density of the fields, $\frac{1}{2}\int \mathbf{E} \cdot \mathbf{D} \, dv$ [see Eq. (5.2b)], or in terms of the energy of the sources, $\frac{1}{2}\int \rho\Phi \, dv$ [see Eq. (5.7)]. In picking the form to be used, we are guided by the symmetry of the expressions to which it leads, as this facilitates systematic computation, and by the rapidity with which it converges to the optimal answer, since, in some cases, we are unable to reach an exact solution. Of course, the solution must also satisfy Poisson's equation and the specified boundary conditions. It is found that a suitable mathematical form is obtained using both energy expressions to arrive at the energy functional:

$$\mathscr{F}(\Phi) = \int \left(\frac{1}{2} D \cdot E - \rho\Phi \right) dS = \int \left(\frac{1}{2} \varepsilon \nabla\Phi \cdot \nabla\Phi - \rho\Phi \right) dS \qquad (8.25)$$

As we are dealing with two-dimensional problems, the third dimension has been eliminated by writing the functional $\mathscr{F}$ as an integral of energy/area (or by considering a unit length in the third direction). Note that at its optimal value, $\mathscr{F}$ is equal to the system energy minus twice the system energy, so that $|\mathscr{F}|$ is still the system energy. Before applying Eq. (8.25), let us establish that it has the desired properties.

Suppose that Φ_0 is the correct solution to the problem under consideration, and that we have arrived at a solution $\Phi_0 + \kappa\phi$, where ϕ is a function that also satisfies the boundary conditions, and κ is a numerical factor that will be small if we are close to the desired answer. Then

$$\mathscr{F}(\Phi_0 + \kappa\phi) = F(\Phi_0) + \kappa\varepsilon \int \nabla\Phi_0 \cdot \nabla\phi \, dS + \frac{1}{2}\kappa^2\varepsilon \int \nabla\phi \cdot \nabla\phi \, dS - \kappa \int \rho\phi \, dS$$

$$(8.26)$$

The second term on the right-hand side can be integrated by parts to give

$$\kappa\varepsilon \int \nabla\Phi_0 \cdot \nabla\phi \, dS = \kappa\varepsilon \int \nabla \cdot (\phi\nabla\Phi_0) \, dS - \kappa\varepsilon \int \phi\nabla^2\Phi_0 \, dS$$

$$= \kappa\varepsilon \oint \phi\nabla\Phi_0 \cdot \mathbf{n} \, d\ell - \kappa\varepsilon \int \phi\left(-\frac{\rho}{\varepsilon} \right) dS \qquad (8.27a)$$

where the divergence theorem has been invoked to obtain an integral along the perimeter of the two-dimensional region (i.e., of unit depth), and use has been made of the condition that Φ_0 satisfies Poisson's equation. The integral

$$\oint \phi\nabla\Phi_0 \cdot \mathbf{n} \, d\ell = \oint \phi\frac{\partial\Phi_0}{\partial n} \, d\ell = 0 \qquad (8.27b)$$

because either ϕ or $\partial\Phi_0/\partial n$ is zero at points on the boundary. The last term of Eq. (8.27a) cancels the last term of Eq. (8.26), leaving

$$\mathcal{F}(\Phi_0 + \kappa\phi) = \mathcal{F}(\Phi_0) + \frac{1}{2}\kappa^2\varepsilon \int |\nabla\phi|^2 \, dS \qquad (8.28)$$

Since the last term on the right-hand side is positive-definite, it is clear that $\mathcal{F}$ is at a minimum when $\phi = 0$, and, from the previous derivation, the minimum solution Φ_0 solves Poisson's equation provided the boundary conditions are satisfied. Furthermore, the error term is proportional to the square of the small factor κ, so that the solution accuracy is high in the vicinity of the correct answer.

In each triangle, $\mathcal{F}$ can now be expressed in terms of the vertex potentials. Within a single triangle, from Eq. (8.22),

$$\nabla\Phi = \mathbf{a}_x \frac{\partial\Phi}{\partial x} + \mathbf{a}_y \frac{\partial\Phi}{\partial y} = \mathbf{a}_x b + \mathbf{a}_y c \qquad (8.29a)$$

so that the first term in Eq. (8.25) is

$$\int \frac{1}{2}\varepsilon\nabla\Phi \cdot \nabla\Phi \, dS = \frac{1}{2}\varepsilon A(b^2 + c^2) \qquad (8.29b)$$

Within a single triangle, the second term in Eq. (8.25), $\int \rho\Phi \, dS$, is the integration of a linear function over the triangular area (remember that ρ is a constant within each triangle). Evaluation of this integral yields the centroid of the vertex values

$$\int \rho\phi \, dS = \frac{1}{3}\rho A(\Phi_1 + \Phi_2 + \Phi_3) \qquad (8.30)$$

While term-by-term integration can be used to establish Eq. (8.30) directly, it is convenient here to justify it through analogy. Consider the one-dimensional case of a linear function $f(x) = a + bx$. Its average value over the range from x_1 to x_2 is

$$\langle f \rangle = \frac{1}{(x_2 - x_1)} \int_{x_1}^{x_2} (a + bx) \, dx = a + \frac{1}{2}b(x_2 + x_1)$$

$$= \frac{1}{2}(f_1 + f_2)$$

At each point, we can write $f = \langle f \rangle + \Delta f$. We then have

$$\int_{x_1}^{x_2} f \, dx = (x_2 - x_1)\langle f \rangle = \frac{1}{2}\ell(f_1 + f_2)$$

because positive and negative deviations, Δf, occur with equal weight about the average, so their integral is zero.

Extrapolation of ℓ to A and of the linear end points to the triangle vertices gives Eq. (8.30).

Substituting Eqs. (8.29b) and (8.30) in Eq. (8.25) and using Eqs. (8.24b) and (8.24c) for a and b gives, for the single triangle,

$$\mathscr{F}_1 = \frac{1}{2}\,\varepsilon A[(b_1^2 + c_1^2)\Phi_1^2 + (b_2^2 + c_2^2)\Phi_2^2 + (b_3^2 + c_3^2)\Phi_3^2$$
$$+ 2(b_1 b_2 + c_1 c_2)\Phi_1 \Phi_2 + 2(b_2 b_3 + c_2 c_3)\Phi_2 \Phi_3$$
$$+ 2(b_1 b_3 + c_1 c_3)\Phi_1 \Phi_3] - \rho A(\Phi_1 + \Phi_2 + \Phi_3) \qquad (8.31)$$

This energy functional can be expressed in matrix form as

$$\mathscr{F}_1 = \frac{1}{2}[\Phi]^t[P][\Phi] - [\Phi]^t[q] \qquad (8.32)$$

where

$$[\Phi] = \begin{bmatrix} \Phi_1 \\ \Phi_2 \\ \Phi_3 \end{bmatrix} \qquad [\Phi]^t = [\Phi_1 \quad \Phi_2 \quad \Phi_3] \qquad [q] = A\rho \begin{bmatrix} \dfrac{1}{3} \\ \dfrac{1}{3} \\ \dfrac{1}{3} \end{bmatrix} \qquad (8.33\text{a, b, c})$$

$$[P] = \varepsilon A \begin{bmatrix} b_1^2 + c_1^2 & b_1 b_2 + c_1 c_2 & b_1 b_3 + c_1 c_3 \\ b_2 b_1 + c_2 c_1 & b_2^2 + c_2^2 & b_2 b_3 + c_2 c_3 \\ b_3 b_1 + c_3 c_1 & b_3 b_2 + c_3 c_2 & b_3^2 + c_3^2 \end{bmatrix} \qquad (8.34)$$

$[P]$ and $[q]$ are referred to as local matrices and their formulation is given in pseudocode in Algorithm 8.3.

8.10 THE GLOBAL MATRICES

Now we come to the key point in finite-element analysis. According to Eq. (8.31), the energy functional $\mathscr{F}_1$ is a second-order polynomial in the unknown node potentials of the single triangle. The total energy functional is the sum of the contributions from all the triangles.

$$\mathscr{F} = \sum_\Delta \mathscr{F}_i \qquad (8.35)$$

Since this must be a minimum, the node potentials must be such as to give the lowest value. By performing the requisite differentiations with respect to Φ_1, for example,

$$\frac{\partial \mathscr{F}}{\partial \Phi_1} = \sum_\Delta \varepsilon A[(b_1^2 + c_1^2)\Phi_1 + (b_1 b_2 + c_1 c_2)\Phi_2 + (b_1 b_3 + c_1 c_3)\Phi_3] - \frac{1}{3}\rho A$$
$$(8.36)$$

which can be recognized as corresponding to the first row of the matrix equation

$$\sum_\Delta ([P][\Phi] - [q]) = 0 \qquad (8.37)$$

As we sum over the triangles, the vectors [Φ] in turn contain the three vertex potentials of the triangle whose contribution is being added. The rules for the summation of the contributions from the various triangles can be discerned by adding the polynomials from two adjacent elements with vertices 1, 2, and 3 and 3, 2, and 4, as shown in Fig. 8.17(a). The first gives a contribution

$$\mathcal{F}_1 = \frac{1}{2}[\Phi]_1^t [P][\Phi]_1 - [\Phi]_1^t [q_1] \tag{8.38a}$$

and the second gives

$$\mathcal{F}_2 = \frac{1}{2}[\Phi]_2^t [M][\Phi]_2 - [\Phi]_2^t [s_2] \tag{8.38b}$$

$[\Phi]_1$ has elements Φ_1, Φ_2, and Φ_3, and $[\Phi]_2$ has elements Φ_3, Φ_2, and Φ_4. $\mathcal{F}_1$ and $\mathcal{F}_2$ can be expanded in the scalar polynomials given by Eq. (8.31) in terms of their respective [Φ] components. Adding them and differentiating $\mathcal{F}_1 + \mathcal{F}_2$ with respect to the four potentials, as was done in Eq. (8.36), gives four linear equations that can be written as a 4 × 4 matrix similar to Eqs. (8.32) to (8.34).

Since each vertex contributes a potential component, we can define the global matrices [G] and [Q] as the square and column matrices, respectively, corresponding to [P] and [q] but of the same size as the total number of vertices of the problem space. Algorithm 8.3 generates the local matrices, and they are placed (added) in the global matrices by Algorithm 8.4.

To understand how this is done, note that Algorithm 8.3 reassigns the vertex numbers 1, 2, and 3 of each triangle in counterclockwise order, as shown in Fig. 8.17(b). To add the adjoining matrices, Algorithm 8.4 recognizes that the vertex points coincide such that the subscript correspondence (second triangle) → (first triangle) is 1 → 2, 2 → 4, and 3 → 3. Thus, for the two triangles shown, the combined matrix equation is

$$\begin{bmatrix} P_{11} & P_{12} & P_{13} & 0 \\ P_{21} & P_{22} + M_{11} & P_{23} + M_{13} & M_{12} \\ P_{31} & P_{32} + M_{31} & P_{33} + M_{33} & M_{32} \\ 0 & M_{21} & M_{23} & M_{22} \end{bmatrix} \begin{bmatrix} \Phi_1 \\ \Phi_2 \\ \Phi_3 \\ \Phi_4 \end{bmatrix} - \begin{bmatrix} q_1 \\ q_2 + s_1 \\ q_3 + s_3 \\ s_2 \end{bmatrix} = 0 \tag{8.39}$$

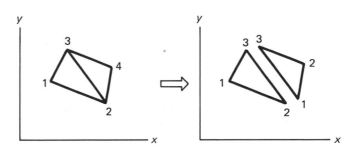

Figure 8.17   Triangle mapping, with order.   (a)   (b)

We therefore have a large matrix equation of the form of Eq. (8.37):

$$[G][\Phi] = [Q] \tag{8.40a}$$

where $[G]$ is the global matrix, which is the sum of the $[P]$, $[M]$, etc. matrices, $[\Phi]$ is the column matrix of all the node potentials of the mesh, and $[Q]$ is the column matrix, which is the sum of all the $[q]$, $[s]$, etc. contributions in their places. Equation (8.40b) displays the full form of Eq. (8.40a):

$$
\begin{bmatrix}
G_{11} & G_{12} & G_{13} & \cdots & G_{1\mathrm{Unk}} & \cdots & G_{1\mathrm{Nodes}} \\
G_{21} & G_{22} & G_{23} & \cdots & G_{2\mathrm{Unk}} & \cdots & G_{2\mathrm{Nodes}} \\
\vdots & \vdots & \vdots & & \vdots & & \vdots \\
G_{\mathrm{Unk}1} & & G_{\mathrm{Unk}2} & & G_{\mathrm{Unk\,Unk}} & & \\
\vdots & \vdots & \vdots & & \vdots & & \vdots \\
G_{\mathrm{Nodes}1} & & & \cdots & & \cdots & G_{\mathrm{Nodes\,Nodes}}
\end{bmatrix}
\begin{bmatrix}
\Phi_1 \\ \Phi_2 \\ \vdots \\ \Phi_{\mathrm{Unk}} \\ \vdots \\ \Phi_{\mathrm{Nodes}}
\end{bmatrix}
=
\begin{bmatrix}
q_1 \\ q_2 \\ \vdots \\ q_{\mathrm{Unk}} \\ \vdots \\ q_{\mathrm{Nodes}}
\end{bmatrix}
$$

$$\tag{8.40b}$$

The solution of Eqs. (8.40) can be found by iteration techniques or from

$$[\Phi] = [G]^{-1}[Q] \tag{8.41}$$

which requires the inverse of $[G]$, followed by multiplication into $[Q]$. Although matrix solution programs are available in most computer facilities, they are very long processes, particularly for the large matrices that are likely to be involved here. Algorithm 8.4, therefore, carefully places the elements so that the matrix size is minimum. For this, the unknown nodes are numbered first, up to N_{Unk}, and the knowns thereafter up to N_{Nodes}. This is the scheme implied in Eq. (8.40b).

Note that since some of the boundary potentials are known, no variations will be made with respect to them, that is, Eq. (8.36) is identically zero, so that the corresponding rows do not appear in the global matrix. With the numbering system just specified, these will all correspond to the last rows of the matrix equation. This means that there will be no elements below the horizontal dashed line in Eq. (8.40b). Furthermore, the terms derived from the $[G]$ components to the right of the vertical dashed line in Eq. (8.40b) are constants, being the products of known G_{jk} with known Φ_k. The algorithm therefore transposes these to the right of the equation, leaving a matrix whose size is $N_{\mathrm{Unk}} \times N_{\mathrm{Unk}}$.

With the Algorithms 8.2 to 8.4, programs can be written to solve complex problems. The data for a finite-element field problem are contained in three files. The first consists of triangle data—a series of triangles, each defined by three vertices and the values of ρ and ε within it. The second gives the node data, characterized by the x and y coordinates and the solution if the node is known. The third gives the number of triangles, the number of nodes, and the number of unknowns. The finite-element program, after initializing the global matrix, reads every triangle and then, from the node numbers, reads the coordinates of the three vertices and forms the corresponding local matrix using Algorithm 8.3. Algorithm 8.4 then places the local matrix in the global matrix before reading the next triangle. Once the global matrix is assembled, a solver is called upon to find the potentials. Once the vertex solutions are identified, standard plotting routines can be imported to plot equipotential lines.

The following example illustrates use of the finite-element program. Since this is intended merely as a test example, the distant field points have been brought unrealistically close to the field source. However, this reduces the number of unknown nodes sufficiently so that the solution matrices can be checked by hand calculation.

Example 8.5

Solve for the magnetic potential in the region around the rectangular current-carrying cable in Fig. 8.18. The shaded area represents the cable. It is only necessary to solve for the potential in the first quadrant and a coarse grid is shown there, with node point numbering ordered as previously described.

Since this is a magnetostatic problem, all of the previous finite-element derivations are appropriate with A in place of Φ, μ^{-1} substituted for ε, and J in place of ρ. Table 8.3 gives the three required data files.

When worked out by hand, it will be seen that the local matrices for the element 1,2,3 are

$$[p] = \mu_0^{-1}\begin{bmatrix} 0.25 & -0.25 & 0 \\ -0.25 & 1.25 & -1.0 \\ 0.0 & -1.0 & 1.0 \end{bmatrix} \qquad [q] = \begin{bmatrix} 1.0 \\ 1.0 \\ 1.0 \end{bmatrix}$$

Since nodes 5 through 9 represent distant points, they have $A = 0$, so there are only four points left: 1 through 4. The final global matrix equation is

$$\begin{bmatrix} 1.25 & -0.25 & 0.0 & -1.0 \\ -0.25 & 2.5 & -2.0 & 0.0 \\ 0.0 & -2.0 & 5.0 & -0.5 \\ -1.0 & 0.0 & -0.5 & 2.5 \end{bmatrix} \begin{bmatrix} A_1 \\ A_2 \\ A_3 \\ A_4 \end{bmatrix} = \mu_0 \begin{bmatrix} 2.0 \\ 1.0 \\ 2.0 \\ 1.0 \end{bmatrix}$$

If the program written by the reader is working correctly, then the final assembled equation should correspond to this matrix. A standard solver can now be called to solve this matrix equation. A plot of the solution is provided in Fig. 8.19. Note that the plotting program has interpo-

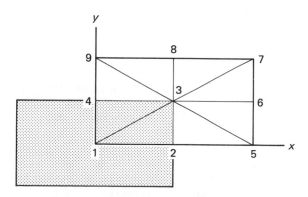

Figure 8.18 Rectangular conductor with coarse solution mesh.

TABLE 8.3. DATA FILES FOR EXAMPLE 8.4

\multicolumn{5}{Triangle data}					Node coordinates		General
v_1	v_2	v_3	ε	ρ_0	X	Y	
							No. of unknowns = 4
1	2	3	1	3	0	0	No. of points = 9
3	4	1	1	3	2	0	No. of triangles = 8
5	2	3	1	0	2	1	
3	6	5	1	0	0	1	
3	4	9	1	0	4	0	
9	8	3	1	0	4	1	
7	8	3	1	0	4	2	
3	6	7	1	0	2	2	
					0	2	

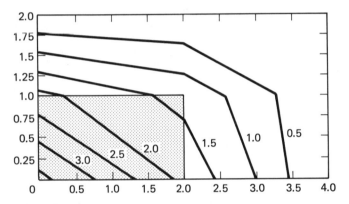

Figure 8.19   Potential contours in first quadrant.

lated potential values between the grid nodes. A Pascal program for this example, involving Algorithms 8.2 to 8.3, is given in an appendix 2 at the end of this text.

Observe that the field lines approach being normal to the two edges on which no conditions were set (x and y axes); for finer meshes, the lines would be more exactly perpendicular. This illustrates the fact, which will not be proven here, that in finite elements, any boundary section on which the potential or its gradient is not set will automatically have a zero normal gradient in the solution. For this reason, the Neumann boundary conditions are called *natural* to the finite-element formulation, so that it is not necessary to specifically set them.

PROGRAMMING ALGORITHMS

Algorithm 8.1 Integrate Filaments

Procedure IntegrateFilaments(Φ, x, y)

```
{Integrates the effects of several charge filamental wires at
a point. Requires a file f containing the coordinates of and
charge in each filament.}
```

```
Begin
    Φ ← 0
    While Not EOF(f) Do
      Read(xf, yf, qf)
      r ← Sqrt(Sqr(x - xf) + Sqr(y - yf))
      Φ ← Φ - qf*ln r
    Φ ← Φ/2π/ε
End
```

Algorithm 8.2 Computing First-Order Matrices *b* and *c*

Procedure Triangle(b, c, A, x, y)

```
{Outputs: b, c: The 1 X 3 first-order matrices defined in
Eqs. (8.24). A = area of triangle.
Inputs: x, y = 3 X 1 vectors containing the coordinates of the
three vertices}
Begin
    Delta ← x[2]*y[3] - x[3]*y[2] + x[3]*y[1] - x[1]*y[3] + x[1]*y[2] -
            x[2]*y[1]
    A ← Abs(Delta)/2                                        {Eq. (8.24d)}
    For i ← 1 To 3 Do
        i1 ← Mod(i,3) + 1 {Or i Mod3 + 1}
        i2 ← Mod(i1,3) + 1
        b[i] ← (y[i1] - y[i2])/Delta
        c[i] ← (x[i2] - x[i1])/Delta
End
```

Algorithm 8.3 Forming Local Matrices for First-Order Triangle

Procedure FirstOrderLocalMats(P, q, x, y, Eps, Rho)

```
{Function: To compute the first-order differentiation matrices
Outputs:
    P = The 3 X 3 first-order Local Element Matrix-Eq. (8.34)
    q = The 3 X 1 local right-hand side vector-Eq. (8.33c)
Inputs:
    x, y = 3 X 1 vectors containing the coordinates of the
            three vertices
    Eps = The constant material value ε in the triangle
    Rho = A constant giving source ρ in the triangle
Required = Procedure Triangle
}
Begin
    Triangle(b, c, A, x, y)
    For i ← 1 To 3 Do
        q[i] ← A*Rho/3
        For j ← 1 To 3 Do
            P[i,j] ← Eps*A*(b[i]*b[j] + c[i]*c[j])        {Eq. (8.34)}
End
```

Algorithm 8.4 Placing a Local Matrix in a Global Matrix

Procedure GlobalPlace(PG, qG, P, q, v, Phi)

```
{Function Adds the local matrices to the global matrices
Outputs
    PG = The global matrix of size NUnk X NUnk
    qG = The global right hand side qG of size NUnk X 1.
Inputs
    P = 3 X 3 local matrix P
    q = 3 X 1 local right-hand side vector
    NUnk = Number of unknown nodes in mesh numbered before the
           known nodes
    v = A 3-vector containing the node numbers of the triangle
    Phi = A vector as long as there are finite nodes. The
          first NUnk elements are unknown and to be determined
          by the finite-element method and the rest of the
          elements are known through the boundary conditions.
}
Begin
    For Row ← 1 To 3 Do
        If v[Row] ≤ NUnk
          Then {Else Row Corresponds to a Known; Ignore it}
              qG[v[Row]] ← qG[v[Row]] + q[Row]
              For Col ← 1 To 3 Do
              If v[Col] ≤ NUnk
                  Then {Add to Global Matrix}
                    PG[v[Row],V[Col]] ← PG[v[Row],v[Col]] + P[Row,Col]
                  Else {Column Corresponds to a Known. Shift Right}
                    qG[v[Row]] ← qG[v[Row]] - P[Row,Col])*Phi[v[Col]]
    End
```

Graphics Program

The Algorithms and programs presented in this chapter write the solution onto a file in the form:

X coordinate	Y coordinate	Potential
. . .	. . .	. . .
. . .	. . .	. . .

```
{The following is a short program that specifies the file name
and plots two-dimensional contours. A graphics plotting
systems must be used.}
read-from fieldplot.dat x,y,z
graphics-mode
plot-type 2-d-contour
plot-range x-min -1 x-max 8 y-min -1 y-max 6
set-parameter contour-increment 0.3
set-parameter contour-label-incr 1
quit
plot x y z
quit
```

EXERCISES

Extensions of the Theory

8.1. Mathematically show that the square finite-difference mesh gives the same solution when used as a finite-element mesh with the diagonal terms drawn at 45° so as to cut each rectangle into two triangles. *Hint:* Consider the local matrix of a right-angled triangle and the contributions to the equation corresponding to a node from the six surrounding triangles. Check the truth of this numerically. What then are the advantages of the finite-element method?

8.2. A field domain R surrounded by a Dirichlet boundary contains a line of symmetry and a mesh has been constructed for a half of the problem. A second mesh is constructed for the whole domain by reflecting the first mesh about the symmetry line. In the first problem, the Neumann condition at the line of symmetry is ignored, whereas in the second problem, symmetry is used to reduce the matrix size by a half by mapping nodes on one half to those on the other half; that is, if a node in the second half has potential Φ_i corresponding to a node in the first half with potential Φ_j, then wherever we have Φ_i, we substitute Φ_j. As a result, both problems have the same matrix size although the larger problem will take twice as long to form the matrix.

Compare the matrix equations that result from both problems in finite-element analysis. From your answer, what can you say about the condition that the results of the problem posed by the half mesh will reflect at the line of symmetry?

8.3. Figure 8.20 is an electric fuse carrying current. When the current exceeds a certain value, the narrow part melts and opens the circuit. The governing equations are

$$\nabla \cdot \mathbf{J} = 0 \qquad \nabla \times \mathbf{E} = 0 \qquad \mathbf{J} = \sigma \mathbf{E}$$

Two approaches are possible to this problem. We can say from the nondivergence of $\mathbf{J}$ that

$$\mathbf{J} = \nabla \times T\mathbf{a}_z$$

where T is the current vector potential, so that, from the irrotationality of $\mathbf{E}$,

$$-\sigma^{-1}\nabla^2 T = 0$$

Alternatively and more conventionally, from the irrotationality of $\mathbf{E}$, we have $\mathbf{E} = -\nabla\Phi$, which gives us

$$-\sigma\nabla^2\Phi = 0$$

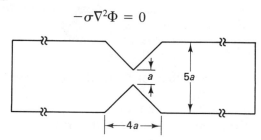

Figure 8.20   For Exercise 8.3.

Choose the most appropriate of the two variables Φ and T to solve for the following cases.

(a) The voltage across the fuse is specified.

(b) The total current through the fuse is specified.

For both these cases, obtain and plot the solution and compute the resistance of the fuse in terms of its conductivity.

8.4. Figure 8.21 (see below) shows an electromagnet used to lift magnetic scrap metal in air. The magnet is 2 m long and the coil about the central portion carries 4 A/m^2. The magnet iron has $\mu = 300\,\mu_0$ and the large piece of scrap has $\mu = 5\mu_0$. Find the force exerted on the scrap piece. *Hint:* Find the energy in the volume that disappears if the magnet and scrap move toward each other.

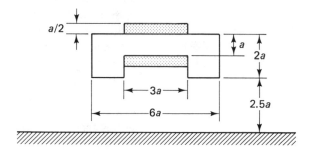

Figure 8.21   For Exercise 8.4

8.5. Under strong magnetic fields, the permeability μ is a function of the magnetic field, so that the associated field problem is nonlinear. Such problems are solved by assuming a certain solution, computing the corresponding permeability, solving for fluxes on the assumption that this permeability is correct, updating the permeability, and repeating the cycle until the fluxes change no more. This procedure, however, is subject to nonconvergence unless we undercorrect the flux changes. That is, if the flux density in one solution is B^i and in the next is B^{i+1}, then when we update the permeability, we slow down the change in B through the instruction

$$B^{i+1} = B^i + 0.1[B^{i+1} - B^i]$$

Repeat the lifting-magnet problem of Fig. 8.21 when the magnetic material is rolled steel with

$$\mu^{-1} = 396.2 + 3.8e^{2.17B^2}$$

Examine what happens to convergence when the current is increased so as to deeply saturate the device.

Problems

8.6. Use the method of finite differences to find the capacitance of two parallel wires separated by a distance that is large compared to their radii; see Fig. 8.22. The lines can be approximated by point lines at potentials $\pm V_0$. Compare your result with the analytic formula $C/\ell = \pi\varepsilon_0/\ln(d/r)$, where d is

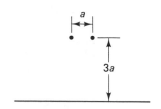

Figure 8.22 For Exercise 8.6.   **Figure 8.23** For Exercise 8.7.

the separation, and r is the wire radius, to determine an effective value of d/r.

8.7. If the lines of Problem 8.6 are assumed to have charges of the same sign, find their combined capacitance to the highly conducting ground plane that is located at three times their separation, as shown in Fig. 8.23.

8.8. The approximation involved in the finite-differences method, of taking the infinitely distant boundaries at a short distance, can be checked to determine the effect it has on the near field distribution. Repeat Problem 8.6, with both the line separation and the contour on which Q is evaluated kept constant, but moving the distant boundaries to a few larger distances. Is the capacitance or the equivalent value of d/r approaching a limit? Note that the field variation at the distant points should not change significantly, so the iteration time and number should also not change greatly. If possible, record these as well.

8.9. Repeat Problem 8.6 if the plane containing both wires separates air (above) from an infinite half space that has $\varepsilon = 6\varepsilon_0$ (below).

8.10. Repeat Problem 8.7 if the ground separation is a and $10a$. Plot the results to see if they are approaching a limit at large values of a.

8.11. A square line with side a carries a linear charge density ρ_ℓ. It is at a distance $2a$ above a conducting plane, as shown in Fig. 8.24. Plot the total charge per length induced on the plane as a function of distance from the foot of the normal from the line charge. Note: $\rho_s = D_n$ on the plane.

8.12. Repeat Problem 8.11 if the plane is not conducting but, instead, has below it a half space with $\varepsilon = 6\varepsilon_0$.

8.13. A line at potential V_0 is at a distance a from each of two conducting planes that intersect at a right angle, as shown in Fig. 8.25. What is its capacitance to the planes? By comparing your ground locations, discuss the relation between this problem and Problem 8.6.

8.14. Find the self-inductance per unit length of two antiparallel current lines, $\pm I$. Recall that Φ can be determined from equation 8.21. Compare your result

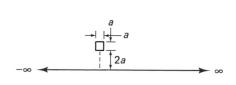

Figure 8.24 For Exercise 8.11.

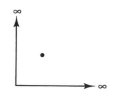

Figure 8.25 For Exercise 8.13.

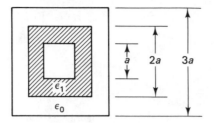

Figure 8.26 For Exercise 8.17.   **Figure 8.27** For Exercise 8.18.

with the analytic formula $L/\ell = (\mu_0/\pi)\ln(d/r)$, where d is the separation and r is the wire radius, to determine an effective value of d/r.

8.15. Repeat Problem 8.14 if the plane containing both wires separates air from an infinite half space that has $\mu = 10\mu_0$.

8.16. Repeat Problem 8.14, taking the distant boundary at 5 times and at 20 times the interwire distance.

8.17. Find the capacitance per length of a transmission line whose conductors and dielectric are coaxial square sections of side a, $2a$, and $3a$, as shown in Fig. 8.26. Compare your result with that of circular conductors to determine the equivalent circular radius ratio ρ_2/ρ_1? Take $\varepsilon_1 = 4\varepsilon_0$.

8.18. Find the capacitance per length of the *butt line* shown in Fig. 8.27.

8.19. Figures 8.28(a) and (b) show a thin and a solid *slot line,* respectively, separated by a distance a.
  (a) Compare their capacitances per meter of length (into the page).
  (b) As a finite width must be selected for summing the charge, take each plate with a width of $5a$, $10a$, and $20a$ and compare the values found.
  Note: At the distant plate ends the influence of the gap is negligible so the potential can be taken to vary linearly from Φ of the plate to zero.

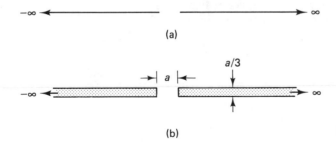

Figure 8.28 For Exercise 8.19.

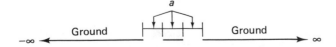

Figure 8.29 For Exercise 8.20.

8.20. The *shielded line* of Fig. 8.29 has a charge Q on the central conductor.
   (a) Assuming Q is uniform on the center conductor, use finite differences to estimate the ground plane width that contains $-0.7Q$. *Hint:* Use trial and error with several widths.
   (b) If the ground planes are truncated so that they are each only $5a$ wide, use the integral technique to find the charge distribution on the central and ground planes.

8.21. Figures 8.30 show a thin and a solid *double stripline,* respectively, separated by a distance a. Compare their capacitances per meter of length.

8.22. Repeat Problem 8.19 if the plane of the lines separates air above from a half space below with dielectric $\varepsilon = 5\varepsilon_0$.

8.23. For the line of Fig. 8.28(b), find c/ℓ if the line thickness is $a/10$ and a.

8.24. A two-wire transmission line is shown in Fig. 8.31. Determine, by integral methods, the electric field above the ground if the lines carry a uniformly distributed charge of $+q$ and $-q$. Find the capacitance of the line.

8.25. **(a)** Repeat Problem 8.24 for magnetic fields for currents of 1 A and -1 A, and determine the external mutual inductance. Draw the equipotentials. Compare your results with the solution to Problem 8.6.
   (b) Find the magnetic-field distribution when the currents are of the same sign.

8.26. A point-like transmission line is 10 m over the earth and at a voltage of 10 kV. A wooden gate under the line has a 4-cm-wide flat conductive rail running horizontally along the plane transverse to the line, as shown in Fig. 8.32. For

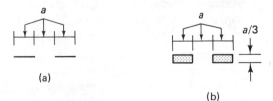

(a)   (b)

Figure 8.30   For Exercise 8.21.

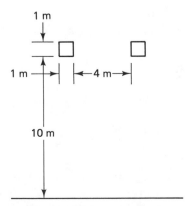

Figure 8.31   For Exercise 8.24.

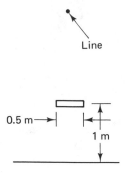

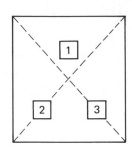

Figure 8.32 For Exercise 8.26.   **Figure 8.33** For Exercise 8.27.

safety reasons, we need to know the voltage induced on the railing. Discuss the appropriateness of integral and differential techniques and solve the problem by the method of your choice.

8.27. Figure 8.33 shows a three-conductor cable in a conducting conduit. Each cable has a square cross section of side a and the conduit is square with side $6a$. The cables are placed as shown, each at a distance a from the nearby conduit walls. Take $V_1 = 100$ V, $V_2 = -73.33$ V, $V_3 = 73.33$ V. Discuss the merit or otherwise of solving for the electric potential in the cable by integral and differential methods. Solve by the method of your choice.

8.28. Determine the capacitance of line 1 of the three-conductor cable system of Problem 8.27. *Note:* For this problem *all* the other conductors are held at ground potential.

8.29. Solve for the vector potential in the current carrying three-conductor cable system of Problem 8.27. *Hint:* At nodes on the edge or corner of the conductor, we can proceed by taking the current density to be, respectively, one-half or one-quarter of the internal density.

8.30. Find the magnetic field within the cable system shown in Fig. 8.34. The conduit is conducting, and has radius $6a$. The conductors have radii a and their centers are $3a$ from the center and are separated by 120°. Take $I_1 = 100$ A, $I_2 = -73.33$ A, $I_3 = 73.33$ A. Compute the magnetic-field strength in every triangular element and produce a plot of arrows with the computer, where the arrow in each triangle represents the flux density. *Hint:* Approximate the circles by high-order polygons.

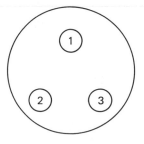

Figure 8.34 For Exercise 8.30.

9

ELECTROMAGNETIC WAVES

Maxwell's equations, in their differential form, were presented at the end of Chapter 2. We then examined the static aspects of the fields they describe and just enough of the time-dependent properties of the equations, so that we could understand the energy of the fields and how the static situation is established. We now turn to a study of the dynamic aspects of electromagnetism; this chapter develops the tools for understanding the time-varying and propagation properties treated in the remainder of the text.

9.1 THE WAVE EQUATION

In the time-varying case, Maxwell's equations involve a mixing of the different field quantities, since a changing magnetic flux induces an electric field and a time-varying electric flux induces a magnetic field. The analysis will be much simpler if the field equations are separated so as to give a complete description of the dynamic behavior of each, for example, of $\mathbf{E}$ or $\mathbf{H}$. To accomplish this, the constitutive equations are used, describing the properties of the medium containing the fields. They are

$$\mathbf{B} = \mu\mathbf{H} \qquad \mathbf{D} = \varepsilon\mathbf{E} \qquad \mathbf{J} = \sigma\mathbf{E} \qquad (9.1)$$

Over most useful ranges, ε, μ, and σ can be regarded as constants and Eqs. (9.1) can be used to simplify Maxwell's equations by eliminating $\mathbf{J}$, $\mathbf{D}$, and $\mathbf{B}$. Thus,

$$\nabla \times \mathbf{H} = \sigma \mathbf{E} + \varepsilon \frac{\partial \mathbf{E}}{\partial t} \tag{9.2a}$$

$$\nabla \times \mathbf{E} = -\mu \frac{\partial \mathbf{H}}{\partial t} \tag{9.2b}$$

$$\nabla \cdot \mathbf{H} = 0 \tag{9.2c}$$

$$\nabla \cdot \mathbf{E} = \frac{\rho}{\varepsilon} \tag{9.2d}$$

A formal procedure to solve for each of the remaining fields is to take the curl of the curl equation, Eq. (9.2b),

$$\nabla \times (\nabla \times \mathbf{E}) = -\mu \frac{\partial}{\partial t} \nabla \times \mathbf{H} \tag{9.3a}$$

The left-hand side of Eq. (9.3a) can be expanded with a vector identity, and the other curl equation, Eq. (9.2a), is substituted on the right-hand side.

$$\nabla(\nabla \cdot \mathbf{E}) - (\nabla \cdot \nabla)\mathbf{E} = -\mu \frac{\partial}{\partial t}\left(\sigma \mathbf{E} + \varepsilon \frac{\partial \mathbf{E}}{\partial t}\right) \tag{9.3b}$$

or, with Eq. (9.2d),

$$-\nabla^2 \mathbf{E} + \mu \varepsilon \frac{\partial^2 \mathbf{E}}{\partial t^2} + \mu \sigma \frac{\partial \mathbf{E}}{\partial t} = -\nabla \frac{\rho}{\varepsilon} \tag{9.4}$$

where $\nabla^2 \mathbf{E} = (\nabla \cdot \nabla)\mathbf{E}$.

In Chapter 1, we introduced $\nabla^2 f$, where f is a scalar function. $\nabla^2 \mathbf{E} = (\nabla \cdot \nabla)\mathbf{E}$ represents the scalar operation applied to each of the components of $\mathbf{E}$. In the coordinate system x_i, x_j, x_k,

$$\nabla^2 \mathbf{E} = \mathbf{a}_i \nabla^2 E_i + \mathbf{a}_j \nabla^2 E_j + \mathbf{a}_k \nabla^2 E_k$$

Equation (9.4) is known as the inhomogeneous wave equation for $\mathbf{E}$ (because not all the terms involve $\mathbf{E}$); it leads to a description of the phenomenon of traveling waves. Starting with Eq. (9.2a) instead of Eq. (9.2b) leads to a similar equation for $\mathbf{H}$.

Let us first investigate the predictions of Eq. (9.4) in some simple cases, so as to build an understanding that can be applied to more complex situations. We will only consider fields in an uncharged medium, so that $\rho = 0$. Except for some special situations, such as propagation through the ionosphere or through ionic solutions, this is the most commonly encountered case. With this understanding, the last term vanishes in Eq. (9.4). Certain other specific assumptions are made now for ease of analysis; they will be relaxed later.

   1. *The Planar Solution.* Many different forms of solution exist for Eq. (9.4), depending on the physical situation of interest. If $\nabla^2 \mathbf{E}$ is written in spherical coordinates, it describes waves radiating in all directions from a source such as an atom or an antenna. This will be examined in a later chapter; for now rectangular, or cartesian, coordinates are employed, so that Eq. (9.4) becomes

$$\frac{\partial^2 \mathbf{E}}{\partial x^2} + \frac{\partial^2 \mathbf{E}}{\partial y^2} + \frac{\partial^2 \mathbf{E}}{\partial z^2} - \mu\varepsilon\frac{\partial^2 \mathbf{E}}{\partial t^2} - \mu\sigma\frac{\partial \mathbf{E}}{\partial t} = 0 \qquad (9.5)$$

This form of the wave equation leads to an important special case that is found by assuming that $\mathbf{E}$ is a function of only a single space variable, say z, as well as of time. Then $\mathbf{E} = \mathbf{E}(z, t)$. This means that $\mathbf{E}$ does not vary with either x or y, so that the magnitude and direction of $\mathbf{E}$ are the same at every point of *each xy* plane; the solution is called a *plane wave*. Note that $\mathbf{E}$ can vary between planes, as shown in Fig. 9.1(a). In Eq. (9.5), the derivatives vanish with respect to x and y, so that

$$\frac{\partial^2 \mathbf{E}}{\partial z^2} - \mu\varepsilon\frac{\partial^2 \mathbf{E}}{\partial t^2} - \mu\sigma\frac{\partial \mathbf{E}}{\partial t} = 0 \qquad (9.6)$$

One consequence of planarity follows immediately from Eq. (9.2d):

$$\nabla \cdot \mathbf{E} = \frac{\partial E_x}{\partial x} + \frac{\partial E_y}{\partial y} + \frac{\partial E_z}{\partial z} = \frac{\partial E_z}{\partial z} = 0 \qquad (9.7)$$

where, again, the second equality results because the components of $\mathbf{E}$ do not vary with x or y and the last equality results from having $\rho = 0$. Equation (9.7) means that E_z, the longitudinal component, does not vary with z either, so that this component is a constant in space. To determine its time variation, take the z component of Eq. (9.6) and, since the first term vanishes,

$$\mu\varepsilon\frac{\partial^2 E_z}{\partial t^2} + \mu\sigma\frac{\partial E_z}{\partial t} = 0 \qquad (9.8a)$$

Either integration or direct substitution shows the solutions to Eq. (9.8a) to be

$$E_z = Ae^{-t/\tau} + B \qquad \tau = \varepsilon/\sigma \qquad (9.8b)$$

or

$$E_z = C \qquad (9.8c)$$

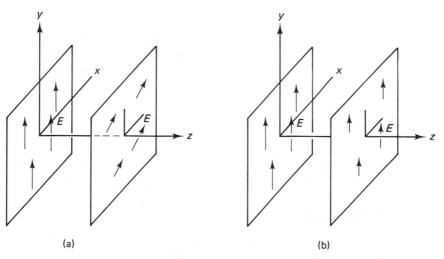

(a)   (b)

Figure 9.1   (a) $\mathbf{E}$ field of a plane wave. (b) $\mathbf{E}$ field of a linearly polarized plane wave.

where A, B, and C are constants. If $\sigma \neq 0$, the time-varying part of E_z vanishes eventually (generally, after a very short time), leaving E_z constant. Not only is a constant field inconsistent with the periodic variation of interest but, in any event, it cannot accompany a plane wave, which has the potential to permeate all space (e.g., radiation from the sun or a distant star), without involving an infinite amount of energy and infinite potentials. Therefore, we must take $E_z = 0$. If $\sigma = 0$, then E_z is constant immediately, and the same conclusion follows. These considerations can be shown to apply also to the longitudinal magnetic field H_z. Both the electric and magnetic fields are, therefore, entirely transverse to the direction of change, that is, they lie entirely in the xy plane; the plane wave is said to be a *transverse wave*.

Note that taking the direction of variation to be along the z axis does not limit the applicability of our development, since it merely corresponds to rotating the coordinate system so that the z axis lies along the particular physical direction in which the wave changes. By doing so, however, the mathematical equations have been greatly simplified so that their consequences are more clearly seen.

2. *Fields in an Insulator.* Consider the field in an insulator, since most media that are commonly used for propagation, such as air, vacuum, or many dielectrics, are close approximations to this case. While one never has a perfect insulator with $\sigma = 0$, we first consider this ideal case. Equation (9.6) then becomes

$$\frac{\partial^2 \mathbf{E}}{\partial z^2} - \mu\varepsilon\frac{\partial^2 \mathbf{E}}{\partial t^2} = 0 \tag{9.9a}$$

3. *Polarization.* Although the electric and magnetic fields lie parallel to the xy plane, their orientations in that plane are yet unspecified. In general, there are both x and y components, so the component equations of Eq. (9.9a) are

$$\frac{\partial^2 E_x}{\partial z^2} - \mu\varepsilon\frac{\partial^2 E_x}{\partial t^2} = 0 \tag{9.9b}$$

$$\frac{\partial^2 E_y}{\partial z^2} - \mu\varepsilon\frac{\partial^2 E_y}{\partial t^2} = 0 \tag{9.9c}$$

We make a further simplifying assumption that one of these fields is zero. Since the $\mathbf{E}$ field is now along a particular line, say the y axis, the plane wave is said to be linearly polarized. The assumption of linear polarization can be regarded as a rotation of our coordinate system about the z axis (which is perpendicular to the plane), so that the direction of the y axis coincides with the $\mathbf{E}$ field.

$$\mathbf{E} = E(z,t)\mathbf{a}_y \tag{9.10}$$

Figure 9.1(b) shows the linearly polarized plane wave as we have it so far. Note that although $\mathbf{E}$ is always parallel to $\mathbf{a}_y$, its magnitude can change as z and t change.

9.2 TRAVELING WAVES

Equation (9.9c) can now be written in its scalar form (since it has only a y component):

$$\frac{\partial^2 E}{\partial z^2} - \mu\varepsilon\frac{\partial^2 E}{\partial t^2} = 0 \tag{9.11}$$

To study the field that satisfies this equation, we first show, from the mathematical form of Eq. (9.11), that E is actually a function of a single variable that is a linear combination of z and t, that is,

$$E = E(\xi) \quad \text{where } \xi = \omega t - \beta z \tag{9.12}$$

The variables z and t always occur together; ω and β are constants. This is established through the chain rule of differentiation:

$$\frac{\partial E}{\partial z} = \frac{dE}{d\xi} \cdot \frac{\partial \xi}{\partial z} = -\beta E' \tag{9.13a}$$

and

$$\frac{\partial^2 E}{\partial z^2} = -\beta \frac{\partial E'}{\partial z} = -\beta \frac{dE'}{d\xi} \cdot \frac{\partial \xi}{\partial z} = \beta^2 E'' \tag{9.13b}$$

Similarly,

$$\frac{\partial^2 E}{\partial t^2} = \omega^2 E'' \tag{9.13c}$$

Here E' and E'' denote the first and second derivatives, respectively, of E with respect to its single variable ξ. Substituting Eqs. (9.13b) and (9.13c) in Eq. (9.11) yields

$$\beta^2 = \mu\varepsilon\omega^2 \qquad \beta = \pm\sqrt{\mu\varepsilon}\,\omega \tag{9.14}$$

as the relation between the constants of Eq. (9.12) and the parameters of the wave equation.

The quantity ξ is difficult to understand as it contains two physical variables, z and t. The physical consequence of having this combination is best illustrated by an example. Consider the impulse function shown in Fig. 9.2(a).

$$f(\xi) = \alpha\xi \qquad 0 \le \xi < 1 \tag{9.15}$$
$$= 0 \qquad \text{otherwise}$$

where α is a constant. Here we will take $\xi = \beta z - \omega t$; reversing the sign doesn't alter the previous proof and it is convenient in the following discussion. The physical behavior of f, in space and time, can be developed by displaying it at several instants. At $t = 0$, we have $f(\beta z)$, which is shown in Fig. 9.2(b); this is a reproduction of $f(\xi)$, with the abscissa scale changed from the generalized coordinate to the specific spatial one. At a later time, $t = t_0$, point A of f, which occurs at $\xi = 0$, is located where $\beta z - \omega t_0 = 0$, or at the point $z = \omega t_0/\beta$. Point B of f occurs at $\xi = \xi_0$, so that at the same later instant, $t = t_0$, this is located where $\beta z - \omega t_0 = \xi_0$, or at the point $z = \xi_0/\beta + \omega t_0/\beta$. Note that the curves $f(t = 0)$ and $f(t = t_0)$ are identical except for being translated through the distance $\omega t_0/\beta$. Therefore, a consequence of the signal being a function of $(\beta z - \omega t)$ [or $(\omega t - \beta z)$] is that it moves in space as time advances. The speed with which it travels is simply the distance traversed divided by the time elapsed:

$$v = \frac{(\omega t_0/\beta)}{t_0} = \frac{\omega}{\beta} = \frac{1}{\sqrt{\mu\varepsilon}} \tag{9.16}$$

where the last equality is a result of Eq. (9.14).

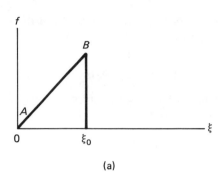

(a)

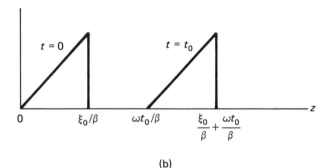

(b)

Figure 9.2   Translating impulse.

A signal that is of particular interest is a sinusoidal variation. In practice, one most frequently deals with sinusoids, but, even when this is not so, the powerful method of Fourier analysis allows analysis of any signal in terms of its sinusoidal components. For the rest of our study, we will therefore emphasize such variations:

$$E = E_0 \cos(\omega t - \beta z) \tag{9.17}$$

For a preliminary consideration, suppose we stand at the origin ($z = 0$) and observe the field variation given by Eq. (9.17):

$$E = E_0 \cos \omega t \qquad z = 0 \tag{9.18a}$$

As shown in Fig. 9.3(a), the field varies cyclically and repeats after a time T, called the *period*, such that the angle $\omega T = 2\pi$, or $\omega = 2\pi/T$. The period of the wave (seconds) is related to its frequency (cycles per second = hertz) by $f = 1/T$. Therefore, $\omega = 2\pi f$ is the number of radians per second, called the *phase frequency* or the *angular frequency*.

Now examine the spatial distribution of the field when $t = 0$:

$$E = E_0 \cos(-\beta z) = E_0 \cos \beta z \qquad t = 0 \tag{9.18b}$$

As shown in Fig. 9.3(b), here again the field varies cyclically (now with position) and repeats after a distance λ, called the *wavelength,* such that the angle $\beta\lambda = 2\pi$, or $\beta = 2\pi/\lambda$. β has the units of radians per meter; it is called the *phase*, or *angular, wave number*.

Figure 9.4 now shows the spacial variation of Eq. (9.17) at successive instants of time, taken at intervals of $T/4$. The first row of this figure has already been dis-

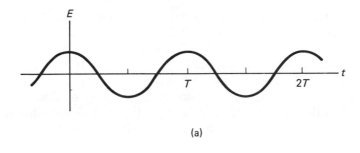

(a)

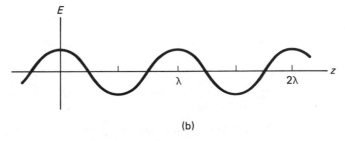

(b)

Figure 9.3   Sinusoidal field variations.

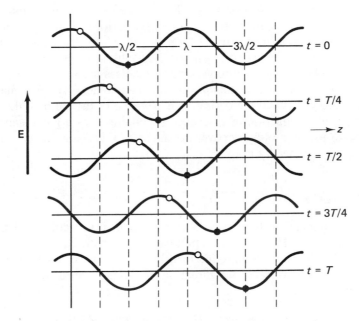

Figure 9.4   Demonstrating the traveling wave.

cussed for the case of $t = 0$ [Fig. 9.3(a)]. In the second row of Fig. 9.4, the time has advanced so that $t = T/4$ and

$$E = E_0 \cos\left(\beta z - \omega \frac{T}{4}\right) = E_0 \cos\left(\beta z - \frac{\pi}{2}\right) \qquad (9.19)$$

where, for convenience, the order of the argument has been reversed. Equation (9.19) is a cosine wave whose maximum value, which occurs when the phase angle $\beta z = \pi/2$, is at $z = \lambda/4$. The third row $(t = T/2)$ has its maximum at $z = \lambda/2$, and so forth. The constant-phase point, $(\omega t - \beta z) = \text{constant} = 0$, in this case, translates to the right as time advances. Once again, the wave is traveling. Other constant-phase points are illustrated in Figs. 9.4 by the closed dots at the phase point π and by the open dots at the point $\pi/6$. Figure 9.4 shows that after a period T, each phase point has moved a distance λ. Its velocity is, therefore,

$$v = \frac{\text{distance}}{\text{time}} = \frac{\lambda}{T} = \frac{2\pi}{\beta}\frac{\omega}{2\pi} = \frac{\omega}{\beta} \tag{9.20}$$

This is the same velocity we found before, for the impulse. In vacuum, $\mu = \mu_0$ and $\varepsilon = \varepsilon_0$, so that

$$v_0 = \frac{1}{\sqrt{\mu_0\varepsilon_0}} = 2.99793 \times 10^8 \approx 3 \times 10^8 \text{ m/sec} \tag{9.21}$$

Historically, when Maxwell first expressed the laws of electrodynamics in their complete dynamical form, he was able to show that a traveling wave resulted from their solution. At that time, the electric units in use had $\varepsilon_0 = 1$ and the magnetic units had $\mu_0 = 1$, so that Maxwell expressed v_0 in terms of the ratio of the units of charge and current in the two systems. He performed an experiment to determine the units ratio and arrived at $v_0 \sim 2.88 \times 10^8$ m/sec. It was apparent that such a huge velocity must correspond to that of light, although, at the time, there was still considerable belief that light traversed all distances instantaneously. Notwithstanding his slightly incorrect value, this prediction made it clear why instantaneous transmission was accepted, since the entire circumference of the earth could be traversed by his radiation in a fraction of a second. It certainly was a great step from Galileo, who tried to measure v_0 by uncovering lanterns to mark the transit time, back and forth, between two nearby hills.

9.3 FIELD VECTORS

Everything said in the previous section about the electric field **E** is also true of the magnetic field **H**. If the derivation had started with Eq. (9.2a) instead of Eq. (9.2b), then **H** would have been found to also obey the wave equation and, in a plane wave, it is also transverse to the direction of propagation of the wave. **H** is, therefore, given by

$$\mathbf{H} = H_0 \cos(\omega t - \beta z)\mathbf{u}_0$$

where $\mathbf{u}_0$ is a unit vector whose direction is still unspecified except that **H**, like **E**, lies in the xy plane. To find the relation between **E** and **H**, take

$$\nabla \times \mathbf{E} = -\mu\frac{\partial \mathbf{H}}{\partial t} \tag{9.22a}$$

or

$$\begin{bmatrix} \mathbf{a}_x & \mathbf{a}_y & \mathbf{a}_z \\ 0 & 0 & \dfrac{\partial}{\partial z} \\ 0 & E_y & 0 \end{bmatrix} = \mu H_0 \omega \, \sin(\omega t - \beta z) \, \mathbf{u}_0 \qquad (9.22\text{b})$$

where zeros have been placed in the determinant to indicate that derivatives vanish with respect to x and y. Using the expression for E_y given by Eqs. (9.10) and (9.17), the left-hand side of Eq. (9.22b) is

$$-\mathbf{a}_x \frac{\partial E_y}{\partial z} = -\mathbf{a}_x \beta E_0 \, \sin(\omega t - \beta z) \qquad (9.22\text{c})$$

Comparison with the right-hand side of Eq. (9.19b) gives $\mathbf{u}_0 = -\mathbf{a}_x$, so that

$$\mathbf{H} = -\mathbf{a}_x H_0 \cos(\omega t - \beta z) \qquad (9.23)$$

and $\beta E_0 = \mu H_0 \omega$, which can be rewritten as

$$\frac{E_0}{H_0} = \frac{\mu \omega}{\beta} = \sqrt{\frac{\mu}{\varepsilon}} = \eta \qquad (9.24\text{a})$$

where η has the units of $E/H = (\text{V/m})/(\text{A/m}) = \text{V/A} = \text{ohms}$. Consequently, η, the ratio of the transverse field magnitudes is called the *transverse wave impedance*. In vacuum (or air), its value is

$$\eta_0 = \sqrt{\frac{\mu_0}{\varepsilon_0}} = 120\pi \approx 377 \text{ ohms} \qquad (9.24\text{b})$$

The orientation of $\mathbf{E}$, $\mathbf{H}$, and the direction of wave motion are shown in Fig. 9.5, where the direction of propagation is parallel to Poynting's vector $\mathscr{P}$:

$$\mathscr{P} = \mathbf{E} \times \mathbf{H} \qquad (9.25)$$

In Chapter 5, it was pointed out that $\mathscr{P}$ gives the power per area flowing in a region in which there are crossed electric and magnetic fields; its units are energy/area/time and it is sometimes referred to as the *wave intensity*.

The meaning of Eqs. (9.24) is not clear, since E and H are different physical quantities and their ratio is a number relating to their magnitudes in V/m and A/m, respectively. Great care must be taken in any comparison of unlike quantities. (The ratio watermelons/apples = 13.4 doesn't indicate what property of these fruits is being compared; it could be mass, volume, water content, number of pits, or it could be a meaningless ratio.) To compare electric and magnetic fields in a plane wave, some property must be considered that is common to both; in this case, it is the energy density of the fields. From Chapter 5, the energy per unit volume of the fields, both in J/m³, is given by

$$U_E = \tfrac{1}{2}\mathbf{D} \cdot \mathbf{E} = \tfrac{1}{2}\varepsilon E^2 \qquad (9.26\text{a})$$

$$U_M = \tfrac{1}{2}\mathbf{B} \cdot \mathbf{H} = \tfrac{1}{2}\mu H^2 \qquad (9.26\text{b})$$

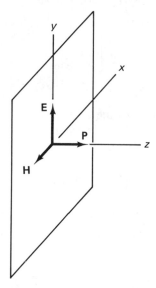

Figure 9.5 Field orientations in a plane wave.

Substituting Eq. (9.24a) into the expression for U_E yields

$$U_E = \tfrac{1}{2}\varepsilon E^2 = \tfrac{1}{2}\varepsilon\left(H\sqrt{\frac{\mu}{\varepsilon}}\right)^2 = \tfrac{1}{2}\mu H^2 = U_M \qquad (9.26c)$$

proving that the electric and magnetic energy densities are equal. In the SI system, E and H are measured in units that make their plane-wave magnitude ratio have the value of 120π ohms. If we measured E in units of millivolts, which we decide to call "potens" (p), then we would have $E/H = 1.2\pi \times 10^5$ p–m/A. The important point is that, regardless of the units chosen to measure E and H, in a plane wave in an insulating dielectric, when E and H are referenced to their common feature — their energy densities — they are equal.

9.4 COMPLEX NUMBERS

In describing sinusoidal field expressions that are functions of both space and time, it is convenient to use a powerful notational technique that is based on an exponential form involving complex variables. Therefore, a brief review of the pertinent properties of complex numbers is now presented.

Figure 9.6(a) shows a line whose every point is represented by a number-location. The line occupies one dimension of the two-dimensional plane, and we wish to find ways of representing all the points of the plane. One way is to refer the location of any point to the horizontal axis shown, and to another that is rotated by $90°$ from it; we normally call these the x and y axes. Instead, let us enter this $90°$ rotation in another way by defining an *operator* j that performs such a rotation. Figure 9.6(b) shows a number $A = a + jb$, that is, a length a along the horizontal axis and a length b along the direction rotated $90°$ counterclockwise from that axis (the positive direction for angles).

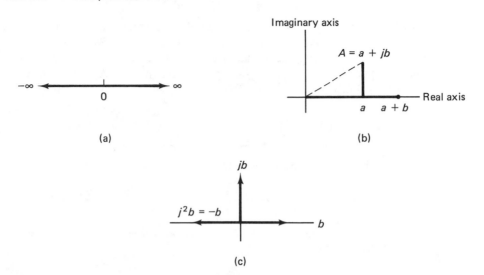

Figure 9.6   Complex numbers.

Figure 9.6(c) shows a point b alone, denoted by the directed arrow. Also shown are point b after a single multiplication by j (i.e., after a single rotation) and $j^2 b$ (i.e., after a double rotation). Note that the double rotation has brought the point back to the axis of real numbers, but to its negative side. Therefore, $j^2 b = -b$, so that multiplication by j^2 is equivalent to multiplying by -1. Since $j^2 = -1$, we can write $j = \sqrt{-1}$. This is a quantity that is not defined in real-number theory; it has no physical existence on the real-number axis. It is therefore referred to as the *imaginary* basis quantity and jb is an *imaginary number*. Previously, we had $a + jb$, a number with a real and an imaginary part; this is called a *complex number*.

It is vitally important to remember that complex numbers are *not* the same as vectors; $A = a + jb$ is *not* equivalent to the vector $\mathbf{A} = \mathbf{a}_x a + \mathbf{a}_y b$, even though they may both be considered to describe a point in the plane with the same coordinates or components. The allowed operations and interpretations are different in the two cases. For example, consider the multiplication of two complex numbers:

$$AB = (a + jb)(c + jd) = (ac - bd) + j(ad + bc) \qquad (9.27a)$$

where term-by-term multiplication has been performed, along with the fact that $j^2 = -1$. Note that this is not a vector dot or cross product. This multiplication is an example of the rule that, in general, operations with complex numbers result in other complex numbers. Another operation that is not defined with vectors is that of division:

$$\frac{B}{A} = \frac{c + jd}{a + jb} = \frac{BA^*}{AA^*} = \frac{c + jd}{a + jb} \cdot \frac{a - jb}{a - jb} = \frac{ac + bd + j(ad - bc)}{a^2 + b^2}$$

$$(9.27b)$$

In the third and fourth parts of Eq. (9.27b), the numerator and denominator of the fraction have been multiplied by A^*, the complex conjugate of the number in the denominator. This is obtained by changing all its factors j to $-j$. Such an operation is equivalent to multiplying the fraction by unity and therefore does not alter its value, but it removes the imaginary quantity from the denominator of the fraction. Note from this that $AA^* = a^2 + b^2 = |A|^2$. The magnitude $|A|$ is shown by the dashed line in Fig. 9.6(b).

Example 9.1

If $A = 3 + 4j$ and $B = 1 + j\sqrt{3}$, find $C = A/B$.

Proceeding as before, we have

$$C = \frac{3 + j4}{1 + j\sqrt{3}} \cdot \frac{1 - j\sqrt{3}}{1 - j\sqrt{3}} = \frac{(3 + 4\sqrt{3}) + j(4 - 3\sqrt{3})}{1 + 3} = 2.48 - 0.30j$$

These complex numbers are shown in the complex plane in Fig. 9.7, with the real ($\mathcal{R}$) and imaginary ($\mathcal{I}$) axes.

In the previous discussion, complex numbers have been described by their real and imaginary components, corresponding to the more usual rectangular coordinates. Complex numbers can be expressed, as well, in terms of polar parameters, the magnitude (or modulus) and angle (or phase), of the number. Figure 9.8 shows this conversion:

$$a = |A| \cos \phi \qquad b = |A| \sin \phi \qquad |A| = \sqrt{a^2 + b^2} \qquad (9.28a)$$

The complex number can now be written as

$$A = a + jb = |A| (\cos \phi + j \sin \phi) = |A|e^{j\phi} \qquad (9.28b)$$

where use has been made of the mathematical identity known as Euler's theorem,

$$e^{j\phi} = \cos \phi + j \sin \phi \qquad (9.29)$$

The exponential representation of complex numbers is a particularly valuable and convenient one for discussion of sinusoidal time- and space-varying functions.

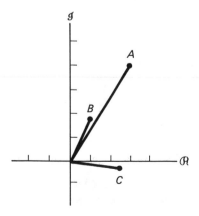

Figure 9.7 Complex numbers.

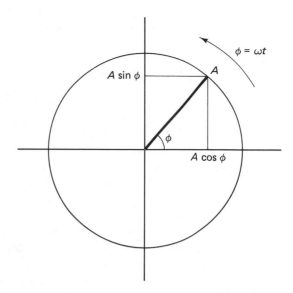

Figure 9.8   Definition of a phasor.

Example 9.1 (cont.)

Repeat Example 9.1 using the polar representation:

$$A = \sqrt{3^2 + 4^2}\, e^{j\arctan 4/3} = 5e^{j53°} \qquad B = 2e^{j60°}$$

$$C = 5e^{j53°}/2e^{j60°} = 2.5e^{-j7°}$$

From Fig. 9.7, it is seen that this result agrees with the previous calculation.

Example 9.2

One particular application of Eq. (9.29) is the polar representation of the imaginary j itself. Since $|j| = 1$ and as $\phi_j = \pi/2$, we have

$$j = e^{j\pi/2} \tag{9.30}$$

9.5 PHASORS

Application of complex variables to the study of electromagnetic fields is through the properties of *phasors*. The discussion of phasors begins by observing that a sinusoid can be regarded as the projection of a uniformly rotating radial line. If $\phi = \omega t$ in Fig. 9.8, then ϕ, the angle of A, increases uniformly in time. A rotates as shown, and its projection on the real (horizontal) axis is $A \cos \phi = A \cos \omega t$. The phasor quantity A can be expressed in terms of its real and imaginary components:

$$A = |A| \cos \phi + j|A| \sin \phi = |A|e^{j\phi} \tag{9.31a}$$

From this, it can easily be shown that

$$|A| \cos \phi = (A + A^*)/2 \qquad |A| \sin \phi = (A - A^*)/2j \tag{9.31b}$$

so that the sinusoids of $\phi = \omega t$ can be written in terms of phasors and their complex conjugates.

For simplicity of mathematical manipulation when using sinusoids, we wish to avoid carrying along both A and A^*, so that it is common to use A to represent only $\cos \phi$. Obviously, this is an incorrect mathematical procedure, but it lends such convenience to algebraic manipulations that it is warranted. One interpretation of this procedure is to understand that one really intends to take only the real part of the phasor A:

$$\cos \phi = \mathbb{R} e^{j\phi} \qquad (9.32)$$

where $\mathbb{R}$ represents the operation of extracting the real part.

To illustrate these operations, consider the example of a sinusoidally varying electric field, that is, one which is given by

$$\mathbf{E} = \mathbf{E}_0 \cos \omega t = \mathbb{R}(\mathbf{E}_0 e^{j\omega t}) \leftrightarrow \mathbf{E}_0 e^{j\omega t} \qquad (9.33)$$

where $\leftrightarrow$ indicates the representation of the cosine function by the exponential; in doing this the operation $\mathbb{R}$ if left unstated, although it is understood to be implied. We will proceed in this way throughout the remainder of the text, omitting both $\mathbb{R}$ and $\leftrightarrow$ unless the discussion specifically warrants including them, and understanding that by e^{jg} we mean $\cos g$.

Since Eq. (9.33) is not mathematically rigorous, its use will give incorrect results in some cases. As long as we add or subtract fields, the real and imaginary components of the fields involved are maintained separately and the operation $\mathbb{R}$ gives correct results. For example,

$$A + B = (a + jb) + (c + jd) = (a + c) + j(b + d) \qquad (9.34a)$$

$$\mathbb{R}(A + B) = a + c = \mathbb{R}A + \mathbb{R}B \qquad (9.34b)$$

However, when multiplications or divisions are involved, the imaginary and real components are intermixed, so that great care must be taken when applying the results. From Eq. (9.27a),

$$\mathbb{R}AB = \mathbb{R}[(ac - bd) - j(ad + bc)] = ac - bd \qquad (9.35a)$$

and

$$\mathbb{R}AB^* = \mathbb{R}[(a + jb)(c - jd)] = ac + bd \qquad (9.35b)$$

The term bd in both of these expressions involves the imaginary parts of A and B. This term is undesirable since we mean to have $AB = ac$, that is, the product of the real parts of the factors. By adding Eqs. (9.35a) and (9.35b), the *correct* real part can be extracted since

$$\tfrac{1}{2}\mathbb{R}(AB + AB^*) = ac = (\mathbb{R}A)(\mathbb{R}B) \qquad (9.35c)$$

9.6 FIELD PRODUCTS

In the present situation, the products of interest are the energy densities $U_e = \mathbf{E} \cdot \mathbf{D}/2$ and $U_m = \mathbf{H} \cdot \mathbf{B}/2$ and Poynting's vector $\mathscr{P} = \mathbf{E} \times \mathbf{H}$. To illustrate the usage, consider

$$A = \mathbf{E} = \mathbf{E}_0 e^{j\theta} \qquad B = \mathbf{H} = \mathbf{H}_0 e^{j\theta} \qquad \theta = \omega t \qquad (9.36)$$

First examine the right-hand side of Eq. (9.35c), which is the physically correct form. By using a trigonometric identity in the vector product,

$$\mathbf{E} \times \mathbf{H} = (\mathbb{R}A) \times (\mathbb{R}B) = \mathbf{E}_0 \cos \omega t \times \mathbf{H}_0 \cos \omega t = \mathbf{E}_0 \times \mathbf{H}_0 \cos^2 \omega t$$
$$= \mathbf{E}_0 \times \mathbf{H}_0 \tfrac{1}{2}(\cos 2\omega t + 1) \tag{9.37a}$$

This indicates that the time variation is given by $2\omega t$, that is, it has twice the frequency of the fields themselves.

$$(\mathbf{E} \times \mathbf{H})_{\text{time dependent}} = \tfrac{1}{2}\mathbf{E}_0 \times \mathbf{H}_0 \cos 2\omega t \tag{9.37b}$$

Furthermore, note that since the term $\cos 2\omega t$ has a zero average value,

$$(\mathbf{E} \times \mathbf{H})_{\text{average}} = \tfrac{1}{2}\mathbf{E}_0 \times \mathbf{H}_0 = E_{\text{rms}} H_{\text{rms}} \tag{9.37c}$$

Now return to the left-hand side of Eq. (9.35c), which consists of two terms. Substituting the first term in the vector product gives

$$\tfrac{1}{2}\mathbb{R}\{A \times B\} = \tfrac{1}{2}\mathbb{R}(\mathbf{E}_0 \times \mathbf{H}_0 e^{j2\omega t}) = \tfrac{1}{2}(\mathbf{E}_0 \times \mathbf{H}_0) \cos 2\omega t \tag{9.38a}$$

giving Eq. (9.37b), which is the time-dependent portion of Eq. (9.37a). For the second term on the left-hand side of Eq. (9.35c),

$$\tfrac{1}{2}\mathbb{R}(A \times B^*) = \tfrac{1}{2}\mathbb{R}(\mathbf{E}_0 \times \mathbf{H}_0) = E_{\text{rms}} H_{\text{rms}} \tag{9.38b}$$

giving Eq. (9.37c), which is the average portion of Eq. (9.37a).

Equations (9.38) combined give Eq. (9.37a), so that an interpretation of the two terms on the left-hand side of Eq. (9.35c) is now at hand: complex vector phasors can be used in vector multiplications if it is remembered that $\tfrac{1}{2}\mathbb{R}\{A \# B\}$ yields the time-dependent portion of the product and that $\tfrac{1}{2}\mathbb{R}\{A \# B^*\}$ gives its average value, where $\#$ denotes either the vector cross product (as when taking $\mathcal{P}$) or the vector dot product (as when considering energy density). As a result, Eq. (9.35c) gives the complete result.

9.7 POLARIZATION

Early in the development of the wave equation, it was proved that the electric and magnetic fields of a plane wave are entirely in the plane perpendicular to the direction of propagation. Furthermore, both planar components E_x and E_y satisfy the same differential equations, Eqs. (9.9). In order to simplify the analysis, the additional assumption was made that $\mathbf{E}$ has only one component, as given by Eq. (9.10). In fact, however, this additional assumption is unnecessary, since, if the electric field is

$$\mathbf{E}_0 = \mathbf{a}_x E_x + \mathbf{a}_y E_y \tag{9.39}$$

then we can consider that there are two plane waves: $\mathbf{E}_0^{(1)} = \mathbf{a}_x E_x$ and $\mathbf{E}_0^{(2)} = \mathbf{a}_y E_y$. Each is of the type that has been considered, with the added condition that they are traveling together. Their relationship, as they travel together, can take several forms. Consider the case that $E_y = \kappa E_x$, where κ is constant. Then Eq. (9.39) describes a vector that is at an angle $\arctan \kappa$ to the x axis. It is still linearly polarized, but the selection of coordinate system has not produced the simplest description, that is, where $\mathbf{E}$ would have only one coordinate component.

By using the fact that complex-phasor notation describes the time variation of vectors, more complicated situations can now be described.

Example 9.3

The field of a wave is described by Eq. (9.39) for the specific case that

$$E_y = -jE_x = -e^{j\pi/2}E_x \tag{9.40a}$$

The meaning of this expression is unclear since it appears to insert an imaginary multiplier between two physical field magnitudes. As j has been associated with phasor descriptions, the interpretation of Eq. (9.40a) must be that it shows a different orientation of E_x and E_y *on the phasor diagram,* that is, it corresponds to a phase difference between E_x and E_y. By combining Eqs. (9.39) and (9.40a),

$$\mathbf{E} = \mathbf{E}_0 e^{j\omega t} = (\mathbf{a}_x E_x + \mathbf{a}_y E_y)e^{j\omega t} = E_x(\mathbf{a}_x - e^{j\pi/2}\mathbf{a}_y)e^{j\omega t} \tag{9.40b}$$

To interpret this further, recall that

$$\mathbf{E} = \mathbb{R}(E_x \mathbf{a}_x e^{j\omega t}) - \mathbb{R}(E_x \mathbf{a}_y e^{j(\omega t + \pi/2)})$$

$$= \mathbf{a}_x E_x \cos \omega t - \mathbf{a}_y E_x \cos\left(\omega t + \frac{\pi}{2}\right)$$

$$= \mathbf{a}_x E_x \cos \omega t + \mathbf{a}_y E_x \sin \omega t \tag{9.40c}$$

Thus, the x and y components do indeed differ in phase, by 90°. Furthermore,

$$|\mathbf{E}| = = (E_x^2 \cos^2\omega t + E_x^2 \sin^2\omega t)^{1/2} = E_x \tag{9.40d}$$

so that $\mathbf{E}$ is a vector of constant magnitude. From the results of Eqs. (9.40c) and (9.40d), this means that the field can be described as a vector of constant magnitude, E_x, rotating at a uniform rate ω, tracing out a circle. This could be represented by a figure similar to Fig. 9.8, where $A = E_x$, except that the axes now represent the physical x and y coordinates instead of the real and imaginary coordinates.

Example 9.3 describes what is called a *circularly polarized field*. Note that as the wave advances along $\mathbf{a}_z$, $\mathbf{E}$ is along $\mathbf{a}_x$ at $t = 0$ and then rotates to be along $\mathbf{a}_y$ when $\omega t = \pi/2$. If the fingers of the right hand curl with the rotation of $\mathbf{E}$, then the thumb is along the direction of propagation. This is said to be a right-hand, or positive, circularly polarized wave. If, Eq. (9.40a) is changed to $E_y = +jE_x$ the direction of rotation will be opposite, relative to the propagation direction, so the same relation will hold with the left hand, describing a left-hand, or negative, circular polarization. If, instead, Eq. (9.40a) is changed to read $E_y = -j\kappa E_x$, then the same sort of result will ensue, except that the $\mathbf{a}_y$ component in Eq. (9.40c) will be multiplied by the constant κ. Since the components are no longer equal, the circle of Fig. 9.8 is stretched into an ellipse, describing a positive elliptically polarized wave. It is left as an exercise to show that a tilted ellipse results if, instead of Eq. (9.40a), we have $E_y = \pm\kappa e^{j\alpha}E_x$.

From this discussion, it is seen that the various polarizations can be considered to consist of two phase-related linearly-polarized waves, so that the simpler solution given by Eq. (9.10) can be used without hesitation. In doing so, we must be careful to avoid confusing the phasor diagram, which shows time-phase relations, and the cartesian plane, which shows physical orientation of the vector fields.

9.8 LOSSY MEDIA

In deriving the wave equation, our second simplifying assumption was that $\sigma = 0$. We now return to Eq. (9.6) and relax this requirement. Note that a linearly polarized plane wave is still assumed. In anticipation of a different result in this more general case, the wave constant is denoted by k rather than by β. In terms of the exponential phasor notation, a trial solution is written in the form

$$\mathbf{E} = \mathbf{a}_y E_0 \cos(\omega t - kz) \leftrightarrow \mathbf{a}_y E_0 e^{j(\omega t - kz)} \tag{9.41}$$

Substituting this in Eq. (9.6) yields

$$k^2 = \omega^2 \mu \varepsilon - j \omega \mu \sigma = \omega^2 \mu \varepsilon \left(1 - j \frac{\sigma}{\omega \varepsilon} \right) \tag{9.42}$$

Since k^2 is complex, k has the general form

$$k = \beta - j\alpha \tag{9.43}$$

Expressions for β and α can be derived from Eq. (9.42), but it is advantageous instead to examine the general nature of this solution. The earlier analysis of insulators took $\sigma = 0$, in which case, Eqs. (9.42) and (9.43) reduce to the previous result that $k = \omega\sqrt{\mu\varepsilon} = \beta$. In the presence of conductivity, there is a new term: $-j\alpha$. Since conductivity leads to ohmic heating, which draws energy from the electromagnetic field, one effect of this new term is to cause a reduction of the fields. This can be verified by substituting Eq. (9.43) into Eq. (9.41), yielding

$$\mathbf{E} = \mathbf{a}_y E_0 e^{-\alpha z} e^{j(\omega t - \beta z)} = \mathbf{a}_y E_0 e^{-z/\delta} e^{j(\omega t - \beta z)} \tag{9.44}$$

The term $e^{j(\omega t - \beta z)} \leftrightarrow \cos(\omega t - \beta z)$ is the same traveling wave studied earlier. The new feature of Eq. (9.44) is that the amplitude of the wave decreases exponentially as the wave advances, expressed by $e^{-\alpha z}$. Figure 9.9 and Eq. (9.44) show that the wave height has decreased to $1/e$ of its value when it has traveled a distance $\delta = 1/\alpha$ in the conducting medium. As a result, the field is strongest near the source, which is likely to be the surface of the material. It will be seen shortly that at high frequencies, the field attenuation is very rapid, so that a significant field exists only in a thin skin at the surface; δ is called the penetration distance, or *skin depth*.

β, α, and δ can be calculated in some simple cases. First, recognize that if

$$\frac{\sigma}{\omega \varepsilon} \ll 1 \qquad \text{(insulator)} \tag{9.45a}$$

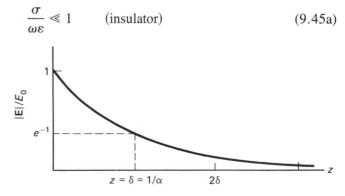

Figure 9.9   Attenuation of a wave moving in a conductor.

this term can be neglected compared to unity in Eq. (9.42), giving $k = \omega\sqrt{\mu\varepsilon} = \beta$. Consequently, the original assumption, that $\sigma = 0$, was more severe than necessary to define an insulator-like behavior; the more proper, less restrictive condition is Eq. (9.45a). In the other extreme,

$$\frac{\sigma}{\omega\varepsilon} \gg 1 \qquad \text{(conductor)} \qquad (9.45b)$$

the medium of propagation can be considered to be a conductor. Whether a material is a conductor or an insulator, according to these classical definitions, depends upon the frequency as well as upon the conductivity.

Figure 9.10 shows representative variations of $\sigma/\omega\varepsilon = 10^s$ for a few materials, over a wide frequency range $f = \omega/2\pi = 10^p$. In this figure, a factor of 100 has been arbitrarily selected to satisfy Eqs. (9.45a) and (9.45b). Therefore, $s > 2$ is the region of conductors, and $s < -2$ is the region of insulators. Note that some materials can display both behaviors at different frequencies. In this figure, it is assumed that σ and ε have their constant low-frequency values over the extended frequency range shown, so that the only frequency dependence is given by the $1/\omega$ factor. In fact, this is not so; variations of σ, ε, and μ with frequency are of great

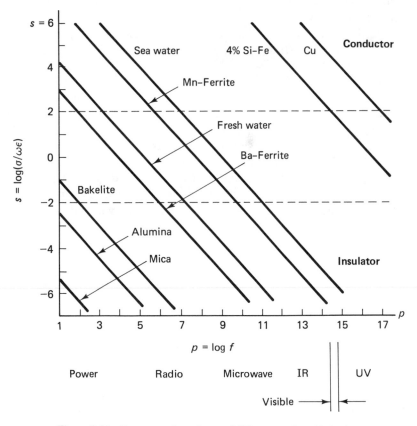

Figure 9.10   Frequency dependence of different modes of behavior.

importance in some cases, and have been studied in detail. Even without this refinement, however, Fig. 9.10 is a useful illustration of the various domains of behavior — the qualitative behaviors predicted there will not be greatly altered by the other analysis.

Example 9.4

Nichrome is an alloy having a resistivity of about 100 microohm–cm at room temperature. Determine the frequency range over which nichrome can be considered to be a conductor, and the range over which it is an insulator.

A resistivity of 100×10^{-6} ohm–cm corresponds to $\sigma = 10^4$ mho/cm $= 10^6$ mho/m $= 10^6$ S/m. In a conductor, $\varepsilon = \varepsilon_0$, so that

$$\sigma/\omega\varepsilon = 10^6/2\pi f\varepsilon_0 = 1.8 \times 10^{16}/f$$

Nichrome will appear to be a conductor at frequencies below $f \approx 10^{15}$ Hz ($\sigma/\omega\varepsilon > 20$); this corresponds to the near ultraviolet. Similarly, nichrome acts like an insulating dielectric above 4×10^{17} Hz. This lies in the X-ray range, where the wavelength is on the order of the atomic separations of matter. Since the classical theory of electromagnetism does not deal with such microscopic phenomena, great care must be taken in drawing any conclusions, other than as an *indication* of the behavioral range.

Example 9.5

Repeat Example 9.4 for intrinsic silicon, which has a resistivity of 2300 ohm–cm and $\varepsilon_r = 11.8$.

Here $\sigma/\omega\varepsilon = 6.6 \times 10^7/f$, so that this silicon acts as a good conductor for $f < 10^6$ Hz and as an insulator for $f > 10^9$ Hz.

9.9 WAVES IN CONDUCTORS

In a good conductor, $\sigma/\omega\varepsilon \gg 1$, so the unity term in Eq. (9.42) can be neglected and

$$k^2 = -j\omega\mu\sigma \tag{9.46a}$$

$$k = (1 - j)\left(\frac{\omega\mu\sigma}{2}\right)^{1/2} = \frac{1 - j}{\delta} \tag{9.46b}$$

Substituting this k into Eq. (9.41) gives

$$\mathbf{E} = \mathbf{a}_y E_0 e^{-z/\delta} e^{j(\omega t - \beta z)} \tag{9.47}$$

where

$$\beta = \frac{2\pi}{\lambda} = \frac{1}{\delta} \tag{9.48}$$

$$\delta = \left(\frac{2}{\omega\mu\sigma}\right)^{1/2} \tag{9.49}$$

This is the same attenuated behavior found earlier, in Eq. (9.44). Now, however, Eq. (9.49) gives a specific expression for δ in a good conductor, and therefore for $\beta = 1/\delta$ and for $\alpha = 1/\delta$. Figure 9.11 shows values of δ for some commonly en-

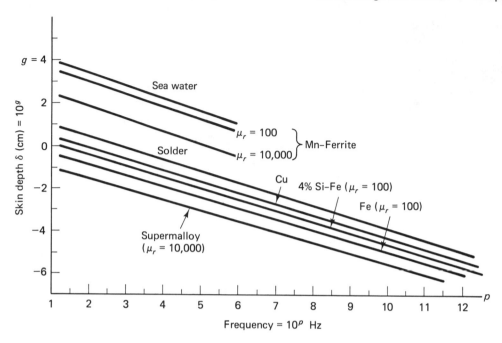

Figure 9.11   Skin depth of conductors.

countered conducting materials. As with Fig. 9.10, it is assumed that μ and σ are constants.

Example 9.6

Calculate the skin depth in copper ($\rho = 2 \times 10^{-6}$ ohm–cm $= 2 \times 10^{-8}$ ohm–m) at 60 Hz (power frequency), at 10^9 Hz (microwave), and at 10^{16} Hz (optical range).

Since $\mu = \mu_0$, we have $\delta = 0.0712/\sqrt{f}$. Therefore,

f (Hz)	60	10^9	10^{16}
δ	0.92 cm	2.25 microns	~ 7 Å

This example shows that the field penetration into a conductor is quite small at high frequencies. Since, from Poynting's vector, the power density transported by the wave is $\mathscr{P} = \mathbf{E} \times \mathbf{H} = E^2/\eta = (E_0^2/\eta)e^{-2\alpha z}$, the energy drops off faster than the amplitude, being multiplied by a factor of $e^{-2} = 0.135$ in one skin depth ($z = \delta = 1/\alpha$) and being reduced to 1.85 percent in 2δ. Advantage is taken of this property, for example, where thin high-conductivity precision coatings can be placed on a more base metal to obtain high-quality waveguides that would otherwise be quite expensive.

Note from Example 9.6 that for frequencies greater than $\sim 10^{13}$ Hz, δ is on the scale of atomic features of a solid. The classical theory of electromagnetics assumes continuous distributions of matter, with the properties of that matter being completely described by the parameters ε, μ, and σ. Consequently, as was pointed out in Example 9.4, at the atomic level, we must be very hesitant to accept any quantitative results.

An important skin-depth limitation occurs in ferromagnetic material. The iron alloy commonly used for quality electrical transformers contains 3.5 percent silicon, which serves to raise its resistivity by a factor of 25 over that of copper, so that eddy-current heating, that is, by induced currents, is reduced. For this alloy, we can often assume that $\mu = 10^4 \mu_0$, so that at 60 Hz, $\delta \approx 0.46$ mm = 0.018 in. Field penetration is therefore effectively limited to approximately this distance; there seems to be little point in making transformer cores thicker, since the field will not penetrate into the bulk and the cost and weight of material serves no useful purpose. As a result, the total flux that can be passed through transformer cores is lowered, and their power-handling capabilities are greatly reduced. To overcome this limitation, the iron cores of most transformers are made of laminations. These are thin sheets, each of which allows nearly complete flux penetration; laminations as thin as 14 mils are in use. The laminations are insulated from each other, so that the bulk effect just described is absent. They are then pressed together and bolted in position to make up the transformer. Insulation between laminations is accomplished in a number of ways, for example, plastic coatings are used where precision and uniformity are demanded. It is not uncommon for the manufacturer to simply expose lamination sheets to the atmosphere and allow the surface to oxidize, so that the oxide serves as an insulating layer. The necessity of using insulating layers reduces the volume percentage of active magnetic material to conduct flux, necessitating larger transformers. This has the side effect of requiring more length of copper windings, with associated losses, and more massive cooling systems for large transformers.

Earlier in this chapter, it was found that the wave impedance is $\eta = E/H = \omega\mu/\beta$. For the insulating dielectric media first considered, $\beta = \omega\sqrt{\mu\varepsilon}$, so that $\eta = \sqrt{\mu/\varepsilon}$. In a conductor, however, β is replaced by k from Eq. (9.46b) resulting in

$$\frac{E}{H} = \frac{\omega\mu}{k} = \frac{\omega\mu\delta}{(1-j)} = \eta_0\sqrt{\frac{\omega\varepsilon}{\sigma}}\,e^{j\pi/4} \tag{9.50}$$

For a good conductor, the square root in the last term is so small that E/H is much smaller than in a dielectric.

Example 9.7

Calculate the ratio of U_E to U_H for a plane wave in copper at 10 GHz.

From the values given for copper in the previous example, we have $E/H = 10.5 \times 10^{-10}\eta_0 f$. For $f = 10^{10}$ this gives $E/H = 10.5 \times 10^{-5}\eta_0 = 0.04$ ohm. Recall that when $E/H = \eta_0 = 377$ ohms, the electric- and magnetic-energy densities are equal. There η is 10^4 times the value found in this example. Since the ratio of energy densities is proportional to the square of the field ratios, for copper at 10 GHz, $U_E/U_H \sim 10^{-8}$. As the parameter values do not vary greatly in other conductors, it is a general conclusion that very little energy is carried in the electric-field wave in a conductor.

$|\mathbf{E}/\mathbf{H}|$ is very small in a conductor (0.04 at 10^{10} Hz and even less at lower frequencies). Since $|\mathbf{E}/\mathbf{H}| = 377$ ohms in air, adjacent to the conductor surface, this ratio changes by a factor of at least 10^4 upon entering the conductor. Therefore, very little error is made when one assumes that $E = 0$ as a boundary condition at a conducting wall. Note that since $|\mathbf{E}/\mathbf{H}| \sim \sqrt{f}$, this reduces to the earlier result that $\mathbf{E} = 0$ in a conductor in the static case.

EXERCISES

Extensions of the Analysis

9.1. In Exercise 5.1, the relation of the magnetic vector potential **A** was introduced with respect to **B** and **E** in the time varying case. In addition, if we specify that it obeys the *Lorentz condition* $\nabla \cdot \mathbf{A} + \mu\varepsilon\, \partial V/\partial t = 0$, show that **A** and V satisfy the wave equation in an insulating dielectric. *Hint*: Use the curl **H** equation and the identity $\nabla \times (\nabla \times \mathbf{H}) = \nabla(\nabla \cdot \mathbf{H}) - (\nabla \cdot \nabla)\mathbf{H}$.

*9.2. It can be verified from Fig. 9.12 that the vector equation for a plane is $\mathbf{r} \cdot \mathbf{n} = d$, where **r** is the vector from the origin to a point on the plane, **n** is the unit normal to the plane, and d is the shortest distance from the origin to the plane

     (a) Using this expression, write the equation of a sinusoidal plane wave traveling in an arbitrary direction **n**. For this, note that the wave vector is defined as $\boldsymbol{\beta} = |\beta|\mathbf{n}$, where $|\beta| = 2\pi/\lambda$ of the wave.

     (b) If $\beta_x = 50 \text{ m}^{-1}$, $\beta_y = 40 \text{ m}^{-1}$, $\beta_z = 0$, $f = 10^9 \text{ Hz}$, and $E_0 = 1 \text{ V/m}$, write a vector expression for the plane wave **E**, and determine its phase velocity.

9.3. A plane wave is given by $E = E_0 \cos(\omega t - \beta z)$.

     (a) Show that the resultant superposition of two waves of this type, of equal amplitude and of angular frequencies $\omega_0 + \Delta\omega$ and $\omega_0 - \Delta\omega$ (and two corresponding wave numbers $\beta_0 + \Delta\beta$ and $\beta_0 - \Delta\beta$), can be expressed as $E = 2E_0 \cos(\omega_0 t - \beta_0 z) \cos(\Delta\omega\, t - \Delta\beta\, z)$.

     (b) If $\Delta\omega/\omega \ll 1$ and $\Delta\beta/\beta \ll 1$, discuss the physical form of the previous expression for E, that is, compare the rates of variation of the two cosine terms.

     (c) Discuss the statement that the group velocity of a wave packet (a group of waves with a small range of ω and β, which move together) in a dielectric medium is given by $v_g = d\omega/d\beta$.

     (d) In most media, the phase velocity $v = \lambda f$ is not a constant. Instead, as frequency changes, so do the wavelength and phase velocity. In general, v increases when λ increases. From this and part (b), show that the group velocity v_g is less than the phase velocity v.

9.4. In a realistic model of dielectric materials, the atomic electrons cannot be assumed to closely follow a rapidly varying field, because of inertia and other

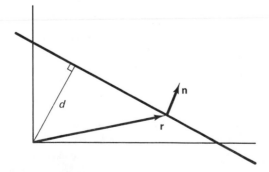

Figure 9.12 For Exercise 9.2.

forces that are present. As a result, the polarization P (which results from the displacement of charge) is out of phase with the electric field. Since $\mathbf{D} = \varepsilon_0\mathbf{E} + \mathbf{P}$, the flux density $\mathbf{D}$ is also out of phase with $\mathbf{E}$. This is expressed by

$$\mathbf{D} = [\varepsilon]\mathbf{E} \quad \text{where } [\varepsilon] = \varepsilon' - j\varepsilon''$$

to denote the phase and magnitude differences between $\mathbf{D}$ and $\mathbf{E}$.

(a) For an *insulating* dielectric of this type, determine the exponential attenuation constant. *Hint:* Derive the wave equation for a plane wave with sinusoidal time variation and *use the fact that, generally, $\varepsilon'' \ll \varepsilon'$, but it is not* negligible.

(b) If a certain material has $\varepsilon' = 4$ and $\varepsilon''/\varepsilon' = 10^{-3}$ at $f = 10^6$ Hz, calculate the attenuation constant. What would have to be the conductivity of a good conductor for it to have this same α?

9.5. Starting with Eqs. (9.42) and (9.43), derive an expression for β and α for a lossy dielectric, that is, one for which $\sigma/\omega\varepsilon \ll 1$, but is not negligible. *Hint:* Use the expansion $(1 + \varepsilon)^n = 1 + n\varepsilon + n(n - 1)\varepsilon^2/2! + \cdots +)$

Problems

9.6. In the text it is shown that E and H vary as

$$e^{j(\omega t - kz)} \quad \text{where } k^2 = \omega^2\mu\varepsilon\left(1 - \frac{j\sigma}{\omega\varepsilon}\right)$$

Explain how this solution describes waves that move in both the negative z and positive z directions.

9.7. (a) Equation 9.4 is the inhomogeneous wave equation for $\mathbf{E}$. Starting from Maxwell's equations, derive the corresponding equation for $\mathbf{H}$. State your assumptions as you proceed.

(b) For an insulating dielectric, state your assumptions and show that

$$\mathbf{H} = \mathbf{H}_0 e^{j(\omega t - \beta z)} = \mathbf{H}_0 \cos(\omega t - \beta z)$$

*9.8. A uniform plane wave is propagating in a material that is characterized by $\varepsilon_r = 2.53$, $\mu_r = 1$, and $\sigma = 0$. If the electric field is given by

$$\mathbf{E} = 10 \cos(10\pi \times 10^9 t - \beta z)\mathbf{a}_y$$

Determine the following:

(a) The phase velocity.
(b) The wavelength.
(c) The magnitude of the magnetic field.
(d) The time-domain expression for the magnetic field.

*9.9. The electric field of a plane wave in a dielectric is given by

$$\mathbf{E} = \mathbf{a}_y 3.77 \times 10^{-2} \cos 2\pi(10^8 t - \tfrac{2}{3}z) \text{ V/m}$$

(a) Prove that $\varepsilon_r = 4$.
(b) Write the expression for the vector magnetic field.
(c) Determine the wave intensity (Poynting's vector, with units).

9.10. For a plane wave, $\mathbf{E} = \mathbf{a}_y e^{j(\omega t - \boldsymbol{\beta} \cdot \mathbf{r})}$, where $\boldsymbol{\beta} \cdot \mathbf{r} = \beta_x x + \beta_y y + \beta_z z$. Show that $\nabla \cdot \mathbf{E} = -j\boldsymbol{\beta} \cdot \mathbf{E}$ and that $\nabla \times \mathbf{E} = -j\boldsymbol{\beta} \times \mathbf{E}$. (See Exercise 9.2.)

*9.11. A uniform plane wave is traveling in the x direction in a lossless medium with the 10 V/m electric field in the z direction. Its wavelength is 25 cm and its velocity of propagation is 2×10^8 m/sec. Determine the frequency of the wave and the relative permittivity of the medium if it is characterized by free-space permeability. Write complete time-domain expressions for the electric and magnetic field vectors.

9.12. (a) A plane wave passes through a dielectric medium whose faces are at $z = 0$ and $z = \ell$. It experiences a phase change that is twice the change it has over the same distance when the dielectric is removed. Find the relative permittivity ε_{r0} of the dielectric.
    (b) The dielectric is gently heated at one face and is constrained from expanding. It is found that $\varepsilon_r = \varepsilon_{r0}(1 + 0.01z/\lambda_0)$, where λ_0 is the free-space wavelength. Find ℓ in terms of λ_0 if the phase shift doubles again.

9.13. The relative dielectric permittivity of a material is given by $\varepsilon_r = 9(1 - \alpha_1 T)^2$, where T is in °C. Its linear expansion is given by $\ell = \ell_0(1 + \alpha_2 T)$. If a plane wave passes through the dielectric, compare its phase change at 400°C to that at 0°C. $\alpha_1 = 2 \times 10^{-3}/$°C and $\alpha_2 = 10^{-3}/$°C.

*9.14. Write the following phasors in time-domain form. The frequency in each case is 10 MHz. Each of the phasors has the form $\mathbf{E} = \mathbf{E}_0 e^{j(\omega t - \beta z)}$.
    (a) $\mathbf{E}_0 = -j30\mathbf{a}_x - \mathbf{a}_y 10/j$
    (b) $\mathbf{E}_0 = 4je^{-j\pi/3}\mathbf{a}_y$
    (c) Find $\langle \mathcal{P} \rangle$, where $\mathbf{E}_0 = 3je^{-2jz}\mathbf{a}_y$, and $\mathbf{H}_0 = (1 + j)e^{-2jz}\mathbf{a}_x$.

9.15. We have seen that when E and H are expressed as phasors, then $\mathcal{P}_1 = \frac{1}{2}\mathbf{E} \times \mathbf{H}^*$ is the average intensity and $\mathcal{P}_2 = \frac{1}{2}\mathbf{E} \times \mathbf{H}$ is the instantaneous intensity. If circuit quantities $V = V_0 \cos \omega t$ and $I = I_0 \cos(\omega t + \theta)$ are expressed as phasors, show that $P_1 = \frac{1}{2}VI^*$ and $P_2 = \frac{1}{2}VI$ give comparable results. Determine θ and calculate P_1 and P_2 for a capacitor and for an inductor.

9.16. Energy moving with a wave can also be understood by taking the electric and magnetic energies in a volume, for example, in a wavelength long cylinder, as translating with the speed of the wave. Show that $\mathcal{P} = U\mathbf{v}_0$, where U is the total energy density of the wave, and $\mathbf{v}_0$ is its phase velocity.

*9.17. (a) A helium–neon laser operates in a single mode at a wavelength $\lambda = 6328$ Å(1 Å $= 10^{-10}$ m). It emits a continuous light beam that is 2 mm in diameter, at a power of 1 mW. Compute the peak values of E and H in free space.
    (b) A pulsed ruby laser emits a pulse of light ($\lambda = 6943$ Å) with the same cross-sectional diameter as the neon laser, in 10^{-8} sec. The pulse has a total energy of 20 J. Compute E and H in the wave train constituting the pulse.
    (c) For how long must the He–Ne laser emit in order to generate the same total energy as one ruby laser pulse?

*9.18. Describe the figure traced in space by the electric field of the wave

$$\mathbf{E} = E_0(1.25\mathbf{a}_x + j\mathbf{a}_y)e^{j(\omega t - \beta z)}$$

*9.19. The electric field intensity of a plane wave

$$\mathbf{E} = (\mathbf{a}_x + \mathbf{a}_y)10^{-3} \cos(2\pi \times 10^7 t - \beta z) \text{ V/m}$$

propagates in ferrite material that has $\mu/\mu_0 = 10^3$, $\varepsilon/\varepsilon_0 = 3$, and $\sigma = 10^{-5}$ S/m.
  (a) Would you classify this material as an insulator, a conductor, or an intermediate type? Why?
  (b) Determine β and λ of the wave.
  (c) What is its average intensity?
  (d) Write the expression for $\mathbf{H}$ of the wave.
  (e) Find the phase difference between the space–time points $(x, y, z, t) = (40, 20, 10, 0)$ and $(41, 21, 11, 0.125)$ [m, m, m, μsec].

9.20. A propagating wave has an electric field strength given by

$$E = (3\mathbf{a}_x + 4e^{j\pi/2}\mathbf{a}_y)e^{j(\omega t - \beta z)} \text{ V/m}$$

Describe and draw the polarization of this wave.

9.21. A plane wave with frequency $f = 5000$ Hz travels in a very good conductor, which has a skin depth $\delta = 1$ mm.
  (a) Determine the phase velocity of the wave in that medium.
  (b) What is the phase difference between two points separated by 2δ along the direction of wave travel? *Note:* A numerical answer requires units.

9.22. In the text, we derived the inhomogeneous wave equation

$$\mu\varepsilon\frac{\partial^2 \mathbf{E}}{\partial t^2} + \mu\sigma\frac{\partial \mathbf{E}}{\partial t} = \nabla^2 \mathbf{E} - \nabla(\nabla \cdot \mathbf{E})$$

and applied it to the case of a plane wave. We divided our analysis into the cases where the medium was an insulating dielectric or a good conductor, depending upon whether the value of $\sigma/\omega\varepsilon$ was small or large relative to unity. Let us now consider the case where $\sigma/\omega\varepsilon = 1$.
  (a) Show that the wavelength and phase velocity are only 10 percent different from that in the corresponding dielectric.
  (b) Over what distance (in terms of the wavelength) does this medium reduce the wave amplitude to $1/e$ of its initial value?

*9.23. In intrinsic germanium, $\sigma = 0.025$ S/m and $\varepsilon = 16\varepsilon_0$.
  (a) Find the frequency range in which a wave will pass through a piece of germanium with almost no attenuation. State your defining condition.
  (b) Find the range in which the wave will be strongly absorbed.
  (c) Over what range can intrinsic germanium be considered a semiconductor?

*9.24. (a) Express Poynting's vector for a plane wave moving through a conductor that has a penetration depth δ. What is happening to the energy?
  (b) Consider a z-directed plane wave with wave intensity $\mathscr{P}$ at a point z_0 in a conductor. What fraction of this intensity is dissipated between z_0 and $z_0 + \delta$? What fraction between $z_0 + \delta$ and $z_0 + 2\delta$?

9.25. Calculate the $1/e$ penetration depth at $f = 3 \times 10^8$ Hz of a lossy dielectric that has $\varepsilon/\varepsilon_0 = 2.6$ and $\sigma = 1.5 \times 10^{-5}$. (See Exercise 9.5.)

9.26. Current density in a conductor with skin depth δ decays exponentially from its surface value J_0. Prove that the total current in the body is equal to $J_0\delta$, that is, as though the current were constant at its surface value for a depth δ and then abruptly drops to zero.

9.27. Using Ampere's law, show that $\mathbf{H} = \mathbf{a}_\phi I\rho/2\pi a^2$ inside a wire of radius a, carrying a steady current I that is uniformly distributed over the wire cross section. At high frequency, the skin depth phenomenon changes this result. If the same wire carries the same total current at frequency ω such that $\delta = a$, find the following.

   (a) Write an expression for the current density $J(\rho)$. (Assume J decays radially from the surface with a skin depth δ.)

   (b) Calculate the new distribution $H(\rho)$.

   (c) If $f = 10^8$ Hz, estimate the value of a for this wire if it is (i) copper, and (ii) iron with $\mu_r = 10^4$.

9.28. In the region close to a short thin-wire antenna (known as an electric dipole antenna) of length ℓ, carrying current $Ie^{j\omega t}$, and oriented along $\theta = 0$ of spherical coordinates, the radiating fields are given by

$$\mathbf{H} = \frac{I\ell}{4\pi r} \sin\theta\, e^{j(\omega t - \beta r)} \mathbf{a}_\phi \left(j\beta + \frac{1}{r} \right)$$

$$\mathbf{E} = \frac{I\ell}{4\pi r j\omega\varepsilon} e^{j(\omega t - \beta r)} \left[\mathbf{a}_r \left(\frac{j\beta}{r} + \frac{1}{r^2} \right) 2\cos\theta - \mathbf{a}_\theta \left(\beta^2 - \frac{j\beta}{r} - \frac{1}{r^2} \right) \sin\theta \right]$$

Note that these fields are functions of θ and r, and in the near-field region $\beta r < 1$.

   (a) Calculate the average intensity.

   (b) Show that only the outgoing power, that is, along $\mathbf{a}_r$, has a nonzero value when averaged (integrated) over the surface of a sphere surrounding the small antenna, that is, over θ.

   (c) Calculate the wave impedance η in the near-field region. Justify the statement that *high-impedance* fields exist near an electric dipole antenna by showing that $\eta > \sqrt{\mu/\varepsilon}$.

10

INTERFACE PHENOMENA

10.1 GENERAL PLANE WAVES

The wave equation for the electric field strength in charge-free media was derived in the previous chapter.

$$\nabla^2 \mathbf{E} = \frac{\partial^2 \mathbf{E}}{\partial x^2} + \frac{\partial^2 \mathbf{E}}{\partial y^2} + \frac{\partial^2 \mathbf{E}}{\partial z^2} = -k^2 \mathbf{E} \qquad (10.1a)$$

where

$$k^2 = \omega^2 \mu \varepsilon \left(1 - j \frac{\sigma}{\omega \varepsilon} \right) \qquad (10.1b)$$

For a linearly polarized plane wave,

$$\mathbf{E} = \mathbf{e}_0 E(x, y, z) e^{j\omega t} \qquad (10.1c)$$

where $\mathbf{e}_0$ is orthogonal to the direction along which $\mathbf{E}$ changes; in Chapter 9, this propagation direction was taken to be along $\mathbf{a}_z$, so that $E(x, y, z) = E(z)$ only. Equation (10.1a) has been written in a more general form to facilitate examining the expression for $\mathbf{E}$ when the wave moves in an arbitrary direction with respect to the coordinate axes, although the physical content of the field must be unchanged. In this case, Eq. (10.1a) can be solved by separating the variables, a technique already used in Chapter 6. Following the same procedure, we take $E(x, y, z) = X(x)Y(y)Z(z)$. Substituting this in Eq. (10.1a) and dividing by XYZ gives

$$\frac{1}{X}\frac{\partial^2 X}{\partial x^2} + \frac{1}{Y}\frac{\partial^2 Y}{\partial y^2} + \frac{1}{Z}\frac{\partial^2 Z}{\partial z^2} = -k^2 \qquad (10.2)$$

Since these terms are independent and their sum is a constant, each is individually equal to a constant, k_x^2, k_y^2, and k_z^2, respectively. The individual equations lead to sinusoidal solutions for X, Y, and Z. Their product can be written

$$\mathbf{E} = \mathbf{e}_0 E_0 e^{j\omega t} e^{-j(k_x x + k_y y + k_z z)} \qquad (10.3a)$$

where

$$k^2 = k_x^2 + k_y^2 + k_z^2 \qquad (10.3b)$$

Since the vector from the origin to a point in space is

$$\mathbf{r} = x\mathbf{a}_x + y\mathbf{a}_y + z\mathbf{a}_z \qquad (10.4a)$$

and we can define the vector

$$\mathbf{k} = k_x\mathbf{a}_x + k_y\mathbf{a}_y + k_z\mathbf{a}_z \qquad |\mathbf{k}|^2 = \mathbf{k} \cdot \mathbf{k} = k^2 \qquad (10.4b)$$

Eq. (10.3a) becomes

$$\mathbf{E} = \mathbf{e}_0 E_0 e^{j(\omega t - \mathbf{k} \cdot \mathbf{r})} \qquad (10.5a)$$

which is the expression for a plane wave moving in the general direction of the vector $\mathbf{k}$, with components given by Eq. (10.4b). This can be seen from Fig. 10.1, displaying a section through the plane whose normal is the unit vector $\mathbf{n}$, and which is a distance d from the origin. The expression for points on this plane is seen to be $\mathbf{r} \cdot \mathbf{n} = d$. Defining $\mathbf{k} = \sqrt{k^2}\,\mathbf{n}$, then Eq. (10.5a) becomes

$$\mathbf{E} = \mathbf{e}_0 E_0 e^{j(\omega t - kd)} \qquad (10.5b)$$

indicating that the distance d is changing in time, that is, the entire plane is translating parallel to $\mathbf{k}$. In a dielectric, $k = \beta = 2\pi/\lambda$ and, as in Chapter 9, $v = \lambda/T = \omega/\beta$.

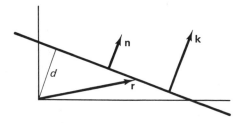

Figure 10.1   General propagating plane.

10.2 SURFACE CURRENTS

In Chapter 9, it was established that an alternating field in a material for which $\sigma \neq 0$ decays into that material with a skin depth δ, which can be quite small at high frequencies and in good conductors. Some attenuation constant $\alpha = 1/\delta$ exists in all real material, although the mechanism producing attenuation can vary. For example, it can be shown that in a dielectric material, one must generally take $\varepsilon = \varepsilon' - j\varepsilon''$, and in a magnetic material, one generally has $\mu = \mu' - j\mu''$. The imaginary components of μ'' and ε'' result from the fact that the response of a physical system lags any excitation. For materials and frequencies of interest, $\varepsilon''/\varepsilon' \ll 1$ and $\mu''/\mu' \ll 1$, generally, so that the electromagnetic behavior is essentially what we have considered heretofore. However, it is straightforward to show that these imaginary components lead to signal attenuation, even in the absence of conductivity. Substituting these μ and ε expressions in Eq. (10.1b),

$$k^2 = \omega^2(\mu' - j\mu')(\varepsilon' - j\varepsilon'') - j\sigma\omega(\mu' - j\mu'')$$

$$= \omega^2\mu'\varepsilon'\left(1 - \frac{\sigma}{\omega\varepsilon'}\frac{\mu''}{\mu'}\right) - j\omega^2\mu'\varepsilon'\left(\frac{\varepsilon''}{\varepsilon'} + \frac{\mu''}{\mu'} + \frac{\sigma}{\omega\varepsilon'}\right) \qquad (10.6a)$$

where second-order terms in the small quantities $\varepsilon''/\varepsilon'$ and μ''/μ' have been dropped. From this it is seen that k^2 is a complex number that can be written as $k^2 = (\beta - j\alpha)^2$. The field variation is, therefore,

$$e^{j(\omega t - \mathbf{k}\cdot\mathbf{r})} = e^{j(\omega t - \boldsymbol{\beta}\cdot\mathbf{r})}e^{-\boldsymbol{\alpha}\cdot\mathbf{r}} = e^{-r/\delta}e^{j(\omega t - \beta r)} \qquad (10.6b)$$

From Eq. (10.6a), there is a value for $\delta = 1/\alpha$ even in the case that $\sigma = 0$.

When $\sigma/\omega\varepsilon \gg 1$, the attenuation is attributable to the generation of conduction current that flows in the conductive (or resistive) medium and converts electromagnetic energy to heat energy, thereby reducing the electromagnetic field. (An alternative explanation is that the electric and magnetic fields from the currents are opposed to the fields that generated them, causing the same reduction.) Now we see that a similar effect can be attributed to the more localized atomic electronic currents whose motions result in dielectric and magnetic properties.

The relations between the induced charge density ρ, the surface current density K, and the magnitudes of the normal (n) and tangential (t) fields in a material are given by the surface boundary conditions found in Chapter 5.

$$D_{2n} - D_{1n} = \rho_s \qquad (10.7a)$$

$$H_{2t} - H_{1t} = K \qquad (10.7b)$$

where subscript 2 denotes the medium in the half space $z > 0$, and subscript 1 denotes the medium in the half space $z < 0$, as shown in Fig. 10.2, where the interface is the plane $z = 0$.

It may seem strange to speak of *surface* charge and *surface* current since, in the world of our common experience, nature does not have any truly two-dimensional, that is, surface, phenomena. Therefore, we recognize that there is actually a physical volume current density **J** near the surface of the medium. The total current flowing parallel to the surface, through the perpendicular area $\Delta x \Delta z$ in Fig. 10.2, is $\Delta \mathbf{I} =$

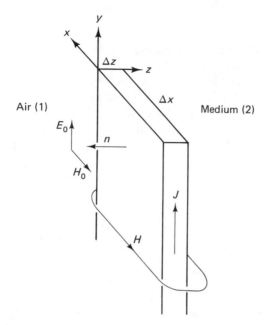

Figure 10.2   Surface fields and current.

$\mathbf{J}\,\Delta x\,\Delta z$. The corresponding surface current density, that is, the current per unit width of surface, is $\mathbf{K} = \Delta \mathbf{I}/\Delta x = \mathbf{J}\Delta z$. Note that the direction of $\mathbf{J}$, $\mathbf{I}$, and $\mathbf{K}$ are parallel to $\mathbf{E}_0$, so that *inside the medium* the magnetic field produced by the current is opposed to the applied magnetic field.

If we let $\Delta z \to 0$ but maintain the finite value of the product $J\Delta z = K$, there will be a *mathematical* surface-current sheet. Furthermore, if the magnitude relation exists that $K = H_0$, then Eq. (10.7b) will be satisfied with $H = 0$ within the medium. If Δz does not go to zero, then after each incremental distance dz, the magnetic field is partly reduced (and, through induction, so is the electric field), producing exponential decay.

From this and Fig. (10.2), it is seen that the magnitude and direction relations between the vector magnetic field and current are satisfied by

$$\mathbf{K} = \mathbf{n} \times \mathbf{H} \tag{10.8}$$

where $\mathbf{n}$ is the unit surface normal directed out of the medium.

A volume-charge density also exists at the medium surface as a result of the law of conservation of charge with the changing currents. By a similar argument to that just presented, and using Eq. (10.7a), it follows that the surface charge density $\rho_s = \varepsilon_0 E_0$ reduces the internal electric field to zero.

Whereas mathematical surface current and charge densities are not strictly obtainable, they can be approached in many cases, and, in any event, they are extremely useful concepts when studying interface phenomena. Recall from Chapter 5 that, when deriving the boundary conditions of Eqs. (10.7a), (10.7b), (10.9a), and (10.9b)

$$E_{t1} = E_{t2} \tag{10.9a}$$

$$B_{n1} = B_{n2} \tag{10.9b}$$

we examined a much enlarged section of the media interface. We made the classical approximation that the surface could be described locally by a mathematical plane. In order to keep this assumption and Eqs. 10.7, it is also necessary to have mathematically consistent surface currents and charge densities.

10.3 WAVES AT AN INTERFACE

The following discussion considers waves at the interface of two dielectric materials. By employing surface current and charge densities, all relevant changes occur *at the surface* of the medium, so that we need not consider attenuation factors; instead, there are only unattenuated waves with wave vectors $\mathbf{k} = \boldsymbol{\beta}$.

Even before electromagnetic theory was applied to the problem, physicists had analyzed wave theories of the reflection and refraction of light at an interface between two media. (This was done, in part, in the context of comparing wave predictions with the Newtonian theory of radiation, in which light consists of a stream of corpuscles that eventually enter and stimulate the eye.) As many properties of light do not depend upon the fact that it is an *electromagnetic* phenomenon, this section and the next considers, merely, that there is a wave amplitude A that must be continuous across the surface of two media. [In electromagnetic theory, this might be the tangential component of the electric or magnetic field strength, according to Eqs. (10.9a) and (10.9b)].

Figure 10.3 shows a wave, denoted by its phase propagation vector β_i, incident to the interface between two media, which is the plane $z = 0$. Also shown are a

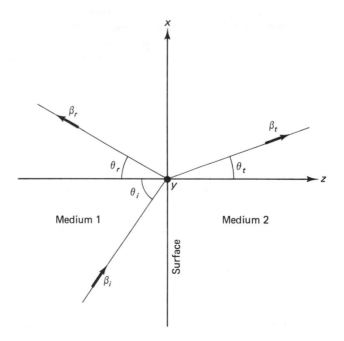

Figure 10.3 Waves at an interface.

reflected ray β_r, and a transmitted ray β_t, all at assumed different angles from the interface normal. For continuity of A across the interface,

$$A_i + A_r = A_t \quad \text{when } z = 0$$

or

$$A_i e^{j\omega t} e^{-j(\beta_{xi}x + \beta_{yi}y)} + A_r e^{j\omega t} e^{-j(\beta_{xr}x + \beta_{yr}y)} = A_t e^{j\omega t} e^{-j(\beta_{xt}x + \beta_{yt}y)} \quad (10.10)$$

where the components of the incident, reflected, and transmitted propagation vectors are carefully distinguished. The frequencies of the waves must be the same if the boundary condition is to be satisfied at all times. Figure 10.3 shows the rays at the origin, but the wavefronts cover the entire xy plane, so that Eq. (10.10) must be satisfied at all points on that plane. The wave phase repeats with periodicity β. Therefore, moving parallel to the x axis or parallel to the y axis, coherence and continuity of A demand that

$$\beta_{xr} = \beta_{xi} = \beta_{xt} \qquad \beta_{yr} = \beta_{yi} = \beta_{yt} \quad (10.11a, b)$$

As the incident wave has been chosen to have $\beta_{yi} = 0$, so must $\beta_{yr} = 0 = \beta_{yt}$, from Eq. (10.11b), so that all three rays remain in the plane of the page; this is called the *plane of incidence*, defined by the incident ray and the surface normal.

In a dielectric, $\beta = \omega/v$, and if we assume that v is a function of the properties of the medium (from electromagnetic theory, we know that $v = 1/\sqrt{\mu\varepsilon}$), then so is β, and $\beta_i = \beta_r = \beta_1$ (i.e., the magnitudes are equal), so that the first equality of Eq. (10.11a) yields

$$\beta_1 \sin \theta_r = \beta_1 \sin \theta_i \quad \text{or} \quad \theta_r = \theta_i \quad (10.12)$$

which is the law of reflection, that is, the incident and reflection angles are equal.

Since v is taken to be a property of the medium, it is convenient to define the index of refraction of a dielectric medium as the ratio of the velocity of light in vacuum to the velocity of light in that medium:

$$n = \frac{v_0}{v} \qquad \left[= \sqrt{\mu_r \varepsilon_r} \text{ in electromagnetic theory}\right] \quad (10.13)$$

With this

$$\beta_t = \beta_2 = \frac{\omega}{v_2} = \left(\frac{\omega}{v_1} \frac{v_1}{v_0} \frac{v_0}{v_2}\right) = \beta_1 \frac{n_2}{n_1} \quad (10.14)$$

so that the second equality in Eq. (10.11a) yields

$$\beta_1 \sin \theta_i = \beta_2 \sin \theta_t = \beta_1 \frac{n_2}{n_1} \sin \theta_t$$

or

$$n_1 \sin \theta_i = n_2 \sin \theta_t \quad (10.15a)$$

which is the law of refraction between two media.

10.4 TOTAL INTERNAL REFLECTION

An important consequence of Eq. (10.15a) can be found by writing it as

$$\sin \theta_t = \frac{n_1}{n_2} \sin \theta_i \qquad (10.15b)$$

A value of θ_t can be found for all θ_i as long as $n_1/n_2 \leq 1$, that is, as long as the wave passes from a less dense to a more dense medium, such as from air to glass. (This statement assumes that n is proportional to medium density, which is generally so. In any event, however, we can speak of an *optical density,* which is defined as proportional to n.) In contrast, if $n_1/n_2 \geq 1$, that is, if the wave passes from a more dense to a less dense medium, then $\theta_t > \theta_i$, and as θ_i increases, a value is reached for which

$$\sin \theta_{i(\text{critical})} = \frac{n_2}{n_1} \qquad (10.16)$$

Beyond this limit, $\sin \theta_t > 1$, meaning that θ_t is undefined. To investigate this situation, we examine the z components of the $\boldsymbol{\beta}$, which describe propagation normal to the interface.

Having dealt with the tangential components of β, it is easily seen, from Eq. (10.12) and Fig. 10.3, that $\beta_{zi} = -\beta_{zr}$. β_{zt}, the transmitted longitudinal component, is found by squaring Eq. (10.14):

$$\beta_{zt}^2 + \beta_{xt}^2 = \left(\frac{n_2}{n_1}\right)^2 \beta_1^2$$

Writing β_{xt} in terms of β_t and θ_t, and substituting Eq. (10.14) for β_t yields

$$\beta_{zt}^2 + \left(\frac{n_2}{n_1}\right)^2 \beta_1^2 \sin^2\theta_t = \left(\frac{n_2}{n_1}\right)^2 \beta_1^2$$

from which

$$\beta_{zt}^2 = \left(\frac{n_2}{n_1}\right)^2 \beta_1^2 (1 - \sin^2\theta_t)$$

and, with Eq. (10.15),

$$\beta_{zt} = \pm\beta_1 \left[\left(\frac{n_2}{n_1}\right)^2 - \sin^2\theta_i\right]^{1/2} \qquad (10.17)$$

Here the $\pm$ refers to the two possible directions of travel. This is a very revealing relation since it divides the transmission properties into two classes of behavior, corroborating the previous distinction.

1. The wave moves from a less dense medium into a more dense medium, that is, $n_1 < n_2$. Then, from Eq. (10.17), β_{zt} is real for all values of θ_i. From Eq. (10.15), it follows that $\theta_t < \theta_i$.

Example 10.1

For many types of optical glass, $n \approx 1.5$. Determine the refracted angle for light incident onto such a plane glass surface from air at an incident angle of 70°.

Here $n_1 = 1$ and $n_2 = 1.5$, so that $\sin \theta_t = (n_1/n_2) \sin \theta_i = 0.9397/1.5 = 0.6265$. $\theta_t = 38.8°$.

2. The wave passes from a more dense to a less dense medium. Then $n_1 > n_2$, and for $\theta_i > \theta_{i(\text{critical})}$ of Eq. (10.16), β_{zt} is imaginary.

$$\beta_{zt} = \pm j\alpha \quad \text{where } \alpha = \beta_i \left[\sin^2\theta_i - \left(\frac{n_2}{n_1}\right)^2 \right]^{1/2} \tag{10.18}$$

This results in a new behavior of the wave in the less dense medium. Since

$$\boldsymbol{\beta}_t = \mathbf{a}_x\beta_{xt} + \mathbf{a}_z\beta_{zt} = \mathbf{a}_x\beta_{xi} \pm \mathbf{a}_z j\alpha \tag{10.19a}$$

then

$$Ae^{j(\omega t - \boldsymbol{\beta} \cdot \mathbf{r})} = Ae^{-\alpha z}e^{j(\omega t - \beta_{xi}x)} \tag{10.19b}$$

where the sign of α has been chosen to give an attenuated rather than a growing wave. Equation (10.19b) describes a wave traveling along the surface of medium 2, that is, in the x direction, but which has its amplitude attenuated inward from the surface of the medium, that is, perpendicular to the direction in which the phase changes. If α is large, the wave has significant amplitude only near $z = 0$; so this can be regarded as a surface wave. Alternatively, it is said to be internally reflected in medium 1, the dense medium.

Example 10.2

Light, with frequency $f = 5 \times 10^{14}$, is incident from the glass side of a glass–air interface at an incident angle of 70°. Take $n_1 = 1.5$ and analyze the surface wave.

In the glass, $\beta_1 = \omega/v_1 = 2\pi f n_1/v_0 = 1.571 \times 10^5 \text{ cm}^{-1}$. Equation (10.18) gives the value $\alpha = 1.571 \times 10^5 (0.883 - 0.444)^{1/2} = 8.5 \times 10^4 \text{ cm}^{-1} = 1/(0.961 \times 10^{-5} \text{ cm})$. Thus, the surface wave attenuates into the air to $1/e = 0.365$ of its value at the glass surface in 961 Å and to 13.5 percent in 1920 Å. Since energy density and Poynting's intensity is proportional to the square of the wave amplitude, it decreases to under 2 percent over the distance 2δ.

Although penetration of the internally reflected wave into the air is very small, it is possible to demonstrate its presence. If a second dielectric surface is carefully placed within a few thousand angstroms of the surface, the weak field in this third medium will propagate and can be sensed. Otherwise, no energy travels away from the first surface; the field in air is said to be *evanescent*.

The idea demonstrated here of total internal reflection in a dense dielectric underlies the propagation of signals along dielectric transmission lines such as optical fibers.

10.5 E IN THE PLANE OF INCIDENCE

The behaviors derived in the previous section do not make specific use of the properties of the electric and magnetic fields of the waves. We now expand our scope by applying the specific electromagnetic boundary conditions of Eqs. (10.9). For this, the orientations of the electric and magnetic fields must be specified, and here two possibilities present themselves for the polarization of the incident wave. $\mathbf{E}_i$ can either lie in the plane of incidence or it can be perpendicular to that plane. In either case, $\mathbf{E}_i$ will be normal to $\boldsymbol{\beta}_i$ and the three vectors $\mathbf{E}_i$, $\mathbf{H}_i$, and $\boldsymbol{\beta}_i$ will form an orthogonal set. Once the orientations of $\mathbf{E}_i$ and $\mathbf{H}_i$ are chosen with respect to the plane of incidence, it is clear that the same orientation must follow for the reflected and transmitted waves if the boundary conditions of continuity of the fields are to be satisfied. In applying the boundary conditions of Eqs. (10.9), use will be made of the relation between E and H of each wave:

$$\frac{E_i}{H_i} = \eta_1 = \frac{E_r}{H_r} \qquad \frac{E_t}{H_t} = \eta_2 \tag{10.20a}$$

where

$$\eta_1 = \sqrt{\frac{\mu_1}{\varepsilon_1}} \qquad \eta_2 = \sqrt{\frac{\mu_2}{\varepsilon_2}} \tag{10.20b}$$

are the transverse-wave impedances of the two media.

It will also be useful to note a relation between the wave impedance η and the index of refraction discussed earlier, $n = \sqrt{\mu_r/\varepsilon_r}$. In most media, $|X_{\mathrm{magnetic}}| \ll 1$, so that $\mu = \mu_0$ to a high degree of accuracy. Even in cases where this is not so at low frequencies, at sufficiently high frequencies, for example, in the infrared or optical range, magnetic changes become very small, so that $\mu \to \mu_0$ and $\mu_r \to 1$. In these cases, we see that $\eta = \eta_0/n$. In the following discussions, we will maintain the more general use of η, but it will be convenient to also examine the results in this *optical* limit, although it is not completely limited to optical frequencies.

Figure 10.4 shows the first case to be considered, that in which $\mathbf{E}$ lies in the plane of incidence. This figure has been drawn with the $\mathbf{H}$ fields of the waves parallel to each other and normal to the incidence plane. It is not necessary to specify whether the $\mathbf{H}$ are parallel or anti-parallel, since an incorrect choice will result in a corrective negative sign in the final solution. The $\mathbf{E}$ vectors are then oriented so that each $\mathbf{E} \times \mathbf{H}$ is parallel to its $\boldsymbol{\beta}$.

Since the tangential $\mathbf{E}$ and $\mathbf{H}$ fields must be continuous across the interface,

$$H_i + H_r = H_t \to \frac{E_i}{\eta_1} + \frac{E_r}{\eta_1} = \frac{E_t}{\eta_2} \tag{10.21a}$$

$$(E_i - E_r) \cos \theta_i = E_t \cos \theta_t \tag{10.21b}$$

where use has been made of the law of reflection, that $\theta_r = \theta_i$. These equations can be solved for the reflection and the transmission coefficients:

$$\rho'_E = \frac{E_r}{E_i} = \frac{\eta_1 \cos \theta_i - \eta_2 \cos \theta_t}{\eta_1 \cos \theta_i + \eta_2 \cos \theta_t} \tag{10.22a}$$

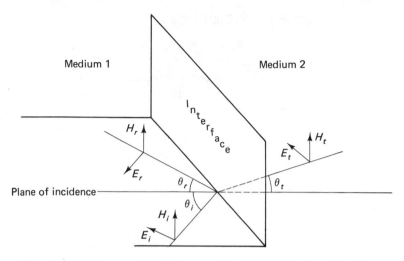

Figure 10.4 Electromagnetic boundary conditions for E in the plane of incidence.

$$\tau'_E = \frac{E_t}{E_i} = \frac{2\eta_2 \cos \theta_i}{\eta_1 \cos \theta_i + \eta_2 \cos \theta_t} \qquad (10.22b)$$

The prime is used to indicate that this result is for the case where E is in the plane of incidence (and H is normal to it), in contrast to the case, considered later, that the field orientations are switched. The subscript E shows that it is the ratio of electric fields that is being expressed. The magnetic-field coefficients can be found by substituting Eq. (10.20a) in these results. This gives

$$\rho'_H = \rho'_E \qquad \tau'_H = \frac{H_t}{H_i} = \frac{2\eta_1 \cos \theta_i}{\eta_1 \cos \theta_i + \eta_2 \cos \theta_t} \qquad (10.23a,\ b)$$

For normal incidence,

$$\rho = \frac{\eta_1 - \eta_2}{\eta_1 + \eta_2} \qquad \tau_E = \frac{2\eta_2}{\eta_1 + \eta_2} \qquad \tau_H = \frac{2\eta_1}{\eta_1 + \eta_2} \quad \text{when } \theta = 0 \qquad (10.24)$$

Example 10.3

Determine the relation of the reflected to the incident wave upon normal incidence from the surface of a very good conductor.

From the discussion in Chapter 9, recall that for a good conductor, $|\eta_2| = (\omega\varepsilon/\sigma)^{1/2} \approx 0$, so that $\rho \approx 1$, $\tau_E \approx 0$, and $\tau_H \approx 2$. Thus, the entire incident signal is reflected. However, a problem seems to be indicated in that while the transmitted E field is almost zero, the transmitted H field is twice that of the incident wave. At the reflecting surface, E_r and E_i are equal and, from Fig. (10.4) and the positive value of ρ, are oppositely directed. As a result, $E = 0$ in medium 1 and, by continuity, inside the conductor on the other side of the interface. However, H_i and H_r add in medium 1, so H must also equal their sum ($= 2H_i$) in medium 2.

By examining the transmitted wave more carefully, Eqs. (10.22b) and (10.23b) indicate that, at $\theta = 0$,

$$\frac{E_t}{H_t} = \frac{\tau_E}{\tau_H} \frac{E_i}{H_i} = \frac{\eta_2}{\eta_1} \eta_1 = \eta_2 = \left(\frac{\omega E}{\sigma}\right)^{1/2} \approx 0$$

as expected. Therefore, E_2 is not zero but merely very small, and the ratio of the fields, equal to the small wave impedance, is still maintained in the conductor. This is consistent with the conclusion at the very end of Chapter 9.

10.6 BREWSTER'S ANGLE

Note from Eqs. (10.22a) and (10.23a) that

$$\rho' = 0 \quad \text{when } \eta_1 \cos \theta_i = \eta_2 \cos \theta_t \tag{10.25}$$

If unpolarized radiation, that is, radiation with components of E both in and normal to the plane of incidence, is reflected at this angle of incidence, called *Brewster's angle,* no component with E parallel to the plane is reflected, so the reflected ray becomes linearly polarized with E normal to the incident plane. This technique has long been used to obtain linearly polarized radiation in the infrared region of the spectrum, where other polarization techniques are not readily available. Reflectors using Brewster's angle are commonly found at the ends of optical lasers to maintain the desirable polarization while building up sufficient intensity in the laser cavity; the radiation that is emitted from the laser is the transmitted component. See Fig. 10.5.

Brewster's angle can be expressed conveniently in the optical case ($\eta = \eta_0/n$) by combining Eqs. (10.25) and (10.15a):

$$\frac{n_2}{n_1} \cos \theta_i - \cos \theta_t = 0 \rightarrow \frac{\sin \theta_i}{\sin \theta_t} \cos \theta_i - \cos \theta_t = 0$$

or

$$\sin 2\theta_i = \sin 2\theta_t$$

This has a solution: $\theta_t = \pi/2 - \theta_i$, that is, at the Brewster angle, θ_i and θ_t are complements. It then follows from Eq. (10.25) that

$$\tan \theta_B = \frac{\sin \theta_i}{\cos \theta_i} = \frac{\cos \theta_t}{\frac{\eta_2}{\eta_1} \cos \theta_t} = \frac{\eta_1}{\eta_2} = \frac{n_2}{n_1} \tag{10.26a}$$

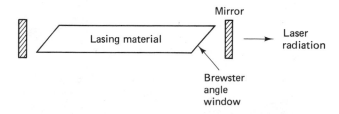

Figure 10.5 Construction of a laser.

It is left as an exercise to show that, in this case,

$$\tau_E' = \frac{\eta_2}{\eta_1} = \frac{n_1}{n_2} \tag{10.26b}$$

Note that Brewster's polarization can occur both for $n_2 > n_1$ and for $n_2 < n_1$.

Example 10.4

The mirror of a gas laser is made of highly polished glass that has $n = 1.5$. What is the transmission coefficient if $\theta_i =$ Brewster's angle? The gas has $n \approx 1$.

Here $n_2/n_1 = 1.5 = \tan \theta_B$, from which $\theta_B = 56.3°$. The law of refraction then yields $\theta_t = 33.7°$. $\tau_E' = 0.667$.

10.7 H IN THE PLANE OF INCIDENCE

Now consider the second polarization case, where the **E** fields are normal to the incidence plane. This is shown in Fig. 10.6. The **E** vectors are taken to be mutually parallel and the **H** vectors are oriented so that each $\mathbf{E} \times \mathbf{H}$ is along its $\boldsymbol{\beta}$.

Applying Eqs. (10.9) and (10.20a) to this figure results in

$$E_i + E_r = E_t \tag{10.27a}$$

$$\frac{(E_i - E_r) \cos \theta_i}{\eta_1} = \frac{E_t \cos \theta_t}{\eta_2} \tag{10.27b}$$

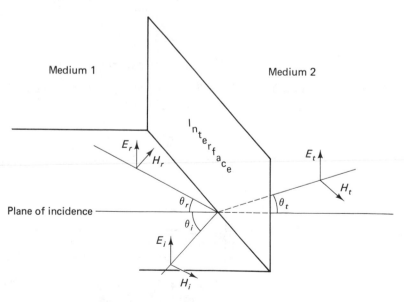

Figure 10.6   *E* normal to the plane of incidence.

From which

$$\rho_E'' = \frac{E_r}{E_i} = \frac{\eta_2 \cos \theta_i - \eta_1 \cos \theta_t}{\eta_2 \cos \theta_i + \eta_1 \cos \theta_t} \qquad (10.28a)$$

$$\tau_E'' = \frac{E_t}{E_i} = \frac{2\eta_2 \cos \theta_i}{\eta_2 \cos \theta_i + \eta_1 \cos \theta_t} \qquad (10.28b)$$

where a double-primed notation distinguishes this case from the orthogonal polarization case, which was considered earlier and indicated with a single prime.

Note that Eqs. (10.28) reduce to Eqs. (10.24) in the case of normal incidence. As with the polarization considered earlier, it is straightforward to find expressions for ρ_H'' and τ_H''.

Figure 10.7 shows the variation of ρ_E and τ_E as functions of the incidence angle at the interface between two dielectrics with $n_2/n_1 = 1.5$. It is seen that over part of the range of θ_i, ρ' is negative, and over the entire range, ρ'' is negative. This indicates a phase reversal, that is, that E_r at the surface is reversed from the direction indicated on Figs. 10.4 and 10.5 and, therefore, so is H_r. It was mentioned earlier that any orientation can be chosen for the field vectors when setting up field configurations, because negative signs enter the final solutions to adjust any incorrect directional assumptions.

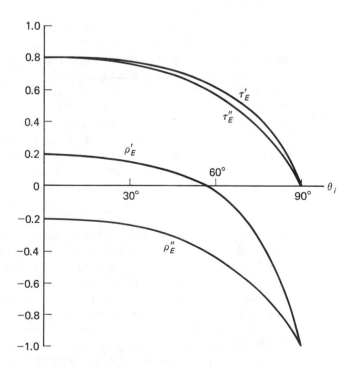

Figure 10.7 Reflection and transmission coefficients; $n_e/n_1 = 1.5$.

10.8 POWER TRANSMISSION

While Eqs. (10.22) and (10.28) give the reflected and transmitted field amplitudes for the two cases of polarization, the intensity ratios, that is, the Poynting vectors $\mathcal{P}$ are also of concern. For this, only the components of $\mathcal{P}$ that are normal to the interface need be considered as these represent the power traveling toward and away from the surface. For the reflected wave, $\theta_i = \theta_r$, so that

$$R = \frac{|\mathcal{P}_r| \cos \theta_r}{|\mathcal{P}_i| \cos \theta_i} = \frac{E_r H_r}{E_i H_i} = \frac{E_r^2/\eta_1}{E_i^2/\eta_1} = \rho^2 \tag{10.29}$$

For the transmitted wave,

$$T = \frac{|\mathcal{P}_t| \cos \theta_t}{|\mathcal{P}_i| \cos \theta_i} = \frac{E_t H_t}{E_i H_i} \frac{\cos \theta_t}{\cos \theta_i} = \frac{E_t^2/\eta_2}{E_i^2/\eta_1} \cdot \frac{\cos \theta_t}{\cos \theta_i}$$

$$= \tau^2 \frac{\eta_1 \cos \theta_t}{\eta_2 \cos \theta_i} = \tau^2 \frac{n_2 \cos \theta_t}{n_1 \cos \theta_i} \tag{10.30}$$

Of course, the power reflection and transmission factors, R and T, respectively, are always positive. It is left as an exercise to show that $T + R = 1$. This expresses the conservation of energy upon reflection and transmission. It is important to avoid a common error by noting that one does *not* necessarily have the sum of ρ and τ equal to unity.

10.9 WAVE MOMENTUM

Among the fundamental revolutions in our understanding of nature, discovered in this century, has been the wave–particle duality, according to which atomic particles sometimes exhibit wave-like properties and waves sometimes exhibit particle-like properties. This duality is often expressed by relationships between the wave properties of wavelength λ and frequency f, and the particle properties of momentum p and energy E. The original proposal by Max Planck was that the radiation interaction with solids could be explained in some detail if radiant energy was absorbed and emitted in multiples of a fundamental energy $E = hf$. This implies that the radiation exists in packets, called *photons,* which behave like particles, and whose energies are proportional to their frequency. Here h is Planck's constant, determined by empirically fitting Planck's radiation law to the known experimental values:

$$h = 6.62620 \times 10^{-34} \text{ J–sec}$$

The implication that radiant energy, that is, light, microwaves, X-rays, etc., is quantized was indeed revolutionary, to the point that Planck himself long denied the conclusion that light had such particle-like properties. This wave–particle correspondence was furthered by de Broglie's proposal that the interaction of atoms and molecules passing through a crystal could be understood if the particles behaved like waves, with wavelengths related to their mechanical momenta by $p = h/\lambda$.

Thus, on an atomic–molecular level, particles seem to display the properties of waves and waves seem to display the properties of particles. If the photons that constitute an electromagnetic wave possess momentum, then, since Newton's law states that $\mathbf{F} = d\mathbf{p}/dt$, the wave must exert a force on a surface on which it impinges. In fact, even before the advent of quantum considerations, it had been known that electromagnetic radiation was capable of exerting a force on a solid (i.e., transferring momentum). In this section, classical electromagnetic considerations are used to derive the relationship between wave momentum and the force it exerts. The wave reflected from a conducting surface at normal incidence will be considered, because this case lends itself to particularly clear and simple explication.

Figure 10.8 shows a conducting surface with the normally incident plane wave. From Eq. (10.24) and Example 10.3, the transmitted fields immediately inside the conductor surface are

$$E_{to} = 2E_i\left(\frac{\varepsilon_0\omega}{\sigma}\right)^{1/2}e^{-j\pi/4} \qquad H_{to} = 2H_i \qquad (10.31)$$

Including attenuation of the wave propagating into a conductor, we have

$$E_t = 2E_i\left(\frac{\varepsilon_0\omega}{\sigma}\right)^{1/2}e^{-z/\delta}e^{j(\omega t - \beta z - \pi/4)} \qquad (10.32\text{a})$$

$$H_t = 2H_i e^{-z/\delta}e^{j(\omega t - \beta z)} \qquad (10.32\text{b})$$

where the skin depth $\delta = (2/\omega\mu\sigma)^{1/2}$. Under the influence of the electric field, a current density $\mathbf{J} = \sigma\mathbf{E}_t$ flows along the surface of the conductor and each infinitesimal current volume experiences a force from its interaction with the transmitted magnetic field. From the discussion of phasor multiplication in Chapter 9, the average force is expressed as

$$d\mathbf{F} = \frac{1}{2}\mathcal{R}(\mathbf{J}\,dv \times \mathbf{B}_t^*) = \frac{1}{2}\mathcal{R}(\sigma\mathbf{E}_t \times \mu_0\mathbf{H}_t^*\,dv) \qquad (10.33\text{a})$$

where $\mathcal{R}$ denotes taking the real part, and * denotes the complex conjugate.

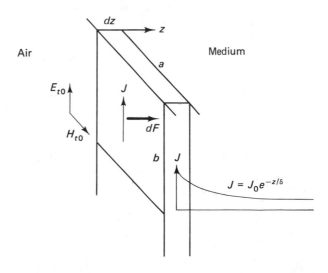

Figure 10.8   Surface force.

As shown in Fig. 10.8, the volume has a surface area ab and extends a distance dz into the metal, so that $dv = ab\,dz$. To find the total force on the projected area ab, $d\mathbf{F}$ must be integrated from the surface $z = 0$ to the distant end of the conductor, here taken to be at $z = \infty$. (While z does not actually extend to infinity, this approximation can be used as long as the conductor thickness is more than several δ, since the fields and currents are negligible beyond that.) Thus,

$$\mathbf{F} = \frac{1}{2}\mathscr{R}\left\{\int \sigma\mu_0\left[2\left(\frac{\varepsilon_0\omega}{\sigma}\right)^{1/2}e^{-z/\delta}e^{j(\omega t - \beta z - \pi/4)}\mathbf{E}_i\right] \times [2e^{-z/\delta}e^{-j(\omega t - \beta z)}\mathbf{H}_i]\,ab\,dz\right\}$$

$$= 2\mu_0(\sigma\varepsilon_0\omega)^{1/2}E_iH_i\mathbf{a}_z ab\,\mathscr{R}(e^{-j\pi/4})\int_0^\infty e^{-2z/\delta}\,dz \qquad (10.33b)$$

It is easily shown that

$$\int_0^\infty e^{-2z/\delta}\,dz = \delta/2 = \left(\frac{1}{2\omega\mu_0\sigma}\right)^{1/2} \quad \text{and} \quad \mathscr{R}\{e^{-j\pi/4}\} = \frac{1}{\sqrt{2}} \qquad (10.33a,\,b)$$

so that the pressure on the surface is

$$\mathbf{P} = \mathbf{F}/ab = \sqrt{\varepsilon_0\mu_0}\,E_iH_i\mathbf{a}_z \qquad (10.34a)$$

Since the Poynting vector of the incident wave is $\mathscr{P} = \mathbf{a}_z\frac{1}{2}E_iH_i$ and the speed of the wave in the air is $v_0 = 1/\sqrt{\mu_0\varepsilon_0}$, this result can be written as

$$\mathbf{P} = \frac{2\mathscr{P}}{v_0} \qquad (10.34b)$$

This is the pressure exerted by the wave upon being reflected from the conductor. As mentioned earlier, the fact that the wave exerts a force per unit area implies that it carries momentum, since Newton's law states that force is equal to the rate of change of momentum. In fact, if the wave has a momentum per unit area M along its direction of motion, then its momentum is reversed on reflection, so that the change of momentum per unit area is $2M$, and the pressure it exerts reflects this factor of 2, as in Eq. (10.34b). Equation (10.34b) can therefore be interpreted in the following way in a traveling electromagnetic wave. On the left-hand side

$$P = \frac{dM}{dt} = \frac{\text{momentum/area}}{\text{time}} = \frac{\text{momentum}}{\text{area-time}}$$

On the right-hand side

$$\frac{\mathscr{P}}{v} = \frac{1}{v}\frac{\text{Power}}{\text{area}} = \frac{1}{v}\frac{\text{energy}}{\text{area-time}}$$

If a wave consists of photons, then these expressions should be equal for each photon. Multiplying by area-time, we have

$$\text{momentum} = \text{energy}/v \quad \text{or} \quad h/\lambda = hf/v \quad \text{or} \quad v = \lambda f \qquad (10.35)$$

which is the basic relationship for traveling waves, thereby illustrating the equivalence of the wave and particle descriptions.

EXERCISES

Extensions of the Theory

10.1. The core of an optical fiber has $n = 2$, and the cladding has $n = 2 - \Delta n$, where $\Delta n/n \ll 1$. If the internal wave is incident to the cladding at an angle just greater than the critical angle for internal reflection, derive an expression for the penetration distance of the evanescent field. Evaluate your expression if $\Delta n/n = 0.05$.

10.2. If $\mathscr{P}'$ and $\mathscr{P}''$ represent the intensities of the orthogonal wave components corresponding to E' and E'', respectively, then the proportion of polarization of a wave is defined as $PP = (\mathscr{P}' - \mathscr{P}'')/(\mathscr{P}' + \mathscr{P}'')$. Thus, a wave that has $E' = E''$ has $PP = 0$. Such a wave is incident from air onto a plane glass surface ($n = 1.6$) at an angle $\theta_i = 50°$.
(a) Find PP of the reflected wave.
(b) Describe the polarization of the reflected wave.

*10.3. When a photon is emitted by an atom, not only does it carry off the energy $E = hf$, but it carries away the momentum $p = h/\lambda$. Conservation of momentum dictates that the atom must recoil with a velocity $v = p/M$, where M is the atomic mass.

The Doppler effect describes, among other features, the fact that a source moving with velocity v emits radiation with wavelength λ', as compared to the same stationary source that emits radiation with wavelength λ_0, where $(\lambda' - \lambda_0)/\lambda_0 = v/v_0$.

Estimate the resultant relative line width $\Delta f/f$ of monochromatic radiation from an atom. This is called the *intrinsic line width*.

Problems

10.4. With reference to the discussion in Exercise 10.3, find the relative line width $\Delta f/f$ of monochromatic radiation from an atom, assuming that the thermal energy of the atom, $3kT/2$, is all kinetic energy.

*10.5. For the plane wave given by Eq. (10.5a), prove that $\nabla \cdot \mathbf{E} = \mathbf{k} \cdot \mathbf{E}$ and that $\nabla \times \mathbf{E} = \mathbf{k} \times \mathbf{E}$.

10.6. For a plane wave in air, with frequency ω and moving parallel to the line $x = y$, write vector expressions for the $\mathbf{E}$ and $\mathbf{H}$ fields.

10.7. An optical signal is undergoing internal reflection. If $n = n_1/n_2$ show that the phase change upon reflection, ζ, is

For E parallel to the incidence plane: $\tan \dfrac{1}{2}\zeta' = n\dfrac{(n^2 \sin^2 \theta_i - 1)^{1/2}}{\cos \theta_i}$

For E normal to the incidence plane: $\tan \dfrac{1}{2}\zeta'' = \dfrac{1}{n}\dfrac{(n^2 \sin^2 \theta_i - 1)^{1/2}}{\cos \theta_i}$

*10.8 For $n = 1.5$, find θ_i from air for which $\rho' = 1/\sqrt{2}$. Also find θ_i for which $\rho'' = 1/\sqrt{2}$.

*10.9. The standing-wave ratio (SWR) is the ratio of E_{max}/E_{min} in a region where a reflected and incident wave interfere. Determine the SWR if a wave with frequency f is normally incident from air onto a surface with $n = 1.25$.

10.10. In the optical limit, show that

$$\rho'_E = \frac{\tan(\theta_i - \theta_t)}{\tan(\theta_i + \theta_t)}$$

and

$$\tau'_E = \frac{2 \sin \theta_t \cos \theta_i}{\cos(\theta_i - \theta_t) \sin(\theta_i + \theta_t)}$$

10.11. In the optical limit, show that

$$\rho''_E = -\frac{\sin(\theta_i - \theta_t)}{\sin(\theta_i + \theta_t)}$$

and

$$\tau''_E = \frac{2 \sin \theta_t \cos \theta_i}{\sin(\theta_i + \theta_t)}$$

10.12. The reciprocity principle states that optical processes, such as reflection, are reversible. With reference to Fig. 10.4, this means that if a wave is incident on the interface from medium 2 at the incident angle θ_t, its transmitted wave is refracted through the angle θ_i, so that it leaves along the formerly incident wave. Furthermore, its reflected wave has $\rho_E = -\rho'_E$. Prove these statements.

10.13. To the lowest order in κ, find the difference of the reflection coefficients for E' and E'' for waves that are incident at the angle $\theta_i = \kappa$, where κ is an extremely small angle (i.e., the waves are nearly normal to the surface).

10.14. In Chapter 9, the wave impedance of a good conductor is given as $\eta = \eta_0(\omega\varepsilon/\sigma)^{1/2}e^{j\pi/4}$, where $\sigma/\omega\varepsilon \gg 1$. To the lowest order in η, write expressions for ρ'_E and ρ''_E.

10.15. Equations (10.28) give ρ'_E and τ'_E. Derive the corresponding equations for ρ''_H and τ''_H.

10.16. Verify Eq. (10.26b) for the transmission coefficient at the Brewster angle.

10.17. Conservation of energy demands that the power-reflection coefficient R and transmission coefficient T add to unity. Explain why this must be so and prove that it is.

*10.18. Find η and n of the medium for which $R = T = \frac{1}{2}$ at normal incidence from air.

10.19. In view of the result of Exercise 10.16, show that energy is conserved at the Brewster angle.

*10.20. Figure 10.9 shows a hypothetical point omnidirectional antenna (i.e., which emits microwave radiation uniformly in all directions) on an aircraft flying over the ocean ($n = 4/3$) at a height h. If the total radiated power is P, compare the intensities of the reflected signals at 30° and at 45° from the normal to the water surface. *Note:* The intensity incident within the cone of width $d\theta$ at the angle θ about h is $dP = (P/4\pi r^2) \cdot 2\pi r \sin \theta \cdot r \, d\theta = P \sin \theta \, d\theta/2$.

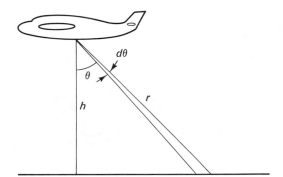

Figure 10.9 For Exercise 10.20.

10.21. One source of the tail of a comet is radiation pressure on its gaseous components. Using the gravitational expression of Exercise 3.16 and the data in
Exercise 14.29, find the ratio of r/m (radius/mass) of a particle on the comet
such that the radiation force just exceeds the gravitational force of the sun.
Assume the radiation is absorbed by the particle and that the comet is at a
distance equal to the earth's orbit. Is your answer consistent with the proposed mechanism? The mass of the sun is 2×10^{30} kg.

11

TRANSMISSION LINES

11.1 FIELD AND CIRCUIT PARAMETERS

Chapter 9 dealt with electromagnetic waves in free space. For purposes of signal delivery and control, it is generally advantageous to conduct the electromagnetic energy along a guiding transmission line. Various kinds of guides have been developed, for example, whose main elements are metals or dielectrics, and whose structures are in the form of rods, fibers, or films. In this chapter and the next, the discussion will be concerned with guiding waves along and within metallic conductors; in a later chapter, we will consider propagation that is at least partially controlled by dielectrics. Even with the restriction of having metallic transmission lines, there are many types of systems in use, several of which are shown in Fig. 11.1. A typical cylindrical coaxial transmisson line is shown in Fig. 11.1(a); this is the workhorse of transmission systems and we will consider it in some detail. Figures 11.1(b) to 11.1(e) are open structures and suffer from allowing radiation at high frequencies. However, their planar geometries lend themselves to easy fabrication, so they are quite popular. Figure 11.1(f) is the common two-wire, or *Lecher,* line, and Fig. 11.1(g) shows a closed rectangular waveguide that will be the topic of the next chapter. With the exception of this last case, all of these systems have separate insulated conductors, so that a voltage difference can be defined between the parts and a distinct current can be measured in each. Such two-conductor systems are our present subject of study.

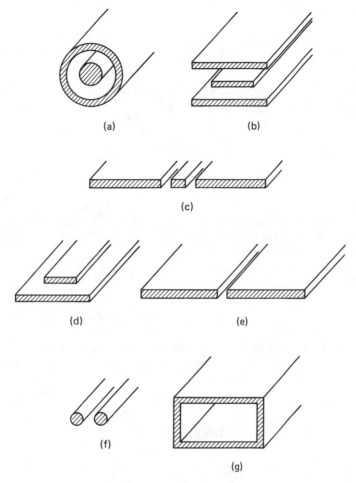

Figure 11.1 Transmission lines.

To illustrate the relationships between field and circuit parameters in describing these transmission lines, we consider, in detail, the coaxial transmission line, which is shown in Fig. 11.2. In many ways, its behavior is representative of the others, and its geometry makes for simple analysis. The figure shows current I flowing out of the center conductor; not shown is the equal current that must be flowing inward on the outer coaxial conductor. If I varies with time and position, then, since $I = dq/dt$, there must be an accompanying charge distribution along the line. The current generates a magnetic field, and charge on the inner and outer conductors are accompanied by an electric field. The charge-carrying coaxial line was discussed in Example 3.3 and the current-carrying line was discussed in Example 5.6, for the cases where the current and charge are constant. In the sense of the quasistationary approximation discussed in Chapter 5, we assume that the result of those examples still apply in each cross section of the present time-varying case.

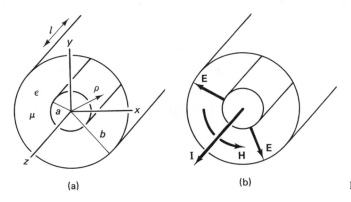

(a)   (b)

Figure 11.2 Coaxial transmission line.

Example 11.1

Figure 11.2(a) shows the line with some of its parameters defined. From the analysis of Examples 3.3 and 5.6, the electric- and magnetic-field intensities in the region between the inner and outer conductors are

$$\mathbf{E} = \mathbf{a}_\rho \frac{\rho_\ell}{2\pi\varepsilon\rho} \tag{11.1}$$

$$\mathbf{H} = \mathbf{a}_\phi \frac{I}{2\pi\rho} \tag{11.2}$$

where ρ_ℓ is the charge per unit length (along z) on each of the conductor surfaces, and I is the current in each. Also, from Example 3.3, we have an expression for the voltage between inner and outer conductors:

$$V = \frac{\rho_\ell}{2\pi\varepsilon} \ln \frac{b}{a} \tag{11.3}$$

Eliminating ρ_ℓ between Eqs. (11.1) and (11.3), we have

$$\mathbf{E} = \mathbf{a}_\rho \frac{V}{\rho \ln \dfrac{b}{a}} \tag{11.4}$$

If I and ρ_ℓ are functions of time and position along the line, then, from Eqs. (11.2) to (11.4), so are V, $\mathbf{E}$, and $\mathbf{H}$. From the field configuration shown in Fig. 11.2(b), it is seen that the Poynting vector, $\mathscr{P} = \mathbf{E} \times \mathbf{H}$, is along $\mathbf{a}_z$, so that power propagates along the line in the form of an electromagnetic wave. $\mathbf{E}$ and $\mathbf{H}$ should therefore carry factors of $e^{j(\omega t - \beta z)}$ and it follows from Eqs. (11.2) and (11.4) that current and voltage waves also propagate along the line. In fact, while there is a geometric dependence of $\mathbf{E}$ and $\mathbf{H}$ on ρ, there is no distinction between the two sets of variables, (V, I) or $(\mathbf{E}, \mathbf{H})$, with respect to z or t dependence.

Example 11.1 (cont.)

This correspondence of (V, I) to $(\mathbf{E}, \mathbf{H})$ is demonstrated by using Eqs. (11.2) and (11.4) in the definition of a quantity Z_0, which is called the characteristic impedance of the transmission line. In terms of the voltage and current traveling waves,

$$Z_0 = \frac{V}{I} = \frac{\ln(b/a)}{2\pi} \frac{|\mathbf{E}|}{|\mathbf{H}|} \tag{11.5}$$

Equations (11.1) and (11.2) show that $|\mathbf{E}|$ and $|\mathbf{H}|$ are both proportional to $1/\rho$ in the transmission line. This is a dependence on location in the plane perpendicular to the direction of wave motion. While it does not alter the axial propagation behavior, it does point out the important difference between transmission-line guided waves and the plane waves studied in the previous chapter: the fields are no longer uniform in the transverse plane. The lines shown in Figs. 11.1(a) to 11.1(f) present even more complex field variations in the transverse plane, making an exact field analysis impossible to carry out. However, since these lines consist of separated metallic conductors, a transverse voltage and a longitudinal current can be defined and an analysis can be carried out using the scalar quantities V and I as general propagation variables, rather than $\mathbf{E}$ and $\mathbf{H}$. Note that the waves are still transverse plane waves in the somewhat limited sense that the phase changes only in the z direction, that is, phase is constant on each xy plane, and the fields lie in the xy plane.

11.2 HIGH-FREQUENCY CIRCUITS

To use the circuit variables V and I in studying transmission lines, it is necessary to know the system circuit parameters. Again, the coaxial transmission line is an illustrative example.

Example 11.2

It was shown in Chapter 5 that the coaxial transmission line of Example 11.1 has an inductance per unit length

$$L = \frac{\mu_0}{2\pi} \ln \frac{b}{a} \tag{11.6a}$$

and a capacitance per unit length

$$C = \frac{2\pi\varepsilon}{\ln \dfrac{b}{a}} \tag{11.6b}$$

Currents flowing in the conductors of a transmission system also experience a series resistance per unit length, R, which we need not calculate at present. Furthermore, particularly at high frequencies, there are losses in the dielectric filling the line because the atomic and molecular electron bonds of most dielectrics have characteristic absorptions, that is, resonant frequencies at which they absorb incident energy and either reradiate it in all directions or convert it to heat. In addition, there can be some transverse current between the conductors, for example, due to ionization of and conduction through the dielectric in the intense electric field of lines that have high voltages or heavy power (in extreme cases, this can lead to dielectric breakdown). These dielectric losses and transverse currents are equivalent to a transverse conductance G per unit length. With these parameters, a length of line can be represented by the equivalent circuit shown in Fig. 11.3(a). Figure 11.3(b) shows the same circuit in a simplified form. With an impressed sinusoidal signal of angular frequency ω, in this figure,

$$Z = R + j\omega L \qquad Y = G + j\omega C \tag{11.7}$$

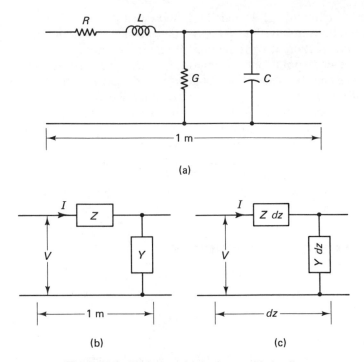

(a)

(b)   (c)

Figure 11.3   Equivalent circuit of a transmission line.

Before using this circuit, it is worthwhile to examine the circuit variables to see if the low-frequency circuit concepts are still applicable at radio frequencies, where the lines are used. Signal wavelengths are given by $\lambda = v_0/f = 3 \times 10^8/f$ meters. At the power frequency of 60 Hz, at the audio frequency of 10^4 Hz, and at the microwave frequency of 10^{10} Hz, this gives

$$\lambda = 3000 \text{ miles} \qquad f = 60 \text{ Hz}$$

$$\lambda = 30 \text{ km} \qquad f = 10^4 \text{ Hz}$$

$$\lambda = 3 \text{ cm} \qquad f = 10^{10} \text{ Hz}$$

The voltage and current reverse their signs over a distance of $\lambda/2$, so that at 10^{10} Hz, a circuit component that is 2 cm long will have voltages of $\pm V$ at different internal points. This is markedly different from the low-frequency cases, where points separated by city blocks can be considered to have the same voltage at any instant. Because of this, an analysis that is to apply to high frequency cannot assume localized, or lumped, parameters, that is, inductors and capacitors that are small compared to the spacial variation of the fields. To assure voltage and current uniformity in the circuit elements of any analysis, a very small element of the transmission line must be used, such as the infinitesimal length shown in Fig. 11.3(c).

11.3 CIRCUIT ANALYSIS

For the infinitesimal circuit of Fig. 11.3(c), ordinary circuit laws give the voltage drop across the series impedance and the leakage current through the parallel admittance. These are

$$dV = -I(Z\,dz) \qquad dI = -V(Y\,dz) \tag{11.8}$$

or equivalently,

$$\frac{dV}{dz} = -IZ \qquad \frac{dI}{dz} = -VY \tag{11.9}$$

The negative signs in Eqs. (11.8) and (11.9) denote the fact that V and I decrease as one moves along the line. V and I in these equations can be separated by taking the derivative of one equation with respect to z, and substituting the other. Thus, for a uniform line (Y and Z are both constant),

$$\frac{d^2V}{dz^2} = YZV \tag{11.10}$$

and an identical equation holds for I. As in the last chapter, a solution to Eq. (11.10) is of the form

$$V = V_0 e^{\gamma z} e^{j\omega t} \tag{11.11}$$

where the factor $e^{j\omega t}$ explicitly gives the sinusoidal time dependence assumed earlier in connection with Eq. (11.7). The factor $e^{\gamma z}$ is a trial solution that can describe traveling and attenuated waves. Substituting Eq. (11.11) in Eq. (11.10) yields

$$\gamma = \pm\sqrt{YZ} \tag{11.12}$$

Before proceeding further, we should note an important practical consideration that affects transmission-line analysis. Currents through the circuit components R and G cause voltage drop, heat loss, and energy dissipation, and result in attenuation and distortion of the propagating voltage and current waves. In practice, a line that has large losses is inefficient; it would be cast aside and not used for signal transmission. As a result, the model of a practical line should assume that the loss-producing elements are very small compared to the reactive components. From Eq. (11.7), this condition can be written as

$$Z = j\omega L\left(1 - j\frac{R}{\omega L}\right) \qquad Y = j\omega C\left(1 - j\frac{G}{\omega C}\right) \tag{11.13}$$

where

$$\frac{R}{\omega L} \ll 1 \qquad \frac{G}{\omega C} \ll 1 \tag{11.14}$$

In this low-loss case, Eqs. (11.13) are substituted in Eq. (11.12) and use is made of the expansion

$$(1 + x)^n = 1 + nx + \frac{1}{2}n(n-1)x^2 + \frac{1}{3!}n(n-1)(n-2)x^3 + \cdots +$$

Retaining only the lowest-order terms in $R/\omega L$ and $G/\omega C$, γ takes the form

$$\gamma = (j^2\omega^2 LC)^{1/2}\left(1 - j\frac{R}{\omega L}\right)^{1/2}\left(1 - j\frac{G}{\omega C}\right)^{1/2}$$

$$= \pm j\omega\sqrt{LC}\left[1 - j\frac{1}{2}\left(\frac{R}{\omega L} + \frac{G}{\omega C}\right) + \cdots\right] \tag{11.15a}$$

or

$$\gamma = \pm(j\beta + \alpha) \qquad \beta = \omega\sqrt{LC} \qquad \alpha = \frac{1}{2}\beta\left(\frac{R}{\omega L} + \frac{G}{\omega C}\right)$$

$$\tag{11.15b}$$

Only the first-order (x^1) terms in the small quantities $x = R/\omega L$ and $x = G/\omega C$ are retained in Eqs. (11.15), since the second- and higher-order terms $(x^2, x^3, \text{etc.})$ are even smaller and can generally be ignored. For example, if $x = 0.05$ and $n = 1/2$, then we retain the term with x, which is 5 percent of the unity term; by ignoring the higher-order terms, we are accepting an error of $\frac{1}{2}n(n - 1)x^2 \approx 0.03$ percent.

Substituting the first of Eqs. (11.15b) in Eq. (11.11) gives the voltage wave

$$V = V_1 e^{-\alpha z}e^{j(\omega t - \beta z)} + V_2 e^{\alpha z}e^{j(\omega t + \beta z)} \tag{11.16}$$

Two solutions are included in Eq. (11.16), corresponding to the plus-or-minus sign in Eq. (11.15b). Since Eq. (11.10) is a second-order differential equation, it has two independent solutions of amplitudes V_1 and V_2 instead of the single amplitude V_0 of Eq. (11.11). The first of these (the V_1 solution) is a wave traveling to the right, with an attenuation constant α. Although we recognize this direction of travel from our previous experience with plane waves, it can also be established by examination of the condition for constant phase, that is, $\Phi = \omega t - \beta z = $ constant. Differentiate this condition with respect to time:

$$\frac{d\Phi}{dt} = \omega - \beta\frac{dz}{dt} = \omega - \beta v = 0$$

or

$$v = +\frac{\omega}{\beta} \tag{11.17a}$$

This is the relation found for a wave traveling to the right, that is, the constant-phase point has positive velocity. The relationship $\beta = 2\pi/\lambda$ still applies.

The second solution in Eq. (11.16) represents a wave traveling to the left, that is, toward decreasing z. The constant-phase point, $\Phi = \omega t + \beta z = $ constant, changes according to

$$\frac{d\Phi}{dt} = \omega + \beta\frac{dz}{dt} = \omega + \beta v = 0$$

or

$$v = -\frac{\omega}{\beta} \tag{11.17b}$$

The negative value of v gives the direction of motion of the constant-phase point. The negatively traveling wave has a positive attenuation factor, indicating that the amplitude decreases as z decreases (in the direction of wave motion). The full solution of the wave equation results in both positively and negatively traveling waves.

Note from Eq. (11.15b) that α is directly proportional to the loss components R and G; because these are small, so is the attenuation. To facilitate understanding the main features of transmission-line behavior, we will further simplify the analysis by moving to the limit of zero losses. In the following discussions, we take $R = 0 = G$, so that $\alpha = 0$. From Eq. (11.15b),

$$\gamma = \pm j\beta = \pm j\omega\sqrt{LC} \qquad \text{(zero loss)} \qquad (11.18)$$

and, from Eq. (11.16),

$$V = V_1 e^{j(\omega t - \beta z)} + V_2 e^{j(\omega t + \beta z)} \qquad (11.19)$$

11.4 BASIC PARAMETERS

An expression for I can be derived by the same steps as those that led to Eq. (11.19), but to find the relationship between V and I, we derive I from V, using Eqs. (11.9), rather than finding it directly. Substituting Eqs. (11.19) and (11.18) in Eq. (11.9) yields

$$I = \frac{V_1}{Z_0} e^{j(\omega t - \beta z)} - \frac{V_2}{Z_0} e^{j(\omega t + \beta z)} \qquad (11.20a)$$

$$Z_0 = \sqrt{\frac{L}{C}} \qquad (11.20b)$$

giving the positively and negatively directed current waves that accompany the corresponding voltage waves. For *each* of these waves taken separately, the ratio $V/I = Z_0$ is called the *characteristic impedance* of the line.

Example 11.3
Substituting Eqs. (11.6) in Eq. (11.20b) for the coaxial transmission line gives

$$Z_0 = \sqrt{\frac{L}{C}} = \frac{1}{2\pi} \sqrt{\frac{\mu}{\varepsilon}} \ln \frac{b}{a} \qquad (11.21a)$$

$$= \frac{60}{\sqrt{\varepsilon_r}} \ln \frac{b}{a} \qquad (11.21b)$$

Z_0 is a function of the properties of the line itself and not of the features of the waves.

Notice from Eq. (11.20a) that the total current is the *difference* of the two currents, whereas in Eq. (11.19), the total voltage is the *sum* of the voltages. This is understandable from Fig. 11.4, where the two potential differences are applied between the conductors of the line; the total is their phasor sum. However, current is the motion of charged particles and it is physically proper that two oppositely directed currents give their difference as the net flow.

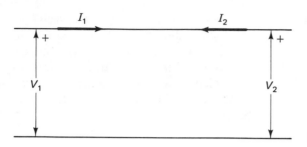

Figure 11.4 Current convention.

Example 11.3 (cont.)

Recall the previous relationship between the circuit and field variables in a coaxial line:

$$Z_0 = \frac{V}{I} = \frac{\ln(b/a)}{2\pi} \frac{|\mathbf{E}|}{|\mathbf{H}|} \tag{11.5}$$

Taken together with Eq. (11.21a), for Z_0, the fields in a coaxial line satisfy

$$\frac{|\mathbf{E}|}{|\mathbf{H}|} = \sqrt{\frac{\mu}{\varepsilon}} \tag{11.22a}$$

which is the same as for plane waves. This correspondence is carried further by calculating the phase velocity of the coaxial transmission wave from Eqs. (11.18) and (11.6):

$$v = \frac{\omega}{\beta} = \frac{1}{\sqrt{LC}} = \frac{1}{\sqrt{\mu\varepsilon}} = v_0 \tag{11.22b}$$

This is the velocity of electromagnetic waves in free space.

The correspondence of the properties of a transmission-line wave and a free-space plane wave is more than a coincidence. It arises from the fact that the analysis of transmission-line theory, *for any type of line,* assumes that the conductors support a difference of potential V in the transverse plane, and that the current I is entirely longitudinal. As a result, $\mathbf{E}$ and $\mathbf{H}$ are entirely transverse to the direction of propagation, a feature of the plane wave. While transmission-line theory assumes that the fields constitute a transverse electric and magnetic (TEM) wave, we will see later, that this is not exactly so.

11.5 IMPEDANCE

In keeping with conventions that are used throughout the engineering communications industry, we make two modifications of the results obtained so far.

1. Figure 11.5 contains a schematic transmission line along the z axis, with positive z measured to the right; this is the coordinate system used up to now. Note that it has not yet been necessary to specify the origin of z, since only differences are important in determining phase changes of the wave. It is purely as a matter of convenience whether the origin of z is located at the generator, at the opposite end of the line, or at a point outside of the line. Also shown in the figure is the variable length ℓ, *with origin at the load* that terminates the line, and with its positive direction

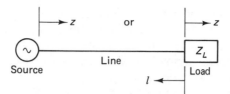

Figure 11.5 Coordinate convention on
transmission lines.

measured to the left. It is customary to use ℓ in discussing transmission-lines properties; the equations used thus far can be properly changed simply by substituting $-\ell$
for z.

   2. The circuit of Fig. 11.5 has an oscillator at its left end, generating a wave
traveling to the right. Since Eqs. (11.19) and (11.20) also have left traveling waves,
and since there is no other signal source on the line, it must be assumed that the reverse wave is produced by reflection from the load. The voltage-reflection coefficient is defined in terms of the waves of Eq. (11.9):

$$\Gamma_L = \frac{V_2}{V_1} \tag{11.23}$$

where the subscript L emphasizes that we are considering the reflection to take place
at the load.

   Using these changes, the expressions for V and for I become

$$V = V_1(e^{j(\omega t + \beta \ell)} + \Gamma_L e^{j(\omega t - \beta \ell)}) \tag{11.24a}$$

$$I = \frac{V_1}{Z_0}(e^{j(\omega t + \beta \ell)} - \Gamma_L e^{j(\omega t - \beta \ell)}) \tag{11.24b}$$

At the load end of the line, $\ell = 0$, so that

$$V_L = V_1(e^{j\omega t} + \Gamma_L e^{j\omega t}) \tag{11.25a}$$

$$I_L = \frac{V_1}{Z_0}(e^{j\omega t} - \Gamma_L e^{j\omega t}) \tag{11.25b}$$

Ohm's impedance law, for the particular impedance placed at the end of the line, is
then

$$Z_L = \frac{V_L}{I_L} = Z_0 \frac{1 + \Gamma_L}{1 - \Gamma_L} \tag{11.26a}$$

This equation gives a relation between the load impedance on a line and its ability to
reflect an incident wave. It can be solved explicitly for the reflection coefficient:

$$\Gamma_L = \frac{Z_L - Z_0}{Z_L + Z_0} = \frac{\eta_L - 1}{\eta_L + 1} \tag{11.26b}$$

where

$$\eta_L = \frac{Z_L}{Z_0} \tag{11.26c}$$

showing that the reflection coefficient depends only on η_L, the ratio of the load impedance to the characteristic impedance of a traveling wave on the line. η_L is called the *normalized* (or *reduced*) *load impedance*. Since Z_L, and therefore η_L, can be a complex number, so also can be Γ_L, indicating that V_2 can differ in phase as well as in magnitude from V_1.

Note that when $Z_L = Z_0$, or $\eta_L = Z_L/Z_0 = 1$, Eq. (11.26b) gives $\Gamma_L = 0$, meaning that there is no reflection from the load; any wave incident upon such a load is completely absorbed. In this case, the line impedance is said to be *matched* by the load.

Z_0 is the impedance (ratio of V/I) for the separate positively or negatively traveling waves, and Z_L is the impedance of the load (also equal to the ratio of V/I at the point $\ell = 0$). We now inquire as to what impedance will be found at other points of the line in the simultaneous presence of *both* traveling waves. At some point, a distance ℓ from the load, the magnitudes and phases of V and of I are measured and their phasor values are divided. From Eqs. (11.24), after canceling the time exponentials, this gives

$$Z = \frac{V}{I} = Z_0 \frac{e^{j\beta\ell} + \Gamma_L e^{-j\beta\ell}}{e^{j\beta\ell} - \Gamma_L e^{-j\beta\ell}} \tag{11.27}$$

A more useful form is obtained by substituting Eq. (11.26b) for Γ_L in terms of η_L. After some manipulation, and using the relations

$$e^{\pm j\theta} = \cos\theta \pm j\sin\theta$$

Eq. (11.27) becomes

$$\eta(\ell) = \frac{Z}{Z_0} = \frac{\eta_L \cos\beta\ell + j\sin\beta\ell}{j\eta_L \sin\beta\ell + \cos\beta\ell} = \frac{\eta_L + j\tan\beta\ell}{1 + j\eta_L \tan\beta\ell} \tag{11.28}$$

From Eq. (11.28), the impedance is seen to vary from point to point on the line, that is, the ratio $Z = V/I$ is not a simple circuit property, as at low frequencies, but varies with location in addition to depending on the load and the wavelength. In general, $Z(\ell)$ does not equal Z_L. In addition, we note that the tangent function has a periodicity of π radians, so that the impedance repeats along the line whenever the angle $\beta\ell$ changes by π. Since $\beta\ell = 2\pi\ell/\lambda$, this occurs when $\Delta\ell = \lambda/2$, that is, every half wavelength.

Example 11.4

Calculate Γ_L and draw $\eta(\ell)$ if $Z_L = 0$ (a short circuit) and if $Z_L = \infty$ (an open circuit).

For $Z_L = 0$, Eq. (11.26c) gives $\eta_L = 0$ and Eq. (11.26b) yields $\Gamma_L = -1$. Then, from Eq. (11.28), we find that $\eta(\ell) = j\tan\beta\ell$. The magnitude and sign of Γ_L indicate that the voltage reflected wave is equal in magnitude and 180° out of phase with the incident wave at the short circuit. We see that $\eta(\ell)$ is purely imaginary, so that it represents a pure reactance at every point on the line. The dependence of $\eta(\ell)$ on position is shown in Fig. 11.6, which is a tangent curve drawn from the right-hand end to indicate that the load is at that point and ℓ is increasing toward the left. The features described are apparent. Particularly, note that the value of the effective impedance at $\ell = \lambda/4$ is infinite, implying that the short-circuit termination is indistinguishable from an open-circuit termination placed a quarter wavelength away. This is also seen by substituting $Z_L = \infty = \eta_L$ in Eqs. (11.26b) and (11.28), yielding $\Gamma_L = 1$ and $\eta = -j/\tan\beta\ell$.

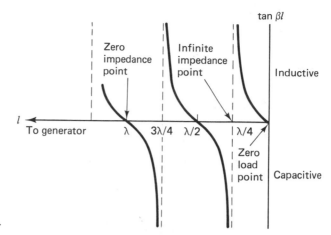

Figure 11.6   Impedance on a shorted line.

One useful result of this example is that an adjustable shorted line can be used to produce a reactive impedance of any magnitude or sign.

The important case where the load matches the line, that is, when $Z_L = Z_0$ so that $\eta_L = 1$, has already been treated. In this case, $\Gamma_L = 0$, which is a most desirable situation from the point of view of power delivery, for example, from a generator to an antenna acting as load. Since $\Gamma_L = 0$ means that there is no reflected wave, all power is absorbed by the load antenna, which then radiates the electromagnetic energy. A matched line is often referred to as being equivalent to an infinite line, since an infinite line has no termination at which to reflect the outgoing wave, so that $\Gamma_L = 0$.

Example 11.5

The normalized impedance at a point on a transmission line is $\eta = W$. Show that its value at a point $\lambda/4$ away is equal to $1/W$.

Take $\eta_L = W$ at $\ell = 0$. Then at $\ell = \lambda/4$, the phase is $\beta\ell = \pi/2$. As ℓ approaches the point $\ell = \lambda/4$, $\tan \beta\ell$ in Eq. (11.28) becomes very large and

$$\eta\left(\ell \to \frac{\lambda}{4}\right) = \frac{j \tan \beta\ell}{j\eta_L \tan \beta\ell} = \frac{1}{\eta_L} = \frac{1}{W}$$

11.6 STANDING WAVES

From Eq. (11.26b), it is seen that for a general complex load, the reflection coefficient is also complex, and can therefore be written

$$\Gamma_L = |\Gamma_L|e^{j\phi} \tag{11.29}$$

Substituting this in Eq. (11.24a) yields

$$V = [1 + |\Gamma_L|e^{j(\phi-2\beta\ell)}]V_1 e^{j(\omega t+\beta\ell)} \tag{11.30a}$$

$$V = \mathcal{V}V_{\text{incident}} \tag{11.30b}$$

where V_{incident} is the right-directed traveling wave (i.e., traveling toward $\ell = 0$), and $\mathscr{V}$ is the voltage at the point ℓ of the transmission line, relative to the incident wave.

$$V_{\text{incident}} = V_1 e^{j(\omega t + \beta \ell)} \qquad \mathscr{V} = 1 + |\Gamma_L| e^{j(\phi - 2\beta \ell)} \qquad (11.30c)$$

In Eq. (11.30a), the voltage configuration on the line can be regarded as the incident wave multiplied by $\mathscr{V}$, which is its complex amplitude. We met a situation like this in the case of an attenuated plane wave moving through a conductor. There, the amplitude variation was visualized by drawing the exponential attenuation term $|\mathbf{E}|/E_0$ (see Fig. 9.9). In this similar situation, the amplitude $\mathscr{V} = V/V_{\text{incident}}$, the total amplitude relative to that of the incident wave, is complex, meaning that it contains phase information. For visualization, it must be drawn in the complex plane.

Figure 11.7(a) shows $\mathscr{V}$, from Eq. (11.30b), at the load ($\ell = 0$). In moving away from the load, the value of $\mathscr{V}$ changes, as shown in Fig. 11.7(b). Note, from Eq. (11.30c), that the change is entirely in the phase angle. The modulus $|\Gamma_L|$ has its origin at the point $(1,0)$ and is constant in magnitude. Figure 11.7(c), therefore, shows that, while moving along the transmission line, the point on the figure traces

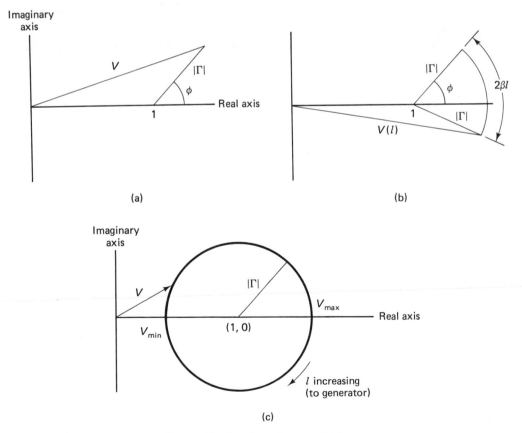

Figure 11.7   Complex-voltage amplitude.

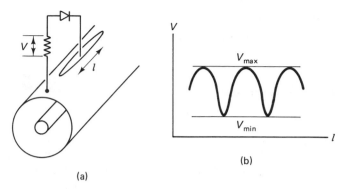

(a)

(b)

Figure 11.8   Measurement of the voltage standing wave.

out a circle of radius $|\Gamma_L|$. At each point (i.e., at each value of ℓ), the phasor from the origin of coordinates to the circle represents $\mathcal{V}$.

The change of $\mathcal{V}$ can be observed experimentally by using a slotted line. This is a section of line with a thin longitudinal slot in the outer shield into which is inserted a probe [see Fig. 11.8(a)]. The probe is thin and the insertion is kept small so the field pattern will not be disturbed. It acts as a minute detecting antenna and the voltage induced on it can be rectified and displayed as the probe moves along the slot. The result is a signal proportional to the magnitude of V, that is, to $|\mathcal{V}|$, while moving around the circle in Fig. 11.7(c). This envelope is shown in Fig. 11.8(b). It represents a voltage wave that is standing on the line, resulting from interference between the incident and reflected waves. This standing wave is fixed in position, with amplitude and location dependent on Z_L and, therefore, on Γ_L. Of course, a similar current standing-wave pattern also exists.

Note that the standing-wave pattern is not a simple sinusoid. This can be seen from Fig. 11.7(c). For a constant change of ℓ, and therefore of the angle of $\mathcal{V}$, the change of $|\mathcal{V}|$ is much smaller at the far side of the transmission-line circle, at $\mathcal{V}_{max}$, than at the near side, at $\mathcal{V}_{min}$. The voltage standing-wave ratio (VSWR) is defined from Fig. 11.8(b) as V_{max}/V_{min}. Using Fig. 11.7(c), this is equal to

$$\text{VSWR} = \frac{V_{max}}{V_{min}} = \frac{1 + |\Gamma_L|}{1 - |\Gamma_L|} \tag{11.31}$$

Example 11.6

A purely resistive load is placed at the end of a 100-ohm line and the measured VSWR on the line is 3. What is the load?

From Eq. (11.31), $|\Gamma_L| = (\text{VSWR} - 1)/(\text{VSWR} + 1) = \frac{1}{2}$. Equation (11.26a), for η_L, requires knowledge of the complex Γ_L, whereas we have only its magnitude. However, this difficulty can be resolved because Eq. (11.26c) indicates that for real Z_L and real Z_0 (we always assume real Z_0; here it is 100 ohms), we have a real η_L. Equation (11.26b) then results in a real value for Γ_L. Taken with the result that $|\Gamma_L| = \frac{1}{2}$, we must have either $\Gamma_L = +\frac{1}{2}$ or $\Gamma_L = -\frac{1}{2}$. By substituting these in Eq. (11.26b), the load is found to be either $\eta_L = 3$ or $\eta_L = \frac{1}{3}$, giving $R_L = Z_0\eta_L = 300$ ohms or $R_L = 33.3$ ohms. Note that the values of η_L are reciprocals.

Before proceeding, note one more aspect of the reflection that results in standing waves. From Eqs. (11.30), a position-dependent reflection coefficient $\Gamma(\ell)$ can be defined:

$$\Gamma(\ell) = \Gamma_L e^{-j2\beta\ell} = |\Gamma_L| e^{j(\phi - 2\beta\ell)} \qquad (11.32)$$

so that Eqs. (11.30) become

$$V = V_{\text{incident}}[1 + \Gamma(\ell)] = V_{\text{incident}} \mathcal{V} \qquad (11.33)$$

Equation (11.33) states that the standing-wave pattern can be generated by reflection at any point on the line, by a load that replaces the remainder of the line and that produces the same effective reflection coefficient $\Gamma(\ell)$ at that point. Equation (11.32) shows that $\Gamma(\ell)$ repeats whenever $2\beta\ell = 2\pi$, or when $\ell = \lambda/2$, that is, every half wavelength. This is consistent with our earlier result that the impedance recurs when ℓ changes by $\lambda/2$. In fact, the relationship between the local impedance and the local reflection coefficient can be found from Eq. (11.27). Factoring $e^{j\beta\ell}$ from the numerator and denominator of that equation gives

$$\eta(\ell) = \frac{1 + \Gamma(\ell)}{1 - \Gamma(\ell)} \qquad (11.34)$$

which is a generalization of Eq. (11.26a) for the case where $\ell \neq 0$. With this understanding, we can refer to Γ in our subsequent discussion rather than $\Gamma(\ell)$ or Γ_L, unless the more detailed notation is specifically needed.

11.7 SMITH CHART

We have seen that every point along a transmission line has a reflection coefficient and a reduced impedance. This means that every point on the circle of radius $|\Gamma|$, of Fig. 11.7(c), has a corresponding reduced impedance. Since $0 \leq |\Gamma| \leq 1$, and can assume any value in this range, depending on the value of the reduced load, there must be an equivalent load impedance value at every point inside the unit circle centered at the origin of Γ, as shown in Fig. 11.9. It will be useful to determine how η varies within this unit reflection-coefficient circle. To do this, take the cartesian form, which is shown in Fig. 11.9,

$$\Gamma = \Gamma_x + j\Gamma_y \qquad (11.35a)$$

along with

$$\eta = \frac{R}{Z_0} + j\frac{X}{Z_0} = \eta_r + j\eta_x \qquad (11.35b)$$

where the subscript r denotes the resistive part and the subscript x denotes the reactive part of η. Substitute these into Eq. (11.26a) for the equivalent load impedance and multiply the numerator and denominator of the resulting equation by the complex conjugate of the denominator, that is, the right-hand side of the equation is multiplied by $(1 - \Gamma^*)/(1 - \Gamma^*)$, where $\Gamma^* = \Gamma_x - j\Gamma_y$ is the complex conjugate of Γ. Cross multiplying and collecting terms then gives

$$(\eta_r + 1)\Gamma_x^2 - 2\eta_r\Gamma_x + (\eta_r + 1)\Gamma_y^2 + j[\eta_x\Gamma_x^2 - 2\eta_x\Gamma_x + \eta_x\Gamma_y^2 - 2\Gamma_y]$$
$$= 1 - \eta_r - j\eta_x$$

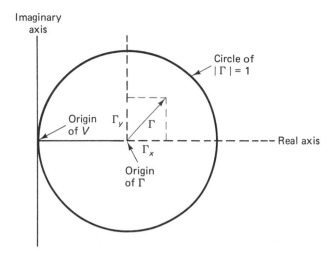

Figure 11.9 Reflection-coefficient representation.

Complex equations comprise two relations that are obtained by equating the real parts and the imaginary parts across the equality. The results are placed in a recognizable form by completing their squares:

$$\left(\Gamma_x - \frac{\eta_r}{1 + \eta_r}\right)^2 + \Gamma_y^2 = \left(\frac{1}{1 + \eta_r}\right)^2 \tag{11.36a}$$

$$(\Gamma_x - 1)^2 + \left(\Gamma_y - \frac{1}{\eta_x}\right)^2 = \left(\frac{1}{\eta_x}\right)^2 \tag{11.36b}$$

> Completing the square of an incomplete quadratic form is a common way of simplifying the algebraic representation. Given $y^2 + ay = b$, we add $a^2/4$ to both sides to have $(y + a/2)^2 = b + (a^2/4)$. This is particularly convenient for placing quadratic expressions in standard form so that they can be recognized as conic sections.

In the $\Gamma_x \Gamma_y$ coordinate system, Eqs. (11.36) give the families of circles shown in Fig. 11.10. For a given value of η_r, Eq. (11.36a) selects one of the circles of Fig. 11.10(a), and for a given value of η_x, Eq. (11.36b) selects one of the circles in Fig. 11.10(b). Figure 11.11 shows both sets of circles combined; every point on this figure represents a different value of $\eta = \eta_r + j\eta_x$. All values of $0 \le \eta_r \le \infty$ and $-\infty \le \eta_x \le +\infty$ are contained within the circle $\eta_r = 0$, $|\Gamma| = 1$. Only parts of the circles of Eq. (11.36b) are shown in Figs. 11.10(b) and 11.11, because only the region within the circle $\eta_r = 0$ is physically meaningful.

Figure 11.11 is called the *Smith chart*. On this chart, the transmission line under study is represented by a circle of constant $|\Gamma|$ centered at the origin; the origin of Γ is at the center of the chart, at the intersection of the circle $\eta_r = 1$ and the horizontal line $\eta_x = 0$. In addition to the curves of constant η_r and constant η_x, several auxiliary scales are shown. Around the perimeter are three circular scales. The innermost is the angle of Γ and the outer two show distance toward and away from the generator, in wavelengths, as described in relation to Fig. 11.7(c). Several linear scales are generally found below the Smith chart on commercially available forms. Among these is a scale for $|\Gamma|$ (VOL REFL COEF) and one for VSWR (VOL RATIO STANDING WAVE). It is easy to see how these scales are derived since $|\Gamma|$ is

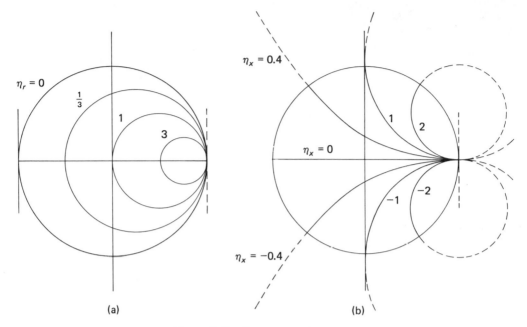

$\eta_r = 0$

$\tfrac{1}{3}$

1

3

$\eta_x = 0.4$

$\eta_x = 0$

1

2

−1

−2

$\eta_x = -0.4$

(a)

(b)

Figure 11.10   Constant η_r and η_x curves.

merely the radial distance from the origin, normalized to the outermost radius of the $\eta_r = 0$ circle, and VSWR is derived from $|\Gamma|$ through Eq. (11.31). We will refer to these scales in the following examples.

It was pointed out earlier that the impedance repeats along a transmission line whenever the angle $\beta\ell$ changes by π. Since $\beta\ell = 2\pi\ell/\lambda$, this occurs when $\Delta\ell = \lambda/2$, that is, every half wavelength. From Fig. 11.7(b) and Eq. (11.30c), it is seen that the Smith chart angle is $2\beta\ell$, so that when $\Delta\ell = \lambda/2$, the chart angle changes by 2π, that is, a complete rotation about the chart back to the starting point. This is also confirmed from the outer circular scales, which show a maximum of 0.50λ about the whole chart.

Example 11.7

A transmission line, carrying a signal at $f = 1$ GHz, has a characteristic impedance of 70 ohms. The line is just 1/2 meter long and is terminated with a load of $105 + 35j$ ohms.

(a) Use the Smith chart to determine the reflection coefficient at the load, the input impedance seen by the generator, and the VSWR on the line.

(b) Find the power delivered to the load if the generator can be represented by a voltage source $V = 60 \cos 2\pi ft$ in series with an internal impedance $Z_{gen} = 60 + 38j$ ohms.

We begin with two, short, preliminary calculations:

(1) The wavelength is $\lambda = v/f = 0.3$ m, so that the length of line corresponds to $0.5/0.3 = 1.667\lambda$.

(2) The load has a value $\eta_L = (105 + 35j)/70 = 1.5 + 0.5j$.

Although the following steps are illustrated in Fig. 11.12, the reader should locate the values on the Smith chart.

Enter the chart at the point corresponding to η_L and draw a line from the origin ($\eta_r = 1$, $\eta_x = 0$) through this point and extended to the perimeter, through the circle scales

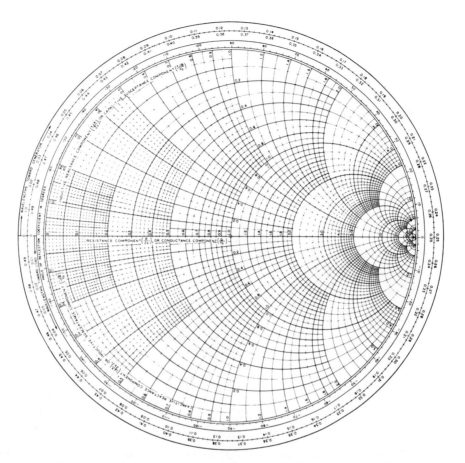

Figure 11.11 Smith chart. Reprinted with permission from *Electronics,* January 1944, pp. 130. Copyright 1944, VNU Business Publications, Inc.

surrounding the chart. Note that it intersects the innermost of these circle scales at the point ≈33.7; on the left side of the chart, this scale is indicated to give the ANGLE OF RE-FLECTION COEFFICIENT IN DEGREES. Use a ruler or a pair of compasses or dividers to measure the radial distance of the load point from the origin (ℓ in Fig. 11.12); transfer the distance to the "VOL REFL COEF" scale below the chart. Measured from the center, this distance corresponds to ≈0.28. From these, the reflection coefficient at the load is $\Gamma_L \approx 0.28e^{j33.7°}$

Transferring the radial distance to the scale marked VOL. RATIO STANDING WAVE, the VSWR value is ≈1.75.

Mark off the circle of constant $|\Gamma|$ (centered at the origin and through the load point) and move along it from the load point *toward the generator* by 1.667λ. Each λ/2 distance represents complete revolution around the circle back to the initial point. Therefore, three such traversals result in the same starting point (actually a different point that has the identical impedance), which must be moved farther through 1.667 − 1.5 = 0.167λ. A radial line through the load intersects the outermost scale (marked, at the left side, WAVE-LENGTHS TOWARD GENERATOR) at the point ≈0.203. Add to this 0.167 to arrive at ≈0.370. Locate this point on the outer circle and draw a line to it from the origin. The

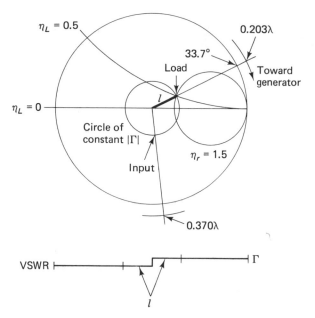

Figure 11.12 Smith-chart solution of Example 11.7.

intersection of this line with the transmission-line circle (constant $|\Gamma|$ circle through η_L) is at the point $\eta_{\text{input}} \approx 0.89 - 0.54j$. Multiplying by $Z_0 = 70$ ohms gives $Z_{\text{input}} \approx 62 - 38j$ ohms as the impedance seen by the generator.

    With the generator input impedance, there is a voltage divider consisting of $Z_{\text{input}} \approx 62 - 38j$ ohms and $Z_{\text{gen}} = 60 + 38j$ ohms, both in series with V_{gen}. The total impedance is 122 ohms. (The input impedance is nearly the complex conjugate of the generator impedance, a condition that leads to optimum power transfer.) Then the power generated by the source is $P_{\text{gen}} = V_{\text{rms}}^2/R = (60/\sqrt{2})^2/122 = 14.75$ W. The power delivered into the line, which must end up in the load since there is no other lossy element in the problem, is $P_L = (62/122) \cdot 14.75 = 7.5$ W.

The previous discussion has used the approximately equal to ($\approx$) sign rather than the equal ($=$) sign in stating values taken from the Smith chart. This indicates that all the values obtained are subject to some error due to inaccuracies in locating points on, and reading values from, the chart. In using graphical methods, we trade precision for speed and ease of use.

    Before proceeding with the next example, recall an important relationship between two points that are separated by $\lambda/4$ on a transmission line. In Example 11.5, it was shown that such points have reduced impedances that are reciprocals, that is, $\eta(\ell + \lambda/4) = 1/\eta(\ell)$. Since separation by $\lambda/4$ corresponds to a half rotation on the Smith chart, points that are on opposite ends of a diameter of the *circuit circle* have reciprocal impedances. Stated alternatively, the reduced impedance η of one is the reduced admittance $\zeta = Y/Y_0 = (1/Z)/(1/Z_0) = Z_0/Z$ of the other. Therefore, in using the Smith chart, it is simple to change between impedances and admittances by moving across a diameter of the circuit circle.

    This also points up that the η_r and η_x curves on the Smith chart can also be regarded as ζ_c and ζ_s curves, where $\zeta = \zeta_c + j\zeta_s$, c indicating the conductance and s the susceptance. In our derivation of transmission-line theory and the Smith chart, we have emphasized the voltage-reflection coefficient and impedances; an equivalent derivation using the current-reflection coefficient would have involved admittances and led to the same Smith chart with the normalized admittance ζ in place of the normalized impedance η.

Example 11.8

A 50-ohm transmission line is terminated with a load $Z_L = 20 + 10j$ ohms. Using the Smith chart, find lengths ℓ_1 and ℓ_2 in Fig. 11.13(a), so that the line appears to be matched from the generator end.

The schematic diagram of Fig. 11.13(a) can be clarified by examination of Fig. 11.13(b), showing the "Tee" at point J, which creates parallel paths through the two lines, and which illustrates how the lengths are changed. Length ℓ_1 contains a sliding section that changes the length of the line without disturbing the load or input to the section. ℓ_1 is a *trombone* section; it is one type of device called a *line stretcher,* which serves to change the phase of a transmitted signal. Such adjustable sections inevitably produce some reflections even though they are intended to present a minimal impedance mismatch. For our analysis, we will ignore this effect. The line ℓ_2 contains a sliding short that contacts the inner and outer line elements through spring contacts and that can be locked in place when properly adjusted.

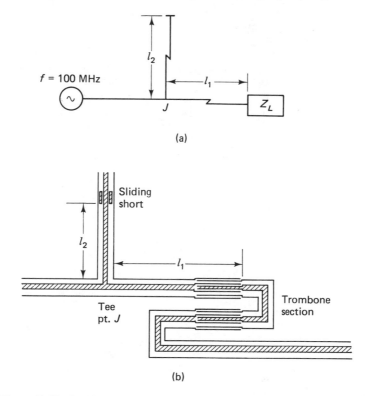

Figure 11.13   Double adjustable-length tuner. (a) Schematic diagram. (b) Structural diagram.

Example 11.8 (cont.)

Before proceeding, it is appropriate to indicate the methodology of matching the line with this double adjustable-length tuner. We wish to adjust ℓ_1 so that the normalized admittance presented by the load, at junction point J, has the form $\zeta_1 = 1 + jQ$, that is, the real part of the admittance is made equal to $\zeta_0 = 1/\eta_0$. We will then adjust ℓ_2 to have an admittance, *from the shorted line* at J, of $\zeta_2 = -jQ$. As these two admittances are in parallel, their sum is $\zeta = \zeta_1 + \zeta_2 = 1$, so that η (from the generator) $= 1/\zeta = 1$, that is, the line is matched. This is possible since it was shown in Example 11.4 that the impedance, and therefore also the admittance, of the shorted line ℓ_2 is purely imaginary, that is, a reactance or a susceptance.

The reduced load is $\eta_L = 0.4 + 0.2j$ and we enter the chart at this point (A') in Fig. 11.14(a). The radius from the origin to A' is Γ and we could read its magnitude and phase from the chart.

The circle of radius $|\Gamma|$ represents points along the transmission line ℓ_1, and if we move across the diameter of the circle from A' to point A, the value read from the chart is the reduced admittance of the load, ζ_L, as previously discussed. From here on, we regard the chart as an admittance chart. We can adjust ℓ_1 so that ζ of A, transferred to point J, is given by the value at either of points B or C. These points are chosen by taking the intersections of the $|\Gamma|$ circle and the $\zeta_0 = 1$ circle (which is also the $\eta = 1$ circle). Thus, ζ(point C) $= 1 + j\zeta_s(C)$. Similarly, ζ(point B) $= 1 + j\zeta_s(B)$. The length of transmission line, read from the perimeter scale of the chart, is ℓ_1(to reach admittance point B) $= 0.05\lambda$ and ℓ_1(to reach admittance point C) $= 0.375\lambda$. Directly from the chart, we find $\zeta(B) = 1 - j$, and $\zeta(C) = 1 + j$.

We must now determine ℓ_2 so that it has $\zeta = \pm j$. Since the short has $Z = 0$, or $Y = \infty$, we can again regard the Smith chart as an admittance diagram and enter it at point D in Fig. 11.14(b) ($\zeta = \infty$). We move toward the generator to point E, where $\zeta = -j$, or to point F, where $\zeta = +j$. From the perimeter scale, $\ell_2(E) = 0.375 - 0.25 = 0.125\lambda$, and $\ell_2(F) = 0.25 + 0.125 = 0.375\lambda$.

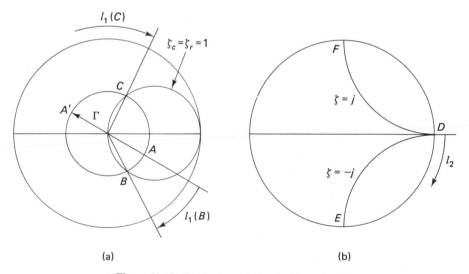

Figure 11.14 Smith-chart solution for Example 11.8.

Finally, for $f = 10^8$ Hz, we have $\lambda = v/f = 3$ m, so that our solutions are

Points B, F $\ell_1 = 0.05\lambda = 0.15$ m, $\ell_2 = 0.375\lambda = 1.125$ m

Points C, E: $\ell_1 = 0.375\lambda = 1.125$ m, $\ell_2 = 0.125\lambda = 0.375$ m

Of course, any of these lengths can be increased by multiples of $\lambda/2 = 1.5$ m, since each such increase takes us once around the chart and back to the same point.

Before leaving this example, it is interesting to note the VSWR on each line section from the scale below the chart. Along line ℓ_1, the VSWR = 2.62; along ℓ_2, the VSWR is infinite; and along the generator line, VSWR = 1. These indicate that there are considerable reflected waves on lines ℓ_1 and ℓ_2, but their phases are such that they cancel at J. No waves return to the generator, and, since we have ignored the possibility of losses along the transmission line, all the energy leaving the generator must find its way to the load.

11.8 ATTENUATION

Initially, in our study of transmission lines, we established the equivalent circuit of Fig. 11.3(a). We then reasoned that the losses in a practical line should be so small that we could neglect the loss components. These are R, the series resistance of the conducting material, and G, the leakage across and energy absorbed by the dielectric medium filling the cable. In fact, while these are small, their effect may not be completely negligible, particularly if long lengths of transmission line are used. In this section, we will investigate the resistive loss arising from R; the effect of the dielectric will not be included, since, at least in air-filled lines, dielectric losses are generally small compared to resistive losses.

The primary effect of the loss is to draw energy from the traveling waves. This is usually expressed through the attenuation constant α, given by Eq. (11.15b), so that the voltage and current of a wave traveling to the right can be written

$$V = V_0 e^{-\alpha z} e^{j(\omega t - \beta z)} \tag{11.37a}$$

$$I = I_0 e^{-\alpha z} e^{j(\omega t - \beta z)} \tag{11.37b}$$

indicating that they decrease exponentially as the wave advances. The average power is

$$P = \tfrac{1}{2} VI^* = \tfrac{1}{2} V_0 I_0 e^{-2\alpha z} = P_0 e^{-2\alpha z} \tag{11.38}$$

From this, the attenuation coefficient can be obtained as

$$\alpha = -\frac{1}{2P} \frac{dP}{dz} \tag{11.39}$$

The power loss per unit length, $-dP/dz$, can be written in terms of $R\,dz$, the series resistance of an infinitesimal length dz. From Fig. 11.3,

$$-dP = I^2 R\,dz \quad \text{or} \quad -\frac{dP}{dz} = I^2 R \tag{11.40}$$

where the negative sign indicates that P decreases as z increases. At each point, the traveling wave has

$$P = VI = Z_0 I^2 \qquad (11.41)$$

so that Eqs. (11.39) through (11.41) can be combined to give

$$\alpha = \frac{R}{2Z_0} \qquad (11.42)$$

This expression could also have been found by combining Eqs. (11.15b) and (11.20b).

Example 11.9

Calculate R and α for a 70-ohm transmission line that is constructed of copper and with a dielectric having $\varepsilon_r = 3$. The inner conductor has a radius of 0.5 cm. The line is excited at $f = 10^9$ Hz.

For this line, the expression for Z_0 gives $b = 7.5a$. Since copper has $\sigma = 5 \times 10^7$ S/m, the skin depth is $\delta = 0.0712/\sqrt{f}$ meters (see Example 9.6).

Figure 11.15 shows a coaxial transmission line, schematically showing the current penetration depths δ. The expression for the resistance per unit length is found in Exercise 11.1 to be

$$R = \frac{1}{2\pi a \sigma \delta} + \frac{1}{2\pi b \sigma \delta} \qquad (11.43)$$

from which $R = 1.01 \times 10^{-5} \sqrt{f} = 0.321$ Ω/m. Equation (11.42) then gives $\alpha = 2.29 \times 10^{-3}$ m^{-1}. Therefore, according to Eq. (11.38), the power decays to 13.5 percent (i.e., to a factor of $1/e^2$) in a distance of $x = 1/\alpha = 437$ m.

Equation (11.38) can provide another expression for α:

$$\alpha = -\frac{1}{2z} \ln \frac{P}{P_0} \text{ nepers/meter} \qquad (11.44)$$

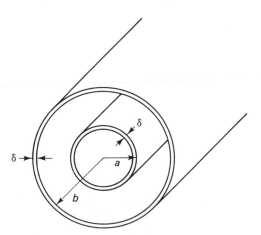

Figure 11.15 Skin depth in a coaxial transmission line.

The neper is actually a unitless quantity denoting that a natural logarithm (i.e., to the base e) is being taken. It is named after John Napier, the Scottish mathematician who first described logarithms in 1614. Note that α is an amplitude attenuation factor. In engineering practice, it is more common to use a power attenuation factor and to be concerned with powers of 10, as described by Henry Briggs in 1617. Thus,

$$\frac{P}{P_0} = 10^{-\alpha'z} \qquad (11.45a)$$

A measure of relative power levels is the *bel*, defined by

$$\text{number of bels} = \log \frac{P}{P_0} \qquad (11.45b)$$

If the power decreases then $P < P_0$ and the logarithm is negative; P is then said to be that number of bels "below" P_0. From Eq. (11.45a) one has

$$\alpha' = -\frac{1}{z} \log \frac{P}{P_0} \text{ bels/meter} \qquad (11.45c)$$

The more standard form is given in tenths of a bel/meter, that is, decibels (dB) per meter, and therefore is 10 times larger. Relating P/P_0 from Eqs. (11.38) and (11.45b), and taking the logarithm (to base 10) of both sides gives

$$\alpha' \text{ (bels)} = 2\alpha \log_{10} e \qquad (11.46a)$$

$$\alpha' \text{ (dB/m)} = 8.686\alpha \text{ (Np/m)} \qquad (11.46b)$$

Signal attenuation results in a loss of signal strength, requiring additional energy input at costly relay stations of a communication system. In addition, from Example 11.9, α increases as $\sqrt{f}$, so that the various frequency components are absorbed differentially and, when the signal components are not uniformly reduced, signal distortion results (e.g., pulse-shape change). For example, consider a square pulse in terms of its Fourier components. The higher-frequency terms define the sharp corners of the pulse, and if these are lost in the course of a transmission that has $\alpha \sim \sqrt{f}$, the pulse becomes rounded. This, together with the overall attenuation, can make it difficult to determine whether the pulse is present or, at least, just when it begins and ends.

At high frequencies, where α is large, losses limit the use of transmission lines. Because the major loss occurs at the inner conductor (88 percent in Example 11.9), this conductor can be removed, thereby producing a hollow system that can still propagate a signal and that has lower attenuation. However, without two separate conductors to define voltage and current, the propagation mode is different and must be analyzed in terms of field theory rather than circuit theory. This introduces the consideration of hollow waveguide structures, which will be covered in the next chapter.

There is a more subtle implication of the existence of attenuation in any real transmission line. Current flowing longitudinally in resistive conducting lines has, according to Ohm's law, $\mathbf{J} = \sigma\mathbf{E}$, an associated parallel electric field. This field lies along the direction of wave propagation, so that the propagation is not entirely in a transverse electric and magnetic field (TEM) mode. As will be shown in the next chapter in a more extreme case, non-TEM modes display dispersion (wavelength and velocity vary with frequency) and have frequency domains where propagation cannot occur. Analysis of signal transmission under such conditions is much more complicated. Fortunately, in most commonly used transmission lines, the metal conductivity σ is quite high, so that the longitudinal field is small, that is, $E_{\text{long.}} \ll E_{\text{trans.}}$, where $E_{\text{trans.}}$ is the transverse TEM field we have been considering up to now. As a result, the departure from a true TEM mode is generally quite small. The actual mode is therefore referred to as being quasi-TEM, and the TEM analysis and properties are assumed, generally without significant error.

EXERCISES

Extensions of the Theory

11.1. The resistance of a conductor is given by $R = \ell/\sigma A$, where ℓ is its length, A is the cross-sectional area through which current flows, and σ is the material conductivity. At high frequencies, the effective cross-sectional area of a conducting line is modified due to the skin-depth phenomenon.

It was shown in Exercise 9.26 that we could consider the current to have its surface value uniformly through a depth δ and to be zero elsewhere. Use this fact to derive the high-frequency resistance of a unit length of a coaxial transmission line [Eq. (11.43) of the text].

11.2. The load determines an apparent impedance at every point of a line.
   (a) Regardless of the load, show that there are points on the line where the impedance has a pure real value R.
   (b) How are the impedances related at two such neighboring points?
   (c) Prove that the VSWR on the line is equal to the value of R at one of these points.

11.3. The reflection and transmission of plane waves such as light and radio waves can be analyzed using transmission-line theory if the transverse wave impedance (E/H) of the wave in the medium is used in place of the characteristic transmission-line impedance. Using this concept, analyze the following in terms of adjoining transmission lines.
   (a) A uniform plane wave traveling in free space is incident normal to the surface of a lossless slab of polyethylene ($\sigma = 0$, $\varepsilon = 2.25\varepsilon_0$, $\mu = \mu_0$) that is standing alone in infinite space ($Z_0 = 377$ ohms). The frequency of the incident wave is 300 MHz and the slab thickness is 45 cm. Determine the E-SWR in front of the slab. How far in front of the slab face is the first E-field maximum located?
   (b) A polished copper block is pressed against the back of the slab. How much does the first E-field maximum move? *Hint:* The copper acts as a short circuit.

11.4. In keeping with the idea of Exercise 11.3, the coating of optical lenses with quarter-wave dielectric layers, as a means to eliminate reflections, can be understood in terms of transmission-line theory if the coating is taken as a $\lambda/4$ length of line between two different and very long lines.

A thick lens with dielectric constant ε_2 is coated with a $\lambda/4$ thickness of dielectric with permittivity ε_1. This intermediate layer (between air and lens) acts as an impedance-matching transformer. Show that $\varepsilon = \sqrt{\varepsilon_0 \varepsilon_2}$ if there is no refection from the lens. For all dielectrics $\mu = \mu_0$.

Problems

*11.5.   Two antennas, each with resistive input impedances of 100 ohms, are fed with three equal lengths of identical $\lambda/2$ transmission line, as shown in Fig. 11.16. Determine the input impedance seen by the source and the phasor voltage at the terminals of one of the antennas.

*11.6.   An air-filled 100-ohm transmission line of unknown length is terminated in an unknown impedance and is operated at a frequency of 1 MHz. The input impedance is measured as $10 - j50$ ohms. An assistant at the other end of the line is told to remove the load, leaving the line open-circuited. If the input impedance to the line is now $-j75$ ohms, determine the unknown impedance.

11.7.   In a slotted-line experiment, with the load replaced with a short circuit, the voltage minima are located at 15, 25, 35 cm, etc. from the load. With the load attached, the voltage minima are located at 8, 18, 28 cm, etc. and the VSWR is 8. If the slotted line is air-filled and has a characteristic resistance of 100 ohms, determine the frequency of operation and the value of the unknown impedance.

*11.8.   An air-filled transmission line has a characteristic impedance of 75 ohms and is terminated with a load R = 225 ohms. It is operated at 450 MHz. The line is 7 m long and has an *additional* input section that is 1 m long and that is identical with the rest of the line except that it is filled with a dielectric having $\varepsilon = 9\varepsilon_0$. What is the input impedance?

*11.9.   The characteristic impedance of an air–dielectric transmission line is 50 ohms. It is conducting a signal at 100 MHz. The line, load, and an adjustable stub tuner shorted at its end are arranged as shown in Fig. 11.17.
(a) Find the VSWR between the stub tuner and the load.
(b) Find the VSWR on the stub-tuner line.

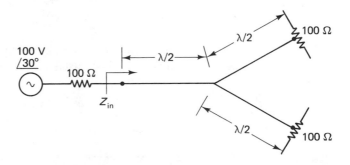

Figure 11.16 For Exercise 11.5.

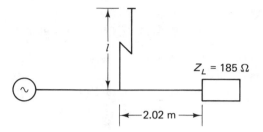

Figure 11.17   For Exercise 11.9.

(c) Find the length of the stub ℓ if the generator line is to be matched.

(d) Find the VSWR between the generator and the stub tuner under the condition of part (c).

*11.10. A transmission line with characteristic impedance Z_0 is terminated with a complex load $Z_L = \sqrt{2}Z_0 e^{j45°}$. At what distances from the load (in wavelengths) is the impedance a pure resistance?

11.11 Determine the terminal impedance on an air–dielectric transmission line that has a VSWR of 2.5, with the $V_{\min}$ 8.75 cm from the load. The line has $Z_0 = 50$ ohms and is operated at 800 MHz.

11.12. A load Z_L terminates a 70-ohm transmission line. At a distance $3\lambda/16$ away, the voltage standing wave has its minimum value and the impedance is 35 ohms. Determine the VSWR on the line and the value of Z_L.

*11.13. Replacing the unknown load of a 100-ohm transmission line with an open circuit yields an input impedance to the line of $-j80$ ohms. With Z_L attached, the input impedance is $30 + j40$ ohms. Determine the value of Z_L.

11.14. A lossless air-filled slotted line with a 100-ohm characteristic impedance has a load that produces a standing wave ratio of 4. The load is replaced with a short-circuit termination and the voltage minimum shifts 2 cm closer to the load end of the line. If the frequency is 1 GHz, find the load impedance.

11.15. A transmission line to be used at 100 MHz, has a characteristic impedance of 120 ohms. It is terminated with a load of $-168j$ ohms. As a circuit designer, you wish to change the load by giving it an additional real part R, so that the VSWR on the line is minimum.

(a) Describe how you arrive at R, and give an appropriate value.

(b) Give the SWR values before and after the change.

(c) Determine the reflection coefficient of the modified load.

(d) If the generator is 1.5 m from the load, determine the input impedance it senses.

11.16. A coaxial transmission line with $Z_0 = 100$ ohms is shown in Fig. 11.18. The load is 25 ohms and ℓ is adjustable. At point P, a cut is made in the center

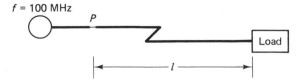

Figure 11.18   For Exercise 11.16.

conductor, resulting in a series capacitance C. Find the length ℓ and the value of C if the line is to appear matched to the line at point P.

11.17. When a load is placed on a transmission line, the VSWR is found to be 2.2 and the minima of the voltage standing wave are at 36, 56, and 76 cm from the load. The real part of the load is known to be 75 ohms.
  (a) Find the characteristic impedance of the line.
  (b) Find the reactance of the load.
  (c) Find the voltage-reflection coefficient at the load.

11.18. A load of $100 - j75$ ohms is placed at the termination of a 50-ohm transmission line.
  (a) What is the SWR on the line?
  (b) Find the locations on the line where the load is purely resistive. What are the resistance values (reduced and absolute)?
  (c) Ideally, the line appears to the generator as though it has a 50-ohm load. From the Smith chart, show where you would place a shorted stub across the line in order to cancel the reactive component of the load.

11.19. A transmission line with characteristic impedance of 75 ohms is terminated with an impedance $Z_L = 131.25 + 37.5j$ ohms, as shown in Fig. 11.19. Three meters from the load, a 75-ohm resistor is inserted in the line. Three meters farther from the load is a shorted stub that is 1.875 m long. Find the input impedance seen by a 20-MHz generator placed at a point 3 m farther from the load.

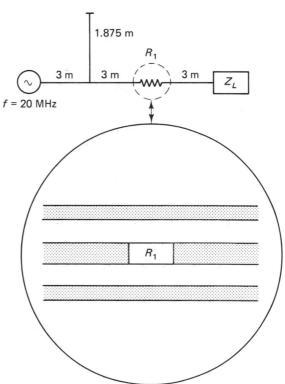

Figure 11.19   For Exercise 11.19.

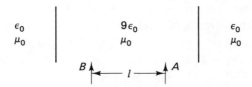

Figure 11.20 For Exercise 11.20.

11.20. In the multiple dielectric structure shown in Fig. 11.20, measurements at points A and B give $Z(A) = (0.4 - 0.4j)Z_0/3$ and $Z(B) = (1.1 + 1.2j)Z_0/3$, where $Z_0 = \sqrt{\mu_0/\varepsilon_0}$. The frequency of the signal is 1 GHz. What is the distance ℓ (see Exercise 11.3)?

11.21. Verify the statement in the text that 88 percent of the power loss in the transmission line of Example 11.9 occurs at the inner conductor.

12

WAVEGUIDES

12.1 NON-TEM PROPAGATION

Transmission systems of the type discussed in Chapter 11 have the great advantage of propagating with electric and magnetic fields in the plane transverse to their direction of propagation; they are referred to as propagating in a transverse electric and magnetic, or TEM, mode. In this, they closely resemble plane waves in an infinite medium, such as were studied in Chapter 9. This property is accompanied by a relative ease in defining the currents in and the potential difference between conducting components. However, at high frequencies, two-conductor systems encounter a number of problems, such as the existence of higher modes, in which the fields do not have the configuration that geometrically simple analysis would lead one to expect. Furthermore, at the end of Chapter 11, it was shown that the coaxial line of Fig. 9.1(a) exhibits large losses, and it was suggested that these can be reduced by removing the center conductor. This produces a single-conductor circular waveguide; the rectangular equivalent is shown in Fig. 12.1. Besides overcoming some of these difficulties, closed systems also have the advantage of allowing no radiation loss at high frequencies and no interference from objects intercepting their fields, as with the open transmission systems illustrated in Fig. 11.1. However, their analysis and use is more sophisticated. For example, a unique transverse voltage cannot be defined, since all parts of the guide cross section are connected by metallic conducting walls and, as will be seen, there are currents in the transverse as well as in

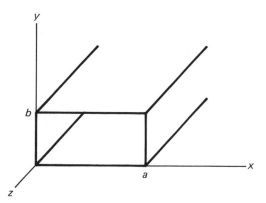

Figure 12.1 Rectangular waveguide.

the longitudinal direction. In addition, at the frequencies where waveguides are used, the wavelength of the wave and the physical size of the guide are comparable, so that there can be a critical dependence of waveguide properties on the physical structure.

For waves traveling along the z axis of the rectangular waveguide of Fig. 12.1, the electric and magnetic fields can be described by

$$\mathbf{E}(x, y, z, t) = \mathbf{E}(x, y)e^{j(\omega t - \beta z)} \tag{12.1a}$$

$$\mathbf{H}(x, y, z, t) = \mathbf{H}(x, y)e^{j(\omega t - \beta z)} \tag{12.1b}$$

A sinusoidal dependence on time and on z has been specified as appropriate for a traveling wave. Note that

$$\frac{\partial\{\,\cdot\,\}}{\partial t} = j\omega\{\,\cdot\,\} \qquad \frac{\partial\{\,\cdot\,\}}{\partial z} = -j\beta\{\,\cdot\,\} \tag{12.2}$$

Where $\{\,\cdot\,\}$ denotes either $\mathbf{E}$ or $\mathbf{H}$. By writing $\mathbf{E}(x, y)$ and $\mathbf{H}(x, y)$, the possibility is admitted that $\mathbf{E}$ and $\mathbf{H}$ vary in the transverse plane. While the phase of a transmission line wave is constant at points of a transverse plane, in, for example, the coaxial transmission line, the fields do vary with ρ [see Eqs. (11.1) and (11.2)]. Such a wave is a nonuniform plane wave; the transverse variations are even more complex in the other systems of Fig. 9.1.

The approach to studying waveguide behavior of closed systems is to describe the electromagnetic field in the dielectric within the hollow interior of the guide. As air is an insulating dielectric, $\mathbf{J} = 0$ and $\rho = 0$. With this and Eq. (12.2), the two Maxwell vector curl equations can be written as the following six component equations:

$$\frac{\partial H_z}{\partial y} + j\beta H_y = j\omega\varepsilon E_x \qquad\qquad \frac{\partial E_z}{\partial y} + j\beta E_y = -j\omega\mu H_x \qquad (12.3a,\ b)$$

$$-\frac{\partial H_z}{\partial x} - j\beta H_x = j\omega\varepsilon E_y \qquad\qquad -\frac{\partial E_z}{\partial x} - j\beta E_x = -j\omega\mu H_y \qquad (12.4a,\ b)$$

$$\frac{\partial H_y}{\partial x} - \frac{\partial H_x}{\partial y} = j\omega\varepsilon E_z \qquad\qquad \frac{\partial E_y}{\partial x} - \frac{\partial E_x}{\partial y} = -j\omega\mu H_z \qquad (12.5a,\ b)$$

Note that Eqs. (12.3a) and (12.4b) are linear in E_x and H_y, so that they can be solved explicitly, yielding, after some algebraic manipulation,

$$E_x = \frac{-j}{\beta_c^2} \left(\omega\mu \frac{\partial H_z}{\partial y} + \beta \frac{\partial E_z}{\partial x} \right) \tag{12.6a}$$

$$H_y = \frac{-j}{\beta_c^2} \left(\beta \frac{\partial H_z}{\partial y} + \omega\varepsilon \frac{\partial E_z}{\partial x} \right) \tag{12.6b}$$

In Eqs. (12.6), an important parameter has been defined.

$$\beta_c^2 = \beta_0^2 - \beta^2 \tag{12.7a}$$

where β_0 is the wave number of a plane wave with the same frequency in an infinite space with the medium filling the guide. From Eq. (9.14), it is given by

$$\beta_0^2 = \omega^2 \mu\varepsilon = \frac{\omega^2}{v_0^2} \tag{12.7b}$$

As the guide is assumed to be air-filled, the appropriate values in Eq. (12.7b) are $\mu = \mu_0$ and $\varepsilon = \varepsilon_0$. From Eqs. (12.1), we see that $\beta = 2\pi/\lambda$ is the wave number of the wave traveling *in the waveguide,* whereas, from Eq. (12.7b), $\beta_0 = 2\pi/\lambda_0$ is the *free-space* wave number of the same frequency. The implication of Eq. (12.7a) is that these wavelengths are not equal. The form of Eq. (12.7a) suggests that β_c can be interpreted as the wave number of a free-space plane wave with frequency.

$$\omega_c = \beta_c v_0 \tag{12.8a}$$

Equation (12.7a) is then

$$\beta^2 = \frac{\omega^2 - \omega_c^2}{v_0^2} \tag{12.8b}$$

Just as Eqs. (12.6) were obtained from Eqs. (12.3a) and (12.4b), so Eqs. (12.3b) and (12.4a) can be used to find

$$E_y = \frac{j}{\beta_c^2} \left(\omega\mu \frac{\partial H_z}{\partial x} - \beta \frac{\partial E_z}{\partial y} \right) \tag{12.9a}$$

$$H_x = \frac{-j}{\beta_c^2} \left(\beta \frac{\partial H_z}{\partial x} - \omega\varepsilon \frac{\partial E_z}{\partial y} \right) \tag{12.9b}$$

The results of Eqs. (12.6) and (12.9) are striking: if the longitudinal fields E_z and H_z are known, then the transverse fields can be determined as well. As a result, it is only necessary to find the two z-directed components to solve the entire problem. In addition, the form of Eqs. (12.6) and (12.9) indicates that the general field is the sum of the fields derived from H_z and from E_z separately. Further analysis can, therefore, be divided into two different problems:

transverse electric (TE) modes, where $E_z = 0$
transverse magnetic (TM) modes, where $H_z = 0$

These are discussed in the following sections.

12.2 FIELD SOLUTIONS

Equations to be satisfied by each of the longitudinal fields can be derived from Maxwell's equations in the same way as was done for the general wave equation. Thus, it was pointed out that Eq. (9.4), the wave equation, applies to **H** as well as to **E**, without the charge-density term. In the air dielectric of the waveguide, we take $\sigma = 0$, so that Eq. (9.4) becomes

$$\nabla^2 \mathbf{H} = -\beta_0^2 \mathbf{H} \tag{12.10a}$$

where, from Eq. (12.2), $\mu\varepsilon\, \partial^2\mathbf{H}/\partial t^2 = -\mu\varepsilon\omega^2\mathbf{H} = -\beta_0^2\mathbf{H}$. As this is a vector equation, it holds for each component of **H**. For H_z in particular,

$$\frac{\partial^2 H_z}{\partial x^2} + \frac{\partial^2 H_z}{\partial y^2} + \frac{\partial^2 H_z}{\partial z^2} = -\beta_0^2 H_z \tag{12.10b}$$

Using Eq. (12.2) gives

$$\frac{\partial^2 H_z}{\partial x^2} + \frac{\partial^2 H_z}{\partial y^2} = -(\beta_0^2 - \beta^2)H_z = -\beta_c^2 H_z \tag{12.10c}$$

This equation can be solved by using the method of the separation of variables, just as was done in Chapter 6 for LaPlace's equation. The difference from that case, imposed by the term on the right-hand side of Eqs. (12.10), makes this solution sufficiently different that it is worth repeating the analysis. As in that case, the trial solution assumes that the x and y dependencies of H_z can be separated into a product of independent functions, such that

$$H_z(x, y) = X(x)Y(y) \tag{12.11}$$

[Note that we have already met functions of this form in the traveling wave, where $H(z,t) = H_0 e^{j(\omega t - \beta z)} = H_0 e^{j\omega t} e^{-j\beta z} = H_0 T(t)Z(z)$.] In substituting Eq. (12.11) into Eq. (12.10c), it must be remembered that X is a constant with regard to y differentiation and Y is a constant with regard to x differentiation. Then

$$Y\frac{d^2 X}{dx^2} + X\frac{d^2 Y}{dy^2} = -\beta_c^2 XY$$

Dividing by XY gives

$$\frac{1}{X}\frac{d^2 X}{dx^2} + \frac{1}{Y}\frac{d^2 Y}{dy^2} = -\beta_c^2 \tag{12.12}$$

Here the first term is a function of x alone and is therefore independent of the second term, which is a function of only y. Since their sum is constant for all values of x and of y, each term separately must be constant. Therefore,

$$\frac{1}{X}\frac{d^2 X}{dx^2} = -\beta_x^2 \tag{12.13a}$$

$$\frac{1}{Y}\frac{d^2 Y}{dy^2} = -\beta_y^2 \tag{12.13b}$$

$$\beta_x^2 + \beta_y^2 = \beta_c^2 \tag{12.13c}$$

At this point, the choice of the separation constants $-\beta_x^2$ and $-\beta_y^2$ as negative squares implies nothing specific about their final form, since the β can be real, imaginary, or complex. While further conditions are necessary to evaluate these β, this particular form is chosen for convenience in a later expression.

As the equations for $X(x)$ and $Y(y)$ are identical in form, the solutions should also have the same form. However, to illustrate two approaches to arrive at these solutions, different, though equivalent, initial expressions will be used. Equations (12.13) are satisfied by

$$X = A \cos \beta_x X + B \sin \beta_x X \tag{12.14a}$$

and

$$Y = Ce^{j\beta_y y} + De^{-j\beta_y y} \tag{12.14b}$$

These field expressions must now be subject to the constraints imposed by the boundaries of the region; here the boundaries are the metal walls of the waveguide. While high-frequency fields penetrate conductor walls to an extent described by the skin-depth phenomenon, if ω is high or the material is a good conductor, this penetration is quite small and can be ignored. (In the limit of a *perfect* conductor or at *very* high frequencies, the skin depth $\delta \to 0$.) Therefore, the electric and magnetic fields can be considered to be zero within the conductor.

In Chapter 5, it was shown that the tangential component of the electric field intensity **E** and normal component of the magnetic flux density **B** do not change at an interface between two media. Since the fields are zero in the conductor, therefore, also in the air dielectric, at the metal walls,

$$B_{\text{normal}} = 0 \qquad E_{\text{tangential}} = 0 \tag{12.15a, b}$$

The normal and tangential directions are with respect to the surface of the waveguide.

12.3 TE MODES

Either condition of Eqs. (12.15), or both, can be applied to obtain the waveguide modes; we will use the electric-field condition of Eq. (12.15b). In terms of Fig. 12.1, this condition becomes

$$E_z(x = 0) = E_z(x = a) = E_z(y = 0) = E_z(y = b) = 0 \tag{12.16}$$

$$E_x(y = 0) = E_x(y = b) = 0 \tag{12.17}$$

$$E_y(x = 0) = E_y(x = a) = 0 \tag{12.18}$$

In considering TE modes, it is immediately clear that Eqs. (12.16) are satisfied since $E_z \equiv 0$. The solution is needed for H_z. The boundary conditions on the transverse electric fields, E_x and E_y, can be obtained from Eqs. (12.11) and (12.14) by applying Eqs. (12.6a) and (12.9a):

$$E_x = \frac{-j\omega\mu}{\beta_c^2} j\beta_y X(x) \left(Ce^{j\beta_y y} - De^{-j\beta_y y}\right) \tag{12.19a}$$

$$E_y = \frac{j\omega\mu}{\beta_c^2} \beta_x Y(y) \left(-A \sin \beta_x x + B \cos \beta_x x\right) \tag{12.19b}$$

The boundary conditions of Eq. (12.17) are independent of x, so they must be imposed by the function of y in Eq. (12.19a). $E_x(y = 0) = 0$ leads to $C - D = 0$, so that $C = D$ and

$$E_x = \frac{\omega\mu\beta_y}{\beta_c^2} X(x)C(e^{j\beta_y y} - e^{-j\beta_y y})$$

$$= \frac{\omega\mu\beta_y}{\beta_c^2} X(x)2jC \sin \beta_y y$$

From $E_x(y = b) = 0$, we have $C \sin \beta_y b = 0$. This could be satisfied by setting the yet undetermined constant $C = 0$, but then E_x would be identically zero. The more general case that E_x does not vanish requires instead that $\sin \beta_y b = 0$, which is true when

$$\beta_y = \frac{n\pi}{b} \qquad n = 0, 1, 2, 3, \ldots \tag{12.20}$$

Negative values of n need not be included because they give identical fields.

The boundary conditions of Eq. (12.18) are independent of y and so must be satisfied by the function of x in Eq. (12.19b). $E_y(x = 0) = 0$ leads to the conclusion that $B = 0$. $E_y(x = a) = 0$ results in having $A \sin \beta_x a = 0$. Reasoning as before, this implies

$$\beta_x = \frac{m\pi}{a} \qquad m = 0, 1, 2, 3, \ldots \tag{12.21}$$

Summarizing these results gives

$$H_z = X(x)Y(y) = AC \cos \beta_x x \, (e^{j\beta_y y} + e^{-j\beta_y y})$$

$$= H_0 \cos \beta_x x \cos \beta_y y \tag{12.22}$$

where $H_0 = 2AC$. Substituting this expression into Eqs. (12.6) and (12.9) yields

$$E_x = \frac{j\omega\mu\beta_y}{\beta_c^2} H_0 \cos \beta_x x \sin \beta_y y$$

$$H_y = \frac{j\beta\beta_y}{\beta_c^2} H_0 \cos \beta_x x \sin \beta_y y$$

$$\tag{12.23}$$

$$E_y = \frac{-j\omega\mu\beta_x}{\beta_c^2} H_0 \sin \beta_x x \cos \beta_y y$$

$$H_x = \frac{j\beta\beta_x}{\beta_c^2} H_0 \sin \beta_x x \cos \beta_y y$$

where

$$\beta^2 = \beta_0^2 - \beta_c^2 \tag{12.24a}$$

$$\beta_c^2 = \beta_x^2 + \beta_y^2 = \left(\frac{m\pi}{a}\right)^2 + \left(\frac{n\pi}{b}\right)^2 \tag{12.24b}$$

12.4 WAVEGUIDE MODE STRUCTURE

Note, from Eqs. (12.22) and (12.23), that the manner of field variation is determined by the choice of m and n, called the *mode numbers*. To examine the physical meaning of these results, consider a specific case, the $TE_{mn} = TE_{10}$ mode, which, as will be seen shortly, is the most important case. For $m = 1$ and $n = 0$,

$$\beta_y = 0 \qquad \beta_c = \beta_x = \frac{\pi}{a} \qquad \beta^2 = \beta_0^2 - \left(\frac{\pi}{a}\right)^2 \qquad (12.25)$$

$$H_z = H_0 \cos \frac{\pi}{a} x$$

$$E_x = 0 = H_y \qquad (12.26)$$

$$E_y = \frac{-j\omega\mu a}{\pi} H_0 \sin \frac{\pi}{a} x$$

$$H_x = \frac{j\beta a}{\pi} H_0 \sin \frac{\pi}{a} x$$

In this case, $\beta_y = 0$ makes $E_x = 0 = H_y$. The remaining field components have only x dependence in the transverse plane. Of course, each of these fields also carries a factor $e^{j(\omega t - \beta z)}$ to describe the traveling wave. By writing this term explicitly, the H-field components are

$$H_z = H_0 \cos \frac{\pi}{a} x \, e^{j(\omega t - \beta z)} \leftrightarrow H_0 \cos \frac{\pi}{a} x \cos(\omega t - \beta z) \qquad (12.27)$$

$$H_x = H_0 \frac{\beta a}{\pi} \sin \frac{\pi}{a} x \, e^{j(\omega t - \beta z + \pi/2)} \leftrightarrow H_0 \frac{\beta a}{\pi} \sin \frac{\pi}{a} x \cos\left[\omega t - \beta\left(z - \frac{\lambda}{4}\right)\right]$$

so that the $j = e^{j\pi/2}$ in the coefficient of H_x causes it to be 90° out of phase with H_z. At any instant of time, the maximum of H_z occurs at a distance $\lambda/4$, along z, from the maximum of H_x. The H-field distribution along the guide, at $t = 0$, is shown in Fig. 12.2(a). There we see that $\mathbf{H} = \mathbf{a}_x H_x$ at $z = \lambda/4, 3\lambda/4, \ldots$, and $\mathbf{H} = \mathbf{a}_z H_z$ at $z = 0, \lambda/2, \lambda, \ldots$. As this field distribution translates down the guide, the $\mathbf{H}$ field at a point appears to rotate in the xz plane. Equations (12.26) indicate that E_y is in phase with H_x (the minus sign merely indicates direction reversal); the $\mathbf{E}$ field shows a sinusoidal variation along the guide, but always along $\mathbf{a}_y$. The transverse-field distributions are shown in Fig. 12.2(b); the arrow lengths indicate the magnitudes of the fields in the xy plane. The central section of Fig. 12.2(b) shows their $\sin \pi x/a$ variation.

Example 12.1

Sketch the transverse field for the TE_{11} mode.

From Eqs. (12.22) and (12.23),

$$H_z = H_0 \cos \frac{\pi}{a} x \cos \frac{\pi}{b} y$$

$$E_x = \frac{j\omega\mu\pi}{b\beta_c^2} H_0 \cos \frac{\pi}{a} x \sin \frac{\pi}{b} y$$

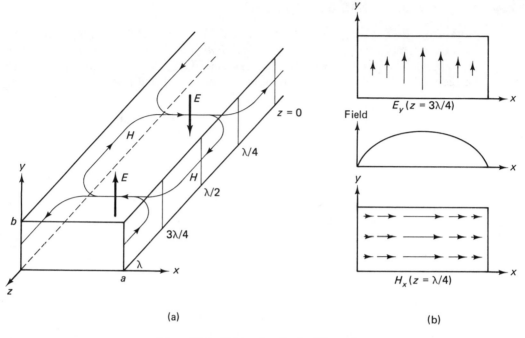

Figure 12.2 Field pattern for the TE$_{10}$ mode.

$$H_y = \frac{j\beta\pi}{b\beta_c^2} H_0 \cos\frac{\pi}{a}x \sin\frac{\pi}{b}y$$

$$E_y = \frac{-j\omega\mu\pi}{a\beta_c^2} H_0 \sin\frac{\pi}{a}x \cos\frac{\pi}{b}y$$

$$H_x = \frac{j\beta\pi}{a\beta_c^2} H_0 \sin\frac{\pi}{a}x \cos\frac{\pi}{b}y$$

Figure 12.3(a) shows schematic representations of $E_x(x, y)$, $E_y(x, y)$, and $\mathbf{E}_{\text{transverse}}(x, y)$. Figure 12.3(b) shows schematic representations of $H_x(x, y)$, $H_y(x, y)$, and $\mathbf{H}_{\text{transverse}}(x, y)$.

Example 12.2

Sketch the transverse fields of the TE$_{20}$ and the TE$_{21}$ modes.

From Eqs. (12.20) through (12.23), it is seen that m and n give the number of half cycles of field variation in the guide, along x and y, respectively. Therefore, the TE$_{20}$ mode has two half cycles of field variation in the x direction, that is, the TE$_{10}$ pattern is repeated twice along the x axis, with a sign reversal arising from the sinusoids in their second half cycle. This is shown in the left-hand side of Fig. 12.4.

The TE$_{21}$ mode repeats the TE$_{11}$ pattern along the x axis with a sign reversal, as shown in Fig. 12.4.

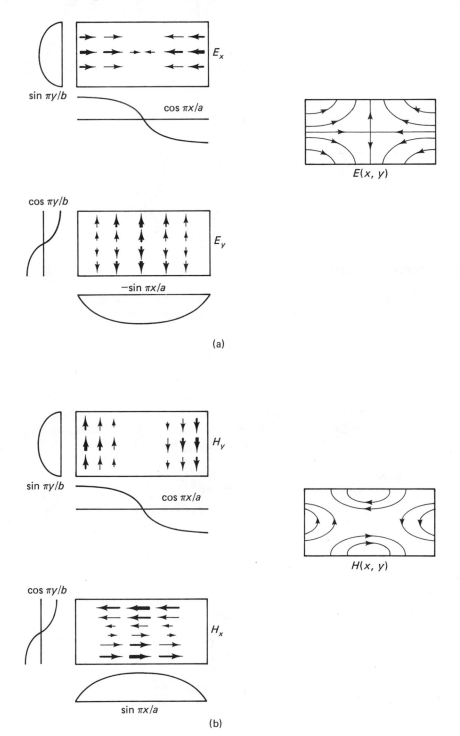

Figure 12.3 TE$_{11}$-mode field patterns.

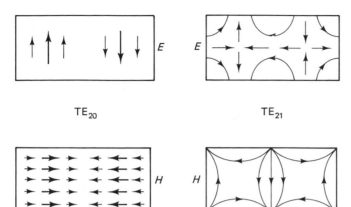

TE$_{20}$ TE$_{21}$

Figure 12.4 Mode field patterns.

12.5 MODE PURITY

Equation (12.24a) provides an expression for the wave number in the waveguide: $\beta^2 = \beta_0^2 - \beta_c^2$, where $\beta = 2\pi/\lambda$. For propagation, β must be real, so it is necessary that $\beta_0 > \beta_c$. From Eqs. (12.7b) and (12.8a), this means that $\omega > \omega_c$ and, with Eq. (12.24b),

$$\omega > \omega_c = v_0\beta_c = v_0\frac{\pi}{a}\left[m^2 + \left(n\frac{a}{b}\right)^2\right]^{1/2} \tag{12.28}$$

Therefore, depending upon the mode numbers m and n, propagation can only occur for frequencies greater than some minimum which is characteristic of that mode. That minimum value is called the *cutoff frequency*.

Example 12.3

To illustrate this, take the specific dimensions of a commonly used waveguide: $a = 0.9$ inch and $b = 0.4$ inch. Then, for propagation, Eq. (12.28) gives

$$f > f_c = 6.56 \times 10^9(m^2 + 5.06n^2)^{1/2}$$

Values of f_c for low-order m and n modes are listed in Table 12.1 and are shown in Fig. 12.5.

At a particular frequency, say 15.0 GHz, the possible propagating modes (for which $f > f_C$) in this waveguide are

10 ($f_c = 6.56$)   20 ($f_c = 13.12$)   01 ($f_c = 14.76$) f_c in GHz

TABLE 12.1.   MODE CUTOFF FREQUENCIES

Mode index m	0	1	0	1	2	2	0	1	3	3	4
Mode index n	0	0	1	1	0	1	2	2	0	1	0
$f_c/6.56 \times 10^9$	0	1	2.25	2.46	2	3.01	4.5	4.61	3	3.7	4

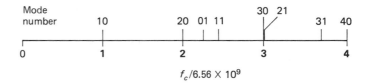

Figure 12.5 Cutoff frequencies.

In general, even if the signal-launching system in the waveguide is set to generate and propagate in only the 10 mode (the method of doing this will be discussed briefly in a later section), it is found that imperfections in the wall, bends in the guide, section connectors, etc. cause scattering into the other allowed modes, so that the field pattern is unstable and signal deterioration can result. However, note from the values in Table 12.1 that there can be only a single mode if this waveguide is operated between the frequencies

$$6.56 \times 10^9 < f < 2 \times 6.56 \times 10^9 = 13.12 \times 10^9$$

In this range, only the 10 mode can exist; the scattering mechanisms mentioned cannot generate propagating signals in the higher modes. To propagate at 15 GHz in the 10 mode only, it will be necessary to use a waveguide with different dimensions.

To assure that only the 10 mode is allowed, a series of waveguides of different physical dimensions is used. Each guide has its own frequency range for propagation. The advantages of using waveguides in this way are a stable and known field configuration, and minimal dispersion (variation of properties, such as phase velocity, with frequency). In addition, knowing the specific field configuration allows development of many useful waveguide components for controlling and manipulating transmitted signals.

The existence of a series of modes and the cutoff of propagation of these modes below their critical frequencies, and the fact that, by limiting the frequency range of use, only one mode can exist in a waveguide, all arise from wave interference within the limited dimensions of the guide. This is a precursor of the phenomenon of quantization found in quantum-wave mechanical systems, where the system dimensions and properties set the condition that only certain specific states (e.g., atomic-energy levels) can exist.

12.6. TM MODES

Now consider those modes for which $H_z = 0$ and $E_z \neq 0$. In these cases, Eqs. (12.10) must be satisfied for E_z instead of for H_z, and the general form of the solution must, therefore, also be that given by Eqs. (12.14). The field configuration can be determined by assuring the boundary condition that the tangent component of E vanishes at the walls. In this case, E_z itself is tangent to the walls, so the solution is particularly simple. The conditions of Eq. (12.16) are sufficient to establish that

$$E_z = E_0 \sin \beta_x x \sin \beta_y y \tag{12.29}$$

where, as before, the β_i are given by Eqs. (12.20) and (12.21), and are related by Eqs. (12.24). The transverse fields can be found from Eqs. (12.6) and (12.9), and, of course, they are different from the TE cases. However, it is clear that the distribution of *modes* is identical to that shown in Fig. 12.5, with the following exception. If $m = 0$ or $n = 0$, we have $\beta_x = 0$ or $\beta_y = 0$, and, in either case, it is seen from Eq. (12.29), that the sine function is zero so that $E_z \equiv 0$. Consequently, the TM mode vanishes if either $m = 0$ or $n = 0$, making the TM_{11} mode the lowest. Since the mode-purity condition established for the TE modes is not altered by the existence of the higher TM modes, the condition still holds that the TE_{10} mode is the leading waveguide mode, and, over the proper range, is the only propagating mode.

Figure 12.5 has not indicated the 00 mode. For this mode, $\beta_x = 0 = \beta_y$ in all cases, and from Eq. (12.23), it is clear that all the transverse fields vanish, leaving only a static field H_z. Equations (9.8) dealt with a similar situation in connection with a possible plane-wave longitudinal field. As in that case, such a field cannot be established and need not be considered, so the 00 mode is nonexistent.

One further distinction between TE and TM modes can be made from Eqs. (12.6) and (12.9). By taking first $H_z = 0$ and then $E_z = 0$, the ratio of the field pairs, E_x/H_y or E_y/H_x, that is, the transverse-wave impedances, can be found. For the TE and TM cases, respectively, these are

$$\eta_{TE} = \frac{|E|}{|H|} = \frac{\omega\mu}{\beta} \qquad E_z = 0 \qquad (12.30a)$$

$$\eta_{TM} = \frac{|E|}{|H|} = \frac{\beta}{\omega\varepsilon} \qquad H_z = 0 \qquad (12.30b)$$

Note that for the 10 mode, there is only one pair of transverse-field components. For the higher modes, this is not the case, but since the wave impedances in each case are the same for both orthogonal pairs of fields, that is, $|E_x/H_y|$ and $|E_y/H_x|$, their values are also for the entire transverse fields.

While $\eta_{TE} \neq \eta_{TM}$, it is interesting to note that $\eta_{TE}\eta_{TM} = \eta_0^2$, where η_0 is the transverse-wave impedance of the plane wave.

12.7 WAVEGUIDE DEVICES

The fact that a rectangular waveguide is used in its single dominant mode (TE_{10}) means that the fields and propagation properties are well-defined, so efficient signal control devices can be designed.

First, however, it is necessary to introduce the signal into the guide. Chapter 14 will discuss antennas, but for now, we need only recall that a current-carrying conductor has an electric field along its length, and, since the tangential E field is continuous across a boundary between two media, the same field exists outside the conductor. If the field is changing rapidly, it radiates from the conductor. Figure 12.6(a) shows a coaxial transmission line carrying a microwave signal and entering a waveguide. The bare center conductor acts as a radiating antenna. Figure 12.6(b) shows more clearly the placement of the antenna in the transverse waveguide section, where it is at the E-field maximum. In the longitudinal section of Fig. 12.6(c), it is placed at a distance $\lambda/4$ from the shorted end, so that the traveling wave is shifted in phase by 90° in traveling to the short, by 180° upon reflection, and by 90° in returning from the short. Thus, the return wave is in phase (360°) with the direct

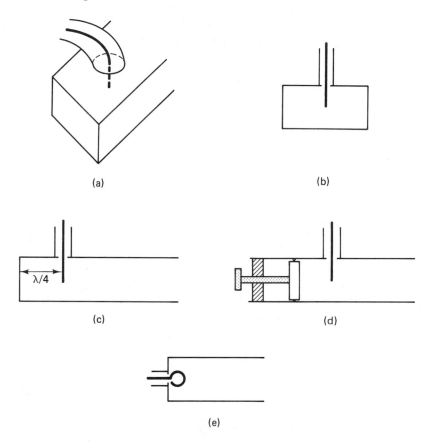

(a)   (b)

(c)   (d)

(e)

Figure 12.6   Waveguide launchers for the TE_{10} mode.

signal from the transmission line and adds constructively to the wave traveling to the right in the figure. In waveguides where precise tuning is required at each frequency of use, an adjustable termination can be used, as shown in Fig. 12.6(d). This description is for coupling by means of an *electric-dipole* antenna. Alternatively, a *magnetic-dipole* antenna could be used, as shown in Fig. 12.6(e). Here a current loop generates the microwave magnetic field that radiates down the guide.

Different excitation means must be used to conform to the field configurations of higher modes, if it is desired to generate these. As an example, Fig. 12.7 shows how one might conceive a TE_{20}-mode launcher. Here the probes' signals are adjusted to differ in phase by 180° and are inserted at the points $x = a/4$ and $x = 3a/4$, corresponding to the E-field maxima and relative orientations.

These generation schemes can also be used to receive waves arriving along the waveguide, since the probe then acts as a receiving antenna that delivers its signal to detection and processing circuits connected to the coaxial cable.

A variable attenuator, constructed by impregnating a dielectric strip with conducting particles, is shown in Fig. 12.8. Currents are induced in the conductor by the tangential microwave electric field, causing heating that draws energy from the field, thereby reducing its amplitude. Turning the external screw causes the slide to

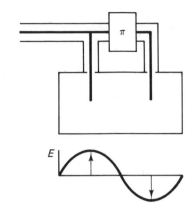

Figure 12.7 Possible TE_{20} launcher.

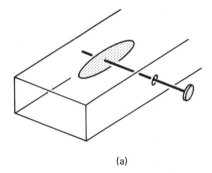

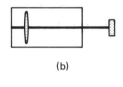

(b)

(a)

Figure 12.8 Variable attenuator.

move across the waveguide, changing the field strength in which the strip resides, and the induced current, and thereby the attenuation. The knob position can be calibrated appropriately. The moveable slide is thin and tapered so as not to interfere with the traveling wave or cause reflections.

It should be recalled, in this regard, that a hollow waveguide is a form of transmission system, with many of the same circuit properties as the transmission line discussed in the previous chapter. Loads, obstructions, and various inserts and structures cause reflections and standing waves. Impedance matching is an important consideration, although it may be analyzed with an electromagnetic field-scattering treatment rather than with the circuit approach used previously. In the coaxial line-to-waveguide coupler shown in Fig. 12.6, the dimensions and coupling hardware of the antenna probe are controlled and adjusted for this purpose.

Tuning devices also exist in waveguides, corresponding to the sliding-section and adjustable stub tuners used with the coaxial line studied earlier. Figure 12.9(a) shows an $E–H$ tuner that generates variable reflections in orthogonal planes of the guide. Figure 12.9(b) represents a different sort of tuner, a slide-screw tuner, consisting of an adjustable probe that is inserted through a thin slot in the waveguide wall. By changing the probe penetration, the amplitude of its reflected wave can be adjusted, and by changing the probe position along the guide, the phase of the reflect. 1 wave is varied.

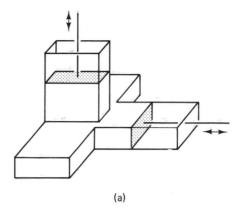

(a)

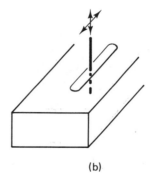

Figure 12.9   Waveguide tuners.   (b)

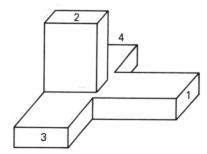

Figure 12.10   Magic T.

These and other fixed inserts in waveguides are used to control the signals being transmitted. They utilize the specific field pattern of the 10 mode and their operation would not be correct if other modes were present. A final example is the *magic T* shown in Fig. 12.10. This is a very sophisticated device that has the following useful properties:

1. A signal into arm 1 divides into two equal signals out of 3 and 4, with no signal from 2.

2. A signal into arm 2 divides into two equal and opposite signals out of 3 and 4, but does not couple to arm 1.

3. If signals are input to arms 3 and 4, their phasor sum appears at 1 and their phasor difference appears at 2.

Such a device is very useful for precisely dividing, adding, and comparing signals with regard to both amplitude and phase.

12.8 PROPAGATION CONSTANT

It has been shown that the critical wave constant β_c depends on the mode and can be determined from Eq. (12.24b). Once β_c is known, the propagation constant β in the waveguide and the guide wavelength $\lambda = 2\pi/\beta$ can be found. From Eq. (12.8b),

$$\lambda = \frac{v_0/f}{\left[1 - \left(\frac{f_c}{f}\right)^2\right]^{1/2}} = \frac{\lambda_0}{\left[1 - \left(\frac{f_c}{f}\right)^2\right]^{1/2}} \tag{12.31}$$

where $\lambda_0 = v_0/f$ is the free-space plane-wave wavelength at the same frequency. This relation is shown in Fig. 12.11, from which it is immediately seen that $\lambda > \lambda_0$ and, when f is close to f_c, we can have $\lambda \gg \lambda_0$.

While Eq. (12.31) is useful, it should be remembered that it is actually β, the phase change per unit distance, that enters into the wave equation. If $\lambda > \lambda_0$, then $\beta < \beta_0$. This is stated quantitatively by Eq. (12.24a), which has the simple geometric representation given by the Pythagorean theorem, as shown in Fig. 12.12. This, in turn, can be interpreted as indicating that a plane wave, with β_0, travels at an angle to the waveguide axis. Its transverse component, that is, across the guide, is β_c, which depends upon m and n, the number of reflections it undergoes at the

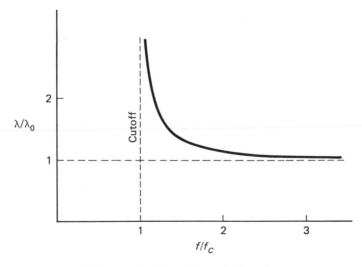

Figure 12.11 Waveguide mode dispersion.

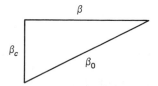

Figure 12.12   Relation of propagation
constants.

various walls. β, the component along the guiding direction, is the propagation
wave number along the guide. If such propagation is to occur, then $\beta \neq 0$, so that
$\beta_0 > \beta_c$, that is, β_c is the minimum allowable value of β_0 for the *mn* mode, corresponding to cutoff.

Just what happens when a signal with $f < f_C$ is injected into a waveguide?
What does it mean to say, "It does not propagate"? If $\omega < \omega_c$ (and, therefore,
$\beta_0 < \beta_c$), Eq. (12.24a) becomes

$$\beta = \pm[-(\beta_c^2 - \beta_0^2)]^{1/2} = \pm\frac{[-(\omega_c^2 - \omega^2)]^{1/2}}{v_0}$$

$$= \pm\frac{j}{v_0}(\omega_c^2 - \omega^2)^{1/2} = \pm j\alpha \qquad (12.32)$$

where α is real. The interpretation of an imaginary β is found by insertion in
Eq. (12.1), giving

$$E \sim E_0 e^{j(\omega t - \beta z)} = E_0 e^{j\omega t}e^{-\alpha z} \leftrightarrow E_0 e^{-\alpha z}\cos \omega t \qquad (12.33)$$

The negative root of Eq. (12.32) is used because it results in an attenuated wave.
The positive root can be neglected because an exponentially increasing field does not
correspond to physical reality. The opposite root would be selected for a wave traveling in the opposite direction. Equation (12.33) shows that there is no longer a traveling wave; instead, the field is stationary and time-varying, with an amplitude that
decays along the guide. No energy is transmitted, although at any instant some is
stored in the fields. When a waveguide is excited with a signal that has $f < f_c$, it is
found that the signal is entirely reflected back to the generator.

Example 12.4

A microwave signal at $f = 10$ GHz is injected into a waveguide with $a = 0.9$ inch and
$b = 0.4$ inch. Describe the propagation if it is excited in a 10 mode and if it is excited in a
20 mode.

From Eq. (12.24b), $\beta_c(10) = \pi/a = 1.374$ cm^{-1}. At $f = 10^{10}$ Hz, $\beta_0 = \omega/v_0 =$
2.096 cm^{-1}. Thus $\beta_0 > \beta_c(10)$, so that $\beta = (\beta_0^2 - \beta_c^2)^{1/2} = 1.583$ cm^{-1}, corresponding to
$\lambda = 3.97$ cm, as compared to $\lambda_0 = 3.0$ cm. Since the walls are assumed to be perfectly
conducting, this mode propagates without any attenuation.

Similarly, $\beta_c(20) = 2.749$ cm^{-1}. From the fact that $\beta_0 < \beta_c(20)$, Eq. (12.32) gives
$\alpha = (\beta_c^2 - \beta_0^2)^{1/2} = 1.78$ cm^{-1} [Np/cm; see Eq. (11.44)]. The fields decay as $e^{-\alpha z}$, so that
the intensity (Poynting vector) or total energy decays as $E^2 \sim e^{-2\alpha z}$. Therefore, the energy
in the field drops to $1/e = 0.368$ of its input value over a distance $z = 1/2\alpha = 0.28$ cm.
This illustrates the general result that waves below the cutoff frequency decay *very* rapidly
from their point of generation.

12.9 CURRENTS

Early in the discussion of waveguides, it was pointed out that because the fields are zero in perfectly conducting walls ($\sigma = \infty$), either condition could be used to determine the field configuration: $B_{normal} = 0$ or $E_{tangential} = 0$ immediately adjacent to the waveguide walls. Now that the fields are known for the case of walls that are perfect conductors, the wall-conductivity condition can be relaxed to the more realistic case that $\sigma/\omega\varepsilon$ is very large, but finite. In this event, $E \neq 0$ at the boundary, but, instead, penetrates into the metal with a skin depth δ. However, the fields in the guide do not depart very much from the case that $\sigma = \infty$, so that the solutions in that case are a good starting point for further study. These fields can be used with the other boundary conditions, $D_{normal} = \rho_s$ and $H_{tangential} = K$, to determine ρ_s and K. Recall that ρ_s is the surface-charge density on the walls (C/m^2) and K is the surface-current density (A/m). ρ_s and K are functions of time and are related by the law of charge conservation. To study waveguide surface currents, the magnetic field interface condition, Eq. (5.40), $H_{2t} - H_{1t} = K$ must be put in the form of a vector relation.

Since **H** decays with distance into the conductor, the field generated by **K** must be in a direction to cancel $H_{tangential}$ inside the conductor. Therefore, from Fig. 12.13, the direction of **K** satisfies

$$\mathbf{K} = \mathbf{n} \times \mathbf{H} \qquad (12.34)$$

where **n** is the unit outward normal from the conductor. Note that whereas $H_{tangential}$ and the **H** arising from **K** cancel inside the conductor, they add outside the surface. For physical reality, we must recognize that $\mathbf{K} = \mathbf{J}\delta$, where δ is the skin depth of penetration. This means that **K** is the surface-current equivalent of the actual physical volume current **J** that flows in the metal. Since J is distributed over a depth $\sim\delta$ in the conductor, its field gradually cancels $H_{tangential}$, causing it to also decay over the same depth. These relations are discussed in more detail in Section 10.2.

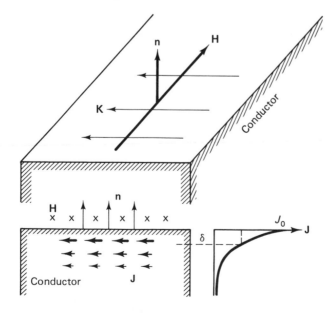

Figure 12.13 Direction and distribution of surface current.

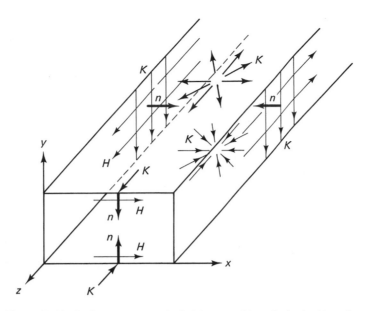

Figure 12.14 Surface currents on the inside waveguide walls, in the 10 mode.

For a waveguide carrying a TE$_{10}$ mode, a half wavelength of **H**-field configuration, modified from Fig. 12.2, is shown in Fig. 12.14. At points along the guide wall, the solution for $\sigma = \infty$ gives a tangential component of **H**. At these points on the boundary, the unit normal **n** (out of the wall) and the current density **K** are constructed. It is seen that the longitudinal currents are oppositely directed on the upper and lower surfaces and that there are transverse currents in the side wall. Using the field expressions of Eq. (12.26) the currents are given by

$$\mathbf{K}_{\text{top surface}} = \mathbf{n} \times \mathbf{H}_{\text{tangential}} = -\mathbf{a}_y \times (H_x \mathbf{a}_x + H_z \mathbf{a}_z)$$

$$= \mathbf{a}_z H_x - \mathbf{a}_x H_z$$

$$= \mathbf{a}_z H_0 \frac{j\beta a}{\pi} \sin \frac{\pi}{a} x - \mathbf{a}_x H_0 \cos \frac{\pi}{a} x \qquad (12.35a)$$

$$\mathbf{K}_{\text{side wall}} = \mathbf{n} \times \mathbf{H}_{\text{tangential}} = \mathbf{a}_x \times \mathbf{a}_z H_z \qquad (\text{at } x = 0)$$

$$= -\mathbf{a}_y H_0 \cos \frac{\pi}{a} x = \mathbf{a}_y H_0 \qquad (12.35b)$$

where we have noted, in Eq. (12.35b), that on the side walls either $x = 0$ or $x = a$, so that the cosine function is unity.

These currents flowing in the conductor cause heating, and the power dissipated comes from the guided wave that is therefore attenuated. Calculation of the attenuation constant α is quite involved; Figure 12.15 shows α as a function of the reduced frequency $\omega/\omega_c = f/f_c$ for a waveguide that has $a = 2b = 2$ cm and copper walls. It is seen that the attenuation rises sharply as f approaches f_c. This limits the range of use of the waveguide at the low-frequency end of its transmission region. Therefore, in Example 12.3, where $f_c = 6.56 \times 10^9$ Hz for the 10 mode, the lower

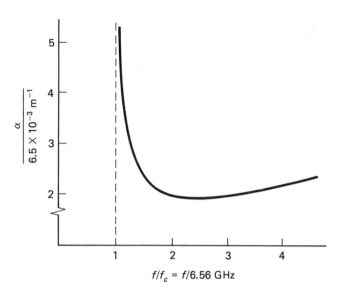

Figure 12.15 Waveguide attenuation for the TE_{10} mode, a = 2b = 2 cm.

limit of use is generally set at 8.4 GHz or $f/f_c = 1.28$, so that the attenuation level is not detrimental to signal propagation. While it is for a different waveguide, it is useful to note that at this value, Fig. 12.15 gives $\alpha = 19 \times 10^{-3}$ m^{-1}; the transmitted power attenuates to 13.5 percent over a distance of about 53 m. While the attenuation in Fig. 12.15 decreases to a minimum at $f/f_c = 2.4$, a waveguide cannot be used at this point because of mode-purity considerations.

Longitudinal (z-directed) currents flow in the conductor walls. This implies, by Ohm's law, that there is also a longitudinal component of electric field. Thus, the imperfectly conducting walls force a condition that the mode is not exactly a transverse electric mode. For walls that are good conductors, this component of E is very small compared to the field at the center of the guide, so that the mode fields are only very slightly altered. However, for this reason, it is not uncommon to refer to the lowest mode in a rectangular waveguide as being a quasi-TE_{10} mode. In Chapter 11, it was mentioned that this same argument holds for transmission lines, and there, also, the true propagation mode is only quasi-TEM.

12.10 FINITE DIFFERENCE ANALYSIS

The numerical-methods approach to electrostatic and magnetostatic problems was described in Chapter 8. These techniques are also applicable in dynamic cases and, although we will not treat them in detail, it is worthwhile to at least indicate some of the methods that can be used.

For this, begin with Eq. (12.10c) for the field H_z of TE modes, and its counterpart for E_z of TM modes:

$$\frac{\partial^2 H_z}{\partial x^2} + \frac{\partial^2 H_z}{\partial y^2} = -\beta_c^2 H_z$$

$$\frac{\partial^2 E_z}{\partial x^2} + \frac{\partial^2 E_z}{\partial y^2} = -\beta_c^2 E_z \qquad (12.36)$$

$$\beta_c^2 = \beta_0^2 - \beta^2$$

Of the methods discussed in Chapter 8, the solution of waveguide problems is well-suited to differential schemes because they have naturally closed boundaries and therefore do not require imposition of an artificial boundary representing points at infinity, on which the unknown (the z-field component in this case) or its normal derivative is set to zero. In a rectangular waveguide carrying TM waves, for example, we have a true close-by boundary on which the component E_z is really zero.

Consider the case where we construct a square finite-difference mesh, as shown in Fig. 12.16, for a rectangular waveguide, with qp unknown points. By following the procedure in Section 8.6, leading to Eq. (8.17), the discretized form of Eq. (12.36) is found to be

$$4E_{z0} - E_{z1} - E_{z2} - E_{z3} - E_{z4} = h^2 \beta_c^2 E_{z0} \qquad (12.37)$$

where the subscripts refer to the points shown on Fig. 8.7. Note that this equation, when assembled as a matrix, takes the form

$$[A][E_z] = \lambda[E_z] \qquad (12.38)$$

where $[A]$ is the square matrix of order qp, constituted of the E_z coefficients, $[E_z]$ is a column matrix, and $\lambda = h^2 \beta_c^2$. Clearly, $[E_z] \equiv 0$ is an allowed but trivial solution. The only possibility for a nontrivial solution is for the secular determinant to be singular, that is,

$$|[A] - \lambda[I]| = 0 \qquad (12.39)$$

where $[I]$ is the unit matrix. This is a classical eigenvalue problem.

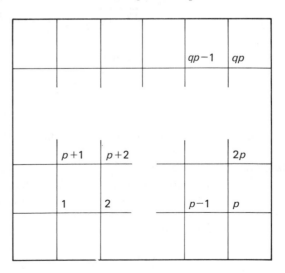

Figure 12.16   Finite-element mesh for waveguide study.

Algorithm 12.1, at the end of the chapter, performs the finite-difference function of assembling the matrix $[A]$ from a file giving the locations of the mesh nodes and a file giving the number of unknown nodes and the known node values. Once this is done, $[A]$ can be passed on to an eigenvalue extraction program from the computer library. This yields values of λ that correspond to nonzero waves. Since the waveguide cutoff of a propagating mode occurs when the propagation constant reduces to $\beta = 0$, then, for each value of λ, the cutoff frequency is

$$\lambda = h^2\beta_c^2 \quad \text{or} \quad \omega_c^2 = \frac{v_0^2\lambda}{h^2} \tag{12.40}$$

The eigenvalues λ, therefore, lead to the spectrum of cutoff frequencies of the allowed modes, generally, in order of increasing ω. The finer the mesh taken, the greater the field resolution and the higher the order of the modes that can be found reliably.

12.11 FINITE ELEMENT ANALYSIS

The finite-element method can also be applied to determine the field lines in waveguide. For this, we must determine the correct functional corresponding to Eqs. (12.36). When dealing with Poisson's and Laplace's equations in Chapter 8, we found that the forms of energy function could be used because if

$$\mathcal{F} = \frac{1}{2}\int(\nabla\Phi)^2\,dv \quad \text{then} \quad \frac{\partial\mathcal{F}}{\partial\Phi} = 0 \quad \text{leads to} \quad \nabla^2\Phi = 0 \tag{12.41a}$$

This follows from the proof given earlier, by setting $\rho = 0$ in Eq. (8.25). We now note that

$$\frac{1}{2}\frac{\partial}{\partial\Phi}(k^2\Phi^2) = k^2\Phi \tag{12.41b}$$

so that minimizing the functional

$$\mathcal{F} = \frac{1}{2}\int[(\nabla\Phi)^2 - k^2\Phi^2]\,dv \tag{12.42}$$

with respect to Φ, leads to the wave equation. It can also be shown that this functional is minimum at the correct solution and leads to second-order errors for small departures from correctness.

The local matrix for the first term in Eq. (12.42) was found in Eq. (8.34). However, with the addition of the second term, complicated matrix analysis is needed to establish the matrix form obtained by minimizing this new $\mathcal{F}$. For this the interested reader is referred to more advanced texts.*

---

*S. R. H. Hoole, *Computer Aided Analysis and Design of Electromagnetic Devices* (New York, NY: Elsevier Science Publishers, 1988).

PROGRAMMING ALGORITHM

Algorithm 12.1 Finite-Difference Matrix Assembly

Procedure FiniteDifference

```
{Function: Assembles the finite-difference matrix equation
Ax=b from data file f1 containing the nodes to the East,
North, West, and South of each node and f2 containing NUnk,
the number of unknowns and the known potentials
}
Begin{Procedure}
     A ← 0
     b ← 0
     Read(f2, NUnk)
     i ← 0
     While NOT EOF(f2) Do
             i ← i + 1
             Read(f2,KnownPot[i])
     For Node ← 1 To NUnk Do
             A[Node,Node] ← A[Node,Node] + 4
             Read(f1,v[1],v[2],v[3],v[4])
             For j ← 1 To 4 Do
                If v[j] ≤ NUnk
                   Then A[Node,v[j]] ← A[Node,v[j]] - 1
                   Else b[Node] ← b[Node] + KnownPot[v[j] - NUnk]
End{Procedure}
```

EXERCISES

Extensions of the Theory

12.1. Equations (12.6) and (12.9) give the transverse-field components in terms of the longitudinal components. Equation (12.10a) is the wave equation for **H**, and there is an identical equation for **E**.

   (a) Starting from the wave equation, use the boundary conditions imposed by a rectangular waveguide to demonstrate how you find the z component of the field for the TM modes, as given by Eq. (12.29).

   (b) Verify that the same distribution of mode cutoff frequencies holds for TM modes as for TE modes.

12.2. For many applications, circular waveguides are used instead of rectangular. For these guides, TE and TM modes also exist.

   (a) Starting from Maxwell's equations in circular cylindrical coordinates, derive the expressions that give the transverse-field components in terms of the longitudinal fields

$$j\beta_c^2 E_\rho = \beta \frac{\partial E_z}{\partial \rho} + \frac{\omega\mu}{\rho} \frac{\partial H_z}{\partial \phi}$$

$$j\beta_c^2 H_\phi = \omega\varepsilon \frac{\partial E_z}{\partial \rho} + \frac{\beta}{\rho} \frac{\partial H_z}{\partial \phi}$$

and similar expressions for E_ϕ and H_ρ.

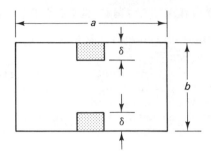

Figure 12.17   For Exercise 12.3.

   (b) Determine the wave impedances for the TE and TM modes in circular
      waveguides.

*12.3. We can modify the short dimension in a rectangular waveguide by inserting
   low ridges of height δ in the center of the guide, as shown in Fig. 12.17.
   Note that the TE_{10} mode has its field maxima at the guide center and, therefore, is very sensitive to this change, whereas the next higher mode, TE_{20},
   has null fields at the center and so is insensitive to it. One effect of the ridges
   can be understood by referring to the cutoff condition of the 10 mode: $\beta_c a =
   \pi$, indicating that the a dimension corresponds to a phase change of π at the
   cutoff frequency.
   (a) If $a = 2.25$ cm, $b = 1$ cm, and $\delta = 0$ (ordinary waveguide), find the
      frequency range for pure TE_{10} propagation.
   (b) Repeat part (a) if $\delta = 0.2$ cm and if the ridges can be considered to add
      a distance δ to the apparent transverse propagation of the TE_{10} mode.

12.4. The TE_{10} mode has components E_y, H_x, and H_z.
   (a) Write the full expression for the Poynting vector.
   (b) Determine the power propagating down the guide in terms of the ampli-
      tude of H_z and of the guide dimensions.
   (c) Show that the transverse power is purely imaginary and find the rms am-
      plitude of the reactive power per unit length of the guide.
   (d) The propagating power crosses the area $a \times b$. Calculate the reactive
      power crossing an area $(\lambda/4) \times b$, where $\lambda/4$ is along the guide. What
      fraction is it of the real power for a waveguide that is 2×1 cm operat-
      ing at 10 GHz?

12.5. For a rectangular waveguide of dimensions $a \times b$, calculate I_{long}, the total
   longitudinal current that flows across a plane normal to the guide axis. (Find
   either the maximum or rms value.) *Hint:* Use Eqs. (12.35).

12.6. A hollow waveguide is a form of transmission line as discussed in Chapter 11. For a guide with dimensions $a \times b$, propagating in the TE_{10} mode,
   (a) Find the effective transmission-line-like voltage V_{trans} defined as the
      transverse voltage between the centers of the horizontal waveguide
      plates (either maximum or rms).
   (b) Using the result of Exercise 12.5, determine the equivalent transmission-
      line characteristic impedance, defined as $Z_0 = V_{\text{trans}}/I_{\text{long}}$.
   (c) Compare Z_0 with η for the TE_{10} mode.

Note that Z_0 is a function of the short guide dimension b, as well as of the long dimension a. Except for this, all the properties we have studied of the TE_{10} mode are independent of b. This exception is important because, for example, it explains the occurrence of reflections when b changes in a waveguide.

12.7. In Exercise 9.3, it was shown that the group velocity (i.e., the effective velocity of a group of waves, e.g., of a broad pulse) is given by $v_g = d\omega/d\beta$. Note that this is different from the phase velocity (i.e., the velocity of a constant-phase point on an infinite wave), $v_p = \omega/\beta$.
 (a) From the expressions in the text for $\beta = \beta(\omega)$, calculate v_g for the waveguide modes.
 (b) Show that $v_g v_p = v_0^2$, where v_0 is the free-space velocity.
 (c) Plot v_{phase} and v_{group} as functions of f/f_c on the same figure.

Problems

12.8. Equations (12.6) and (12.9) give the transverse-field components in terms of the longitudinal components. It appears from these equations that when both $H_z = 0$ and $E_z = 0$, all the transverse fields vanish, so that there is no field solution at all. However, we recognize that this TEM case is just the plane wave studied in Chapter 9. Since the derivations of Eqs. (12.6) and (12.9) are quite general, they must also apply to the TEM wave.
 (a) Examine the derivation of these equations and find the point where a step is taken that is improper for the TEM case.
 (b) From the answer to (a), show that $H_z = 0 = E_z$ and that $\beta^2 = \omega^2\mu\varepsilon$ for TEM propagation.

12.9. Draw the fields for the TM_{11} mode in a cross-sectional view of the waveguide.

12.10. A magnetic circulator can be constructed by placing a small ferromagnetic sample off center in a waveguide. The operation of such a device depends upon rotation of the magnetization of the sample, driven by the microwave magnetic field of the TE_{10} mode. Show that this field traces an ellipse, and determine the axis ratio of this ellipse.

*12.11. Determine the transverse wave impedance, that is, the ratio $\eta = |E/H|$ of the transverse components, for the TE_{10} and the TE_{01} modes in an air-filled waveguide with dimensions $a = 0.9$ inch and $b = 0.4$ inch. Assume $f = 10^{10}$ Hz.

12.12. Repeat Exercise 12.11 for the TM_{11} mode.

*12.13. As a microwave designer, you wish to use the cutoff property of waveguide to make a filter that will not pass a signal with $f_1 = 19$ GHz, but that will pass one with $f_2 = 21$ GHz.
 (a) Specify the transverse dimensions of the waveguide.
 (b) Determine the attenuation of the f_1 signal over 2 cm.
 (c) In terms of Fig. 12.15, discuss whether this is an effective way to filter these particular frequencies.

12.14. Repeat Exercise 12.13 if $f_1 = 6$ GHz and $f_2 = 8.2$ GHz.

*12.15. A microwave engineer transmits a signal at 16 GHz over a distance of 5 cm along a rectangular waveguide that has inside dimensions of 2×1 cm. The input signal has an amplitude of 100 mV/cm. Determine the amplitude and the phase change of the output signal if it propagates in the TM_{11} mode and if it propagates in the TE_{20} mode.

12.16. A waveguide with a square cross section is said to be doubly degenerate in every mode because for every mode field pattern there is another with the same value of β.
   (a) How are these mode pairs geometrically related in the guide?
   (b) For a signal frequency of 3 GHz, determine the range of waveguide dimensions so that only the lowest TE modes propagate.

   By carefully selecting the launching method, we can attempt to have only one of this degenerate pair. Further down the guide, however, waveguide imperfections, bends, junctions, etc. cause energy to couple to the other mode, so that it also propagates, with a mixing of the signals. This is the reason for making the two guide dimensions unequal.
   (c) If the guide has one dimension of 5.5 cm, find the maximum value of the other dimension if one of the degenerate modes is to be cut off. Take $f = 3$ GHz.

*12.17. A waveguide is to propagate in the TE_{10} mode at $f = 25$ GHz. Since attenuation increases near cutoff, we wish to have $f > 1.2 f_{c, 10}$, and to avoid any possibility of mode contamination, we wish to have $f < 0.8 f_{c, mn}$, where mn is the next mode. Find the dimensions of such a waveguide.

*12.18. Given an air-filled waveguide with dimensions $a = 1$ cm and $b = 0.3$ cm, determine the propagating modes for a frequency of 40 GHz. For each of these modes find:
   (a) the cutoff frequency
   (b) the propagation constant β
   (c) the wavelength in the guide
   (d) the phase velocity.

12.19. For the waveguide of Exercise 12.18, find the attenuation constant of the first nonpropagating mode. By what factor is its *power* attenuated over the distance of one wavelength of the TE_{10} mode?

12.20. Repeat Exercises 12.18 and 12.19 for a guide filled with a dielectric having $\varepsilon_r = 4$.

12.21. A student designs a waveguide high-frequency pass filter for the TE_{10} mode, by inserting a 2-cm-long section with dimensions 1×0.5 cm in a guide that is 2×1 cm. Two equal-amplitude signals propagate along the guide with $f_1 = 12.5$ GHz and $f_2 = 17.5$ GHz. For this problem, ignore reflections at the waveguide discontinuities.
   (a) Find the ratio of the output to the input amplitudes of the filter section at the two frequencies.
   (b) What is the minimum length of the reduced section if the f_1 signal is to have its output intensity reduced to 1 percent of the f_2 signal?

12.22. A 2.25 × 1 cm waveguide is filled with a dielectric that has $\varepsilon_r = 4$. It propagates two-equal amplitude TE_{10} signals in the positive z direction with $f_1 = 5$ GHz and $f_2 = 7.5$ GHz. If the dielectric is abruptly terminated at $z = 0$, determine the relative amplitudes of the two signals at $z = 1$, 5, and 20 cm. Ignore reflections at the dielectric discontinuity.

12.23. (a) From the relationships derived in the text, draw the graph of $|\beta_{mn}/\beta_0|$ as a function of $f_{c,mn}/f$, both in the range of propagation and below cutoff.
(b) Repeat part (a) for ω/ω_c as a function of β/β_c.
(c) From these figures, discuss the behaviors and values of β_{mn} at very low and very high frequencies and near cutoff.

12.24. The relation $\beta_0^2 = \beta_c^2 + \beta^2$ has been interpreted as showing that a waveguide mode is a result of having a plane wave, with β_0, reflecting off the guide walls in such a way that the transverse phase change is described by β_c and the longitudinal phase change by β. If this is so, the energy properties of the plane wave must hold. Show that the total electric and total magnetic energies, in a representative length of waveguide, e.g., $\lambda/4$, are equal for any TE mode and for any TM mode.

12.25. Given a waveguide with dimensions $a = 1$ cm, $b = 0.3$ cm, and filled with a dielectric that has $\varepsilon_r = 4$.
(a) For a frequency of 20 GHz, determine the propagating modes. For each of these, find the cutoff frequencies.
(b) Determine the attenuation constant of the first mode that is nonpropagating. By what factor is its *power* attenuated in the distance of one wavelength of the TE_{10} mode?

12.26. (a) Find the fraction of energy that is carried in the transverse electric field, the transverse magnetic field, and in the longitudinal field of the TE_{10} mode. Assume the dimensions $a/b = 2$. *Hint:* Integrate the energy densities over a representative waveguide volume.
(b) Show that the electric- and magnetic-field energies are equal.

12.27. (a) Find the total positive charge Q on the waveguide walls in a one-half wavelength section of a rectangular waveguide with dimensions $a \times b$ in the TE_{10} mode.
(b) Show that there is an equal negative charge.
(c) Since this charge reverses in a half period, you can estimate the effective current $I = \Delta Q/\Delta t$. Compare this value with the current calculated in Exercise 12.5.

12.28. A rectangular waveguide of size $a \times b$ carries a TM wave defined by Eq. (12.29). For a 2 × 3 cm guide, construct a general mesh with $(p + 2) \times (q + 2)$ nodes (as on Fig. 12.16). Relate the eigenvalues to n and m, and obtain them for $(p, q) = (4, 6)$, $(8, 12)$, $(12, 16)$, and $(16, 24)$. What connection do you see between mesh size, the number of eigenvalues and accuracy of the eigenvalues? Plot E_z corresponding to the dominant mode.

12.29. The rectangular waveguide of Exercise 12.28 carries a TE wave characterized by Eq. (12.22).
(a) Show that the zero tangential value of the electric field along the

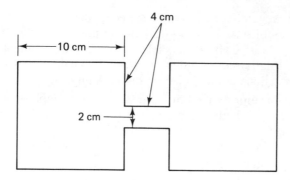

Figure 12.18 For Exercise 12.30.

boundaries of the guide translates into the condition that H_z does not change normal to the walls.

(b) Use this to analyze the guide numerically and compare your results with the analytical solution.

12.30. The ridge-shaped waveguide of Fig. 12.18 is used to increase the capacitance of the line. Obtain the cutoff frequency of the dominant mode of the guide when carrying TE waves. Produce a computer plot of the dominant solution.

12.31. Repeat Exercise 12.30 if the guide carries TM waves.

13

STRIPLINES AND MICROSTRIP

13.1 INTRODUCTION

Figure 11.1 showed seven different types of transmission systems, some of which are now familiar to us: Fig. 11.1(a), coaxial line, was discussed in Chapter 11; Fig. 11.1(g), rectangular waveguide, was the subject of Chapter 12; Fig. 11.1(f), the two-wire system, is familiar from low-frequency usage. The remaining transmission lines are distinguished by having planar conducting elements: Fig. 11.1(b) is known as *stripline*, Fig. 11.1(c) is called *flatline,* Fig. 11.1(d) is *microstrip,* and Fig. 11.1(e) is *slotline*.

The main impetus for the development of these planar lines has been their ease of fabrication, for example, by hybrid- or integrated-circuit techniques, and their wide effective bandwidth, combined with greatly reduced bulk and cost. These factors are of great importance in modern applications that demand large scale and high speed, such as computers and airborne systems. For test purposes, complex planar circuits can be accurately and quickly made to scale using metal-clad circuit boards or printed-circuit techniques.

The theoretical situation with these lines is mixed. The ideal lines are simple to analyze and, in fact, the basic relations have already been studied in Chapters 9, 11, and 12. Real lines, however, present problems that are not insignificant. For example, the conductors of planar lines must be supported; this is generally done by using dielectric posts or sheets, such as those shown in Fig. 13.1. Having both air-filled and dielectric-filled regions creates a condition of inhomogeneity that does not lend

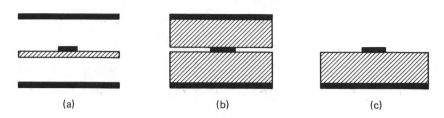

Figure 13.1   (a) and (b) Inhomogeneous strip lines. (c) Microstrip line.

itself to exact analytical solution. In addition, as with the coaxial line and the hollow waveguide, finite conductivity of the metal guiding system and dielectric losses introduce field differences from the ideal that generally require analysis using approximation techniques.

In this chapter, the pertinent features of transmission-line theory will be reviewed, models for strip and microstrip lines will be developed, and some of their properties and limitations will be discussed.

13.2 ELECTROMAGNETIC PROPAGATION THEORY

A review of the discussion of Chapter 11 will reveal that the analysis done there is quite general; the particular properties of the coaxial transmission line are inserted only in the examples. Otherwise, most of the properties described are equally applicable to any transmission system consisting of separated metallic conductors that carry signals along their lengths, such as the systems under consideration here.

Although two-conductor transmission lines propagate along their lengths, they have a transverse interelectrode voltage. Corresponding to this transverse voltage is a transverse electric field. In addition, associated with the longitudinal conduction current is a transverse magnetic field. Therefore, the transmission theory of Chapter 11 applies to systems that have transverse electric and magnetic fields; systems that propagate in a TEM mode.

The development in Section 12.1 is also quite general, and its results are also applicable here. Equations (12.6) and (12.9) express the transverse fields of any propagating signal (regardless of whether it is or is not TEM) as derivatives of the longitudinal fields. However, we have just seen that systems that are also suitable for description by transmission-line theory have only transverse fields. If E_z and H_z, the longitudinal fields, are set equal to zero in the numerators of Eqs. (12.6) and (12.9), we find that the transverse fields also vanish unless the denominators of those equations are zero as well. Both denominators are equal to $\beta_c^2 = \beta_0^2 - \beta^2$, so that, for the general transmission line, $\beta^2 = \beta_0^2$ or $\beta^2 = \omega^2/v_0^2$, where $v_0 = 1/\sqrt{\mu\varepsilon}$. This is the relationship for an electromagnetic plane wave in an unbounded medium, a *free* wave. It can therefore be concluded that signals on transmission lines travel with the free-wave phase velocity v_0, and have the free-wave wavelength and wave number, appropriate to the medium that occupies the space of the transmission line.

Dispersion describes the property of some systems in which phase velocity changes with frequency. Some dispersion occurs if the magnitudes of ε and μ are frequency dependent, but we will ignore this case. Transmission systems display dispersion if their propagation mode is not TEM; this was the case for the waveguides described in Chapter 12. To the extent that a system conforms to transmission-line theory, its phase velocity is v_0 and it is dispersionless.

13.3 LIMITATIONS OF GEOMETRY

The wave equation is derived at the beginning of Chapters 9 and 12 for the case of sinusoidal fields in an insulating dielectric medium. This can be stated for either the electric or magnetic field; for **E**, it is

$$\nabla^2 \mathbf{E} = -\beta_0^2 \mathbf{E} \tag{13.1a}$$

For a wave traveling in the z direction,

$$\mathbf{E}(x, y, z, t) = \mathbf{E}(x, y)e^{j(\omega t - \beta z)} \tag{13.1b}$$

Equation (13.1a) becomes

$$\frac{\partial^2 \mathbf{E}}{\partial x^2} + \frac{\partial^2 \mathbf{E}}{\partial y^2} = (\beta^2 - \beta_0^2)\mathbf{E} = 0 \tag{13.2a}$$

where the last equality arises because, as just shown, $\beta^2 = \beta_0^2$. Note that Eq. (13.2a) resembles Laplace's equation in two dimensions (see Chapter 6), and, in fact, substituting $\mathbf{E} = -\nabla V$ and interchanging the order of taking derivatives yields

$$\nabla\left(\frac{\partial^2 V}{\partial x^2} + \frac{\partial^2 V}{\partial y^2}\right) = 0 \tag{13.2b}$$

This means that the field configuration can be determined by having V and the transverse fields satisfy the appropriate geometric boundary conditions of the guiding structure, using the techniques developed in Chapter 6.

In attempting to find the solutions to Eq. (13.2b), we would soon discover that there are only a few geometries for which the analytic problem can be solved. In most cases, the configuration of the conductors does not lend itself to direct analysis. The configurations of Fig. 13.1, and most of those in Fig. 11.1, have fields in both air- and dielectric-filled spacial regions. This creates a condition of inhomogeneity that also makes it impossible to obtain an exact analytical solution. The lack of a simple analytic solution does not mean that propagation cannot occur in these cases, but only that a more detailed analysis is necessary to describe the actual propagation, that is, approximation methods must be used.

13.4 LIMITATIONS OF MATERIAL

An objection can be raised that the current that generates the transverse magnetic field flows longitudinally in the conductors, and since $\mathbf{J} = \sigma\mathbf{E}$, this implies the existence of a longitudinal electric field, which implies that **E** is not completely transverse. However, in the analytical limit of *ideal* conductors, that is, as $\sigma \to \infty$, the finite current and magnetic field can be maintained only if $E \to 0$ inside the conductor. While this ideal case does not exist, the materials used in transmission lines are chosen because they are good conductors, and it is hoped that the approximation is a reasonable one; in some cases, corrections must be made.

Also recognize that the real dielectrics used in transmission systems cannot be simply described in terms of a single electric permittivity ε. Atomic systems have inertia and do not respond instantaneously to high-frequency fields. This will not be discussed here, but the main result can be inferred, that there is a delay between the signal and the material response. This causes the polarization **P** to be out of phase with the field **E**, so that the material must be described as having a complex dielectric constant $\varepsilon = \varepsilon' - j\varepsilon''$. The (generally) small imaginary component manifests itself as being equivalent to the dielectric showing losses, as though it had a conductivity (see Exercise 9.4 and Section 10.2). It can be visualized as describing high-frequency currents that flow in the dielectric, and it appears as a transverse conductance between the transmission-line conductors, causing the dissipation of energy.

Incommensurate geometric boundaries, inhomogeneity of the dielectric medium, and the effect of the wall and dielectric currents imply that one does not have a true TEM configuration. If the differences are small, the propagation mode is said to be quasi-TEM. In many cases, the ease and advantages of using idealized transmission-line theory is so great that it is customary to assume a perfect TEM case and the applicability of transmission-line theory. The primary effect of the disturbing factors is to cause energy loss and attenuation of the propagating wave. If the line is a good one, the attenuation is small and it can be included as a first-order correction to the ideal behavior. For some applications, this may not be sufficient and second-order corrections for dispersion must also be included.

13.5 TRANSMISSION LINE THEORY

In Chapter 11, the propagation properties of transmission lines were discussed in terms of the circuit properties per unit length of the line. R, L, G, and C are the resistance, inductance, conductance, and capacitance, respectively, per meter of line length; the circuit model of a transmission line was shown in Fig. 11.3. In analyzing the behavior of this model, two parameters of importance are found. These are the propagation constant γ, where the wave response is taken to be $V \sim V_0 e^{\pm \gamma z + j\omega t}$, and the characteristic impedance Z_0, where $V_0/I_0 = Z_0$ for the traveling waves. Equations (11.15) and (11.20) give their values to be

$$\gamma = j\beta + \alpha = j\omega\sqrt{LC} + \tfrac{1}{2}\omega\sqrt{LC}\left(\frac{R}{\omega L} + \frac{G}{\omega C}\right) \qquad (13.3a)$$

$$Z_0 = \sqrt{\frac{L}{C}} \qquad (13.3b)$$

For a low-loss line, α can be neglected. From the imaginary part of Eq. (13.3a), we see that $v = \omega/\beta = 1/\sqrt{LC}$, and since it has already been established that transmission lines propagate with velocity v_0, this means that

$$v_0 = \frac{1}{\sqrt{LC}} \tag{13.4}$$

Combining Eqs. (13.3b) and (13.4) gives

$$Z_0 = \frac{1}{v_0 C} \tag{13.5}$$

In the most common case, where the transmission line contains a dielectric, recall that $v_0 \propto 1/\sqrt{\varepsilon_r}$ and that $C \propto \varepsilon_r$, so that

$$Z_0 \sqrt{\varepsilon_r} = \frac{1}{v_{0\,(\text{vacuum})}\, C_0} \tag{13.6}$$

where C_0 is the capacitance of the identical air-filled capacitor. Thus, the problem of finding Z_0 reduces to that of determining the static capacitance per meter of length of the transmission line. This is the connection to Eq. (13.2b), indicating that only the electrostatic field problem need be solved. The derivation of Eq. (13.6) has been based on the properties of TEM propagation. Therefore, if the system accommodates TEM waves, then Eqs. (13.3) to (13.6) apply. However, it is important to note that the ability to solve the electrostatic problem, to obtain C_0, does not guarantee that TEM propagation actually occurs. This solution often serves as a useful and convenient approximation to the true field configuration, but it may not be the actual dynamic case.

13.6 STRIPLINE IMPEDANCE

A simple approximation that gives physical insight in the case of stripline is shown in Fig. 13.2(a). Here the capacitance has been divided into that due to the parallel plate construction, C_{pp}, and that due to the fringing field, C_f. All the capacitors shown are in a parallel arrangement between the ground plates and the central conducting strip. From the dimensions in Fig. 13.2(b), and using Eq. (5.4) for the ca-

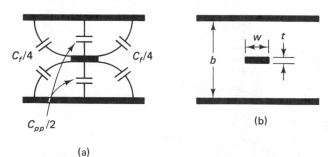

Figure 13.2 Strip line: (a) interelectrode capacitances, and (b) component dimensions.

pacitance of a parallel-plate capacitor, we see that $C_{pp}/2 = \varepsilon_0 w/[(b - t)/2]$. The total capacitance per unit length is

$$C_0 = 2\frac{C_{pp}}{2} + 4\frac{C_f}{4} = \frac{4\varepsilon_0 w}{b - t} + C_f \tag{13.7a}$$

Equation (13.6) then becomes

$$Z_0\sqrt{\varepsilon_r} = \frac{\left(1 - \dfrac{t}{b}\right)\bigg/v_0}{4\varepsilon_0\dfrac{w}{b} + C_f\left(1 - \dfrac{t}{b}\right)} \tag{13.7b}$$

Note that the separation of C into C_{pp} and C_f puts the effect of the lateral extent of the ground or shield planes, beyond the center conductor, entirely into the determination of C_f. In some calculations, it has been shown that there is little change in properties as this extent increases beyond $w + \frac{1}{2}b$.

Most lines in use have $t/b \ll 1$, so that Eq. (13.7b) could be further reduced. Instead, note that for $w/b \gg 1$, the effect of fringing becomes small relative to the parallel-plate capacitance. In this case, Z_0 is inversely proportional to w/b, so that Z_0 is small; if $Z_0 < 25$ ohms, it is a frequently used engineering approximation to neglect C_f.

At the other extreme, for small w/b, corresponding to $Z_0 > 100$ ohms, the line has a small center conductor between two ground planes. This is equivalent to saying that the fringing fields overlap, so that Eqs. (13.7) are inadequate.

Example 13.1

Consider the case where the center conductor is very small. We wish to estimate the capacitance when $w/b \ll 1$ and $t/b \ll 1$. Since details of the conductor shape are not important in this limit, consider that the center conductor is approximated by a cylindrical line charge of radius $t/2$. If the lateral extent of the ground planes exceeds b, then most of the fringing field lines terminate on the inner ground surfaces and the ground planes can be regarded as almost completely enclosing the line charge. The capacitance then should not be sensitive to details of their extent or shape, so that, for the purpose of estimation, we replace the ground planes by a coaxial cylinder of radius $b'/2$, where b' is some effective coupling radius greater than b; a reasonable guess might place $b \le b' \le 2b$.

This approximation structure reduces to the coaxial line shown in Figs. 5.3 and 11.2, and its capacitance per unit length is given by Eqs. (5.5) and (11.6b):

$$C_0 = \frac{2\pi\varepsilon_0}{\ln\dfrac{b'}{t}}$$

With Eq. (13.6), this gives

$$Z_0\sqrt{\varepsilon_r} = 60\ln\frac{b'}{t}$$

A more complete analysis demands the general solution of Laplace's equation, Eq. (13.2b), subject to the boundary conditions of Fig. 13.2(b). If the central con-

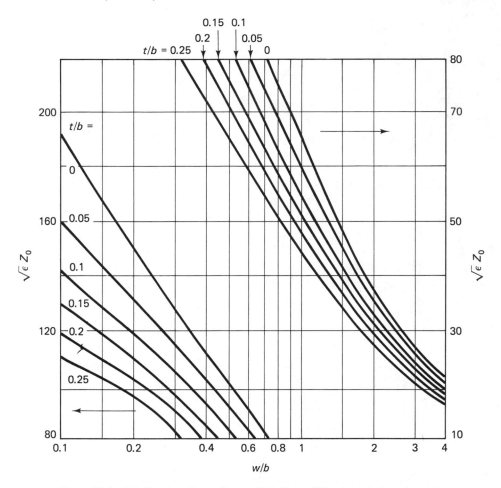

Figure 13.3   Strip-line impedance. *Source:* S.B. Cohn, "Characteristic Impedance of The Shielded-Strip Transmission Line," *Trans. IRE*, 2 (1955), 52-©IRE (now IEEE).

ductor has zero thickness ($t = 0$), a technique called Conformal Mapping has been used to obtain an exact analytic solution for C_0. Using approximation methods, this solution has been extended to lines for which $t \neq 0$. These results are shown in Fig. 13.3. It is left as an exercise to show that Eq. (13.7b) gives reasonable agreement with Fig. 13.3 at the right edge, where $w/b \gg 1$, and that the result of Example 13.1 gives reasonable agreement at the left edge of the figure, where $w/b \ll 1$.

Using the numerical methods described in Chapter 8, the capacitance of any conductor geometry can be found and, by adjusting the problem parameters, the properties of any family of transmission systems can be studied. Many of the exercises at the end of Chapter 8 are, in fact, geometries of planar transmission lines. Their solution gives C_0, and, therefore, Z_0.

13.7 ATTENUATION

Losses in striplines can be examined by expanding Eq. (13.3a) and substituting Eq. (13.3b) to obtain, for the attenuation coefficient,

$$\alpha = \frac{1}{2}\left(\frac{R}{Z_0} + GZ_0\right) \tag{13.8}$$

where the first term represents conductor losses, and the second is due to the dielectric. R, the longitudinal resistance per unit length of line, can be estimated in analogy with Eq. (4.5): $R = 1/\sigma A$ where A is the cross-sectional area traversed by the conduction current [note the difference from Eq. (4.5), that here R is in units of ohms/meter, whereas R is in ohms in Chapter 4]. At high frequencies, the skin-depth phenomenon is significant, so that current only penetrates a distance $\delta = (2/\omega\mu\sigma)^{1/2}$.

On the ground or shield planes, we expect the greatest current density to be in the center, although it is not restricted to the width w. Denoting the effective conduction width by $w + w'$, as shown in Fig. 13.4, the resistance of one shield is

$$R = \frac{1}{\sigma\delta(w + w')} = \frac{\sqrt{\omega\mu/2\sigma}}{w(1 + w'/w)} = \frac{\sqrt{\omega\mu/2\sigma}}{w(1 + C_f/C_{pp})} \tag{13.9}$$

where the ratio w'/w has been taken to be the same as the ratio of the fringing field to the central field, which is equal to C_f/C_{pp}. The resistance of the central strip also depends upon its current distribution, which cannot be uniform along the width w because of fringing fields and edge currents. A numerical determination of the distribution might be made, similar to that done in Example 8.2, but for approximation purposes, we assume that $R_{strip} \approx R_{shield}$. Then, $\frac{1}{2}R_{strip}$ is in series with $\frac{1}{2}R$ of Eq. (13.9); the factors $\frac{1}{2}$ arise because the two ground planes are in parallel and the two strip halves are in parallel. Consequently, Eq. (13.9) represents the full resistance.

The second term in Eq. (13.8) gives the dielectric loss in terms of the admittance G. This can be treated by considering that propagation proceeds according to the wave form $e^{j\omega t - kz}$, where $k = j\omega\sqrt{\mu\varepsilon}$. Recall the earlier comments that in a real dielectric,

$$\varepsilon = \varepsilon' - j\varepsilon'' \quad \text{then} \quad k = j\omega\sqrt{\mu\varepsilon'}\left(1 - j\frac{\varepsilon''}{\varepsilon'}\right)^{1/2} = j\beta(1 - \tfrac{1}{2}j\tan\delta) \tag{13.10a}$$

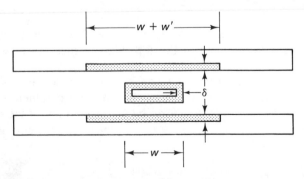

Figure 13.4   Current distribution on a strip line.

where $\beta = \omega\sqrt{\mu\varepsilon'}$ is the normal propagation number, and the term $(1 - j\varepsilon''/\varepsilon')^{1/2}$ has been expanded in powers of $\varepsilon''/\varepsilon'$. Only the leading term of this expansion has to be retained since the loss tangent, $\tan\delta = \varepsilon''/\varepsilon' \ll 1$ for most useful material. Inserting this value of k results in the waveform $e^{-\alpha_d z}e^{j(\omega t - \beta z)}$, where the amplitude attenuation factor is

$$\alpha_{\text{dielectric}} = \frac{\beta}{2}\tan\delta = \frac{\omega\sqrt{\varepsilon_r}\,\tan\delta}{2v_{0\,(\text{vacuum})}}\ \text{n/m}$$

$$= \frac{27.3f\sqrt{\varepsilon_r}\,\tan\delta}{v_{0\,(\text{vacuum})}}\ \text{dB/m}$$

(13.10b)

The conversion from nepers to decibels is taken from Eq. (11.46b).

$\alpha_{\text{dielectric}}$ has been derived from the expression for k of a plane wave. While it is appropriate to use it for the transmission-line TEM wave, recall that this, like the expression for R, is only an approximation. For example, note that Eq. (13.10b) is independent of the geometry of the stripline. The actual fields are more concentrated in the region of the center conductor, so that dielectric losses are enhanced there. More complete analyses, including such geometric effects, give $\alpha_{\text{dielectric}}$ as being more than four times greater than Eq. (13.10b).

There is another attenuation term for a stripline because its open sides allow radiation from the rapidly varying fringe fields. However, because of the typically small line spacing, these fields are quite small; for air lines with $Z_0 < 100$ ohms and whose ground planes are wider than $w + 3b$, the radiation loss can usually be taken as negligible compared to the other loss terms.

13.8 MICROSTRIP

Figure 13.1(c) shows an asymmetric microstrip structure that is commonly used. It consists of a relatively large ground plane and a smaller conducting strip, separated by a thin layer of dielectric, such as a thin board of glass–epoxy, or a layer of alumina, or of silicon dioxide. Unfortunately, the asymmetry of the composite dielectric structure guarantees that a TEM mode cannot exist. In spite of this, and while more exact (though cumbersome) analyses of propagation have been carried out, it is still common to take the convenient approach of using transmission-line concepts. Corrections are then made in greater detail.

The dielectric layer in a microstrip tends to concentrate and localize the fields in the region between the conducting strip and the ground plane, that is, in the dielectric. As a result, the usual approximation assumes that the line is entirely embedded within a uniform dielectric, and an effective dielectric permittivity is used, deriving its value from estimates of the fractional air-to-dielectric propagation. This estimation can be simple or very sophisticated; for our purposes, we merely note that since the half space between the two conductors is dielectric-filled, at least half of the field resides in the dielectric, so we expect that $\frac{1}{2} < \varepsilon_r/\varepsilon_{r\,(\text{dielectric})} < 1$.

Example 13.2

Figure 13.5(a) shows a microstrip line for which the dielectric ($\varepsilon_r = 5$) occupies only the region under the central conductor. This line is equivalent to the capacitor analyzed numeri-

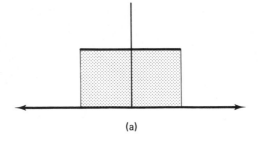

(a)

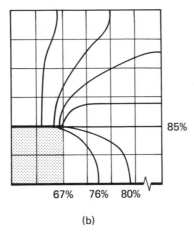

85%

67%   76%   80%

(b)

Figure 13.5   (a) Microstrip-line geometry.
(b) Field lines, as solved in Chapter 8.
Figures show percentage of total energy.

cally in Example 8.3 (That example had two conductors with a central electrical ground plane, which we take here as the actual ground.) Those fringing-field calculations result in values of the potential at nodes of the spacial mesh, and from these the field lines can be drawn and field values determined. The energy in each volume of space can then be determined. Figure 13.5(b) shows the percentage of the total energy contained within a few field lines. The horizontal plane through the upper conductor encloses 85 percent of the field energy.

The approximations involved in solving this problem in Chapter 8 (primarily, moving the infinitely distant points to a finite distance) lead to departures from the true solution, such as the line terminations on the outer boundaries and an intensification of the local fields. However, even so, we see that ~67 percent of the power flows in the dielectric beneath the center strip. If the dielectric extends beyond the central plate, and if the field-concentrating effect of the dielectric is taken into account, the percentage will be significantly greater. This reasoning rationalizes the approximation that the entire field lies in the (or a closely related) dielectric.

In this case, from the energy flow, we might estimate that $\varepsilon_{\text{effective}} \approx 0.67\varepsilon_r\varepsilon_0 + 0.33\varepsilon_0 = 3.67\varepsilon_0$.

After calculating the capacitance per unit length of a uniform microstrip, Eq. (13.6) gives the characteristic impedance. As in the case of the doubly grounded stripline, Conformal Mapping can also give values of C for microstrip. This method has been applied in different ways in the literature; one such result* for the case that $t = 0$ reduces to the form

$$Z_0\sqrt{\varepsilon_r} = 50\frac{b}{w}\left(6 - \frac{\ln\dfrac{11b}{w}}{\ln 2}\right) \qquad t = 0 \tag{13.11}$$

The calculation of ohmic and dielectric attenuation for a microstrip is similar to that for stripline, as previously given, with modification for the different geometry. In this case, however, the open structure means that radiation losses cannot be ignored, although small line-to-ground spacings and high values of the dielectric permittivity tend to hold them to tolerable limits. In practice, microstrip lines tend to be used in relatively short lengths, so it is somewhat inappropriate to calculate an attenuation coefficient for radiation losses from a long uniform line. Instead, the total radiation from a comparable two-wire line has been found, in which the entire transmission line is regarded as an antenna whose length $\ell \gg b$, the strip separation, and wavelength $\lambda \gg b$. For a matched line, that is, with only a forward traveling wave, the radiated power has been found to be[†]

$$P_{\text{rad}} = 160I^2\left(\frac{\pi b}{\lambda}\right)^2 \text{ watts} \tag{13.12}$$

Example 13.3

It is useful to make a comparison of the radiation loss with the copper and dielectric losses of a microstrip line. For these, we use the expressions for the attenuation coefficients of a stripline, Eq. (13.10b) for $\alpha_{\text{dielectric}}$ and Eqs. (13.8) and (13.9) for $\alpha_{\text{conduction}}$, except that it is necessary to double $\alpha_{\text{conduction}}$ to correct for the fact that only half the conduction area is available when calculating R for a microstrip line. For comparison with Eq. (13.12), it is necessary to obtain the total loss due to each of these attenuation coefficients. The line is assumed to have a length $dz = 25$ cm ($\frac{1}{4}$ m) in Eq. (11.39).

$$P_{\text{cond}} = dP = 2\alpha_{\text{cond}}P\,dz = 0.5\alpha_{\text{cond}}I_{\text{rms}}^2 Z_0$$

$$P_{\text{diel}} = dP = 2\alpha_{\text{diel}}P\,dz = 0.5\alpha_{\text{diel}}I_{\text{rms}}^2 Z_0$$

where the transmitted power of a traveling wave on the line is $I_{\text{rms}}^2 Z_0$.

For a line with $w = b = 5$ mm and $\varepsilon = 6$, Eq. (13.11) gives $Z_0 = 51$ ohms. Furthermore, if $\sigma = 5 \times 10^7$ S/m (copper), $\tan \delta = 5 \times 10^{-4}$, and $C_f/C_{pp} = 0.25$, then, for $f = 10^9$ Hz, we find

$$P_{\text{radiation}} : P_{\text{dielectric}} : P_{\text{conduction}} = 0.44 : 0.32 : 0.54$$

Within the variability of the parameters chosen, and the approximations of the various loss terms, this indicates that each of the three mechanisms produce about the same loss for this line and at this frequency.

*J. D. R. McQuillan, "Design Problems of a Megabit Storage Matrix for Use in a High Speed Computer," *IRE Trans. on Elect. Comp.*, EC 11 (1962), 390.

[†]E. J. Sterba and C. B. Feldman, "Transmission Lines for Short-Wave Radio Systems," *Proc. IRE, 20* (1932), 1163.

Here it is important to note, from Eqs. (13.8), (13.9), (13.10b), and (13.12), that

$$P_{\text{radiation}} : P_{\text{dielectric}} : P_{\text{conduction}} \sim \omega^2 : \omega : \omega^{1/2}$$

Therefore, for the same line, but lowering the frequency to 10^7 Hz,

$$P_{\text{radiation}} : P_{\text{dielectric}} : P_{\text{conduction}} = 10^{-2}(0.0044 : 0.32 : 5.4)$$

While all the absolute losses are reduced, their relative roles have changed, showing that conduction loss dominates at low frequencies.

Raising the frequency to 10^{10} Hz yields

$$P_{\text{radiation}} : P_{\text{dielectric}} : P_{\text{conduction}} = 44 : 3.2 : 1.7$$

Here the absolute losses are increased, and, again, their relative importances have shifted, showing that radiation dominates at high frequencies.

Although dielectric dissipation is significant throughout, it is generally not the major loss mechanism except in the vicinity of specific molecular dielectric absorption frequencies. These are scattered through the millimeter wavelength range and at higher frequencies, so that transmission in narrow absorption bands may be limited.

Example 13.3 indicates that radiation from the open structure of microstrip lines becomes quite large as the frequency is raised. More important than the signal attenuation and distortion that results from the loss of energy is the interference problem. In any system where microstrip circuits are used, the radiation from one part is intercepted by other parts, either as magnetic inductive pickup or as electric pickup. This is a situation of self-induced RFI (radio-frequency interference). The result is called *crosstalk* between various parts of the same or different circuits, with the appearance of noise, spurious signals, component overload, etc. Because of the convenience of using microstrip lines, the limits of tolerance of these signals have been expanded, but it has been pointed out that there is an ultimate frequency limit for the use of microstrip circuits because of these effects.

13.9 CIRCUIT ELEMENTS

A wide variety of passive circuit elements are relatively easy to make using stripline and microstrip construction. Thus, a series capacitance can be made by overlapping a section of insulated central conductor or by cutting a gap in the central strip. A series inductance is generated by making a hole in the center conductor. Variable or fixed attenuators can be constructed by pivoting or placing lossy media in the field region.

Couplers can be produced by running the central conductors of two parallel lines in close proximity for a distance of from a quarter to several wavelengths, as shown in the top of Fig. 13.6(a). Another method is to supply high-impedance bridges between the central conductors, as shown in the bottom of Fig. 13.6(a). This construction uses the phase and amplitude changes along the multiple bridges as a means of providing directionality and attenuation. In these cases, the coupling is sensitive to the interline spacing and the bridge dimensions; such couplers are often strongly frequency-dependent. Overlapping two lines and cutting periodic holes in their common ground provides a more easily controlled degree of coupling that can be made less frequency dependent.

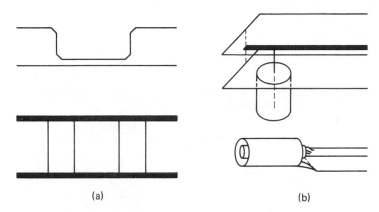

Figure 13.6   (a) Interline couplers. (b) Coax to strip- and microstrip-line connections.

Transitions to coaxial cable can be constructed by inserting the cable normal to the plane of the planar line and connecting the corresponding leads, as shown in the top of Fig. 13.6(b). Alternatively, with the coax running parallel to the strip plane, the outer coax conductor is extended to the strip-line ground and the center conductor is flattened into the strip, as shown in the bottom of Fig. 13.6(b). In such cases, a more gradual transition produces lower reflections, facilitating impedance matching between lines.

Waveguide transitions can involve a center conducting strip in the waveguide, tapering down to the flat strip of the planar line. Alternatively, with either wave-guide or coaxial line, a detector-signal launcher can be placed in one system and the generated signal then used to excite the other system.

One of the advantages offered by the planar waveguide systems is the ease with which these lines can be connected to the leads and pads of modern integrated circuits and elements.

13.10 MODE PURITY

Modern fast-switching elements generate nanosecond and picosecond pulses that must be propagated with fidelity to distant points. A single square pulse of duration τ has a continuous frequency content, with 99 percent of its energy contained in the band below $3/\tau$. For a 30 picosecond pulse, the line must be capable of propagating frequencies up to 100 GHz.

No distortion would result if the transmission system could operate in the true TEM mode, since all component frequencies would then travel with the same phase velocity. However, even for such systems, it is *possible* to excite higher non-TEM modes at sufficiently high frequencies. In that case part of the signal propagates in a TEM-like mode and part in a higher mode and the resulting loss of phase coherence causes distortion in the transmitted signal. The analysis of waveguides in Chapter 12 shows that the lowest non-TEM mode propagates when it is above its cutoff, which occurs when the frequency is greater than that for which $\lambda/2$ equals the transverse guide size. Although more precise cutoff limits have been established in each case,

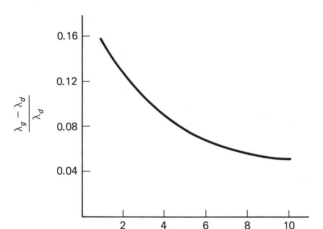

Figure 13.7 Microstrip dispersion ($t = 0$).

we can estimate the cutoff limits for planar lines if we assume that the waveguide criterion applies. Then higher modes are possible when $\lambda/2 < w + b$. For a dielectric loaded line ($\varepsilon_r = 4$) with $w + b = 1$ cm, this means that higher modes can be generated for frequencies greater than ~ 15 GHz.

Early in the discussion, it was indicated that the actual propagation mode in microstrip is not even quasi-TEM. The true lowest mode is a TE mode, and, as is the case for waveguide, this results in a difference between the guide wavelength and the wavelength of a TEM wave at the same frequency. Figure 13.7 shows this difference for a microstrip line having $t = 0$ and a large ground plane width ($>3w$). In the figure, λ_g is the wavelength along the microstrip line and λ_d denotes the wavelength in the corresponding infinite dielectric.*

EXERCISES

13.1. In deriving the transmission-line equations in Chapter 11, it was assumed that $R/\omega L \ll 1$ and $G/\omega C \ll 1$.

    (a) In this case, show that attenuation is a first-order correction to the $R = 0 = G$ solution and verify Eq. (13.8).

    (b) Show that frequency dispersion is a second-order correction and derive its form.

13.2. Consider an air-filled stripline that has $t \approx 0$ and $w/b = 1$. Use numerical programming to find C_0 for the lateral ground plane extent ranging from w to $10w$. Does there seem to be some point beyond which you can consider that fringing has no further effect? What are the values of Z_0 over this range?

*A. F. Harvey, "Parallel Plate Transmission Systems for Microwave Frequencies," *Proc. IEE,* 106B (1939), 129; F. Assadourian and E. Rimai, "Simplified Theory of Microstrip Transmission Systems," *Proc. IRE,* 40 (1952), 1651.

13.3. From Exercise 13.2, examine the power radiated from the edges of the ground planes when these planes have widths ranging from w to $10w$ and if this radiation is proportional to the square of the fringing field at the open ends. Determine whether there is a point beyond which further extension of the ground planes becomes an inefficient way to reduce radiation.

13.4. In discussing radiation from a microstrip, it was noted that this leads to distortion of the microstrip signal. Explain why distortion occurs, as opposed to signal attenuation alone. Consider these arguments with respect to conduction losses.

13.5. Using the integral method of numerical analysis discussed in Chapter 8, find the current distributions on the shield plates and strip of Fig. 13.4. What is the 90 percent value of w'?

13.6. Imagine that a microstrip is approximated by two coaxial semicircular cylindrical surfaces of radii r_1 and r_2 (half of a coaxial line). Neglect fringing and find Z_0. By making a correspondence of r_1, r_2, and the dimensions of a flat microstrip, compare your result to Eq. (13.11).

13.7. Read values from Fig. 13.3 at $w/b = 0.1$ and plot $\sqrt{\varepsilon}Z_0$ as a function of t/b. Compare these values with the prediction of Example 13.1.

13.8. Carefully read values from Fig. 13.3 at $w/b = 4$ and plot $\sqrt{\varepsilon}Z_0$ as a function of t/b. Compare these values with the prediction of Eq. (13.7b).

13.9. Using numerical methods, repeat the field plot of Fig. 13.5 for a microstrip with w/b ranging from 0.1 through 10. Assume an infinite ground plane. Compare the impedance values with Fig. 13.3 and Eq. (13.11).

13.10. Verify the 67 percent and 85 percent energy contours on Fig. 13.5(b). *Hint:* See Fig. 8.10.

13.11. Verify the values of $P_{\text{radiation}} : P_{\text{dielectric}} : P_{\text{conduction}}$ in Example 13.3 at 10^9 Hz.

13.12. While inductive effects cannot be ignored at high frequencies, it is tempting to analyze the coupler in the top of Fig. 13.6(a) in terms of only its capacitive coupling. Consider two very thin parallel lines that are separated by the distance s over a length of one wavelength of the propagating signals.
   (a) Estimate their capacitance.
   (b) Since a harmonic current in one line corresponds to a charge distribution $q = I/j\omega$, find the current that is induced in the other line.
   (c) State the coupling factor in dB as a function of s.

13.13. A square pulse of 0.1-nanosecond duration is injected onto the microstrip line of Example 13.3.
   (a) What is the Fourier harmonic description of the initial signal?
   (b) How has this description changed after traveling 25 cm?
   (c) Estimate how this affects the shape and magnitude of the output signal.

13.14. Imagine that a stripline or microstrip line is operating at a frequency f_0, where propagation can occur in either the quasi-TEM or in the next higher mode. The TEM mode has $f_c = 0$ and the next higher mode has its cutoff at $f_c = 0.8 f_0$. Because propagation in the various modes follows the rule $\beta^2 = \beta_0^2 - \beta_c^2$, find the fractional difference of wavelength and phase velocity of the signal in the two modes.

13.15. Use your answer to Exercise 13.1 and Eq. (13.9) for R of a stripline to find the dispersion of a stripline. For a 50-ohm stripline with $t \approx 0$ and $\varepsilon_{\text{effective}} = 6\varepsilon_0$, plot the dispersion between 10^8 and 10^{11} Hz. Neglect dielectric loss.

13.16. Repeat Problem 13.15 for the microstrip line of Example 13.3.

14

RADIATING SYSTEMS

14.1 ELECTRIC DIPOLE ANTENNAS

Introduction of the time-dependent terms in Maxwell's equations leads to field solutions that represent traveling waves, that is, in which electromagnetic energy travels through space, away from its source. This is different from the static case, where the electric or magnetic field is unchanging with respect to its charge or current source. Although the properties of plane waves were examined in Chapters 9 and 10, the origin of the waves was not considered. When wave-conducting systems were studied in Chapters 11 to 13, the location of the wave source was of interest, but concern was primarily with the mode of propagation and location of the load. In this chapter, the focus is specifically on the source of radiated electromagnetic energy.

First, consider the small time-varying current element shown in Fig. 14.1(a), specified in terms of its electrical and physical magnitudes and its direction by $I(t)\,d\ell$ An oscillating electric dipole, known as a Hertzian dipole, is shown in Fig. 14.1(b). This is specified in terms of its dipole moment $\mathbf{p}(t) = q\boldsymbol{\ell}(t)$. The current element and dipole are seen to be equivalent since $I = dq/dt$, so that $I\,d\ell = d(q\ell)/dt = d\mathbf{p}/dt$. Partly because of this equivalence, the current element of Fig. 14.1(a) is referred to as an *electric-dipole antenna*.

The method of actually exciting such a dipole radiator, by means of shielded leads, is shown in Fig. 14.1(c). Here the antenna is said to be center-fed. If the current is constant along the length, as shown in Fig. 14.1(a), then it is clear that posi-

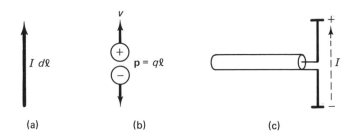

(a)   (b)   (c)   **Figure 14.1**   Antenna representations.

tive charge must leave the bottom half of the antenna and accumulate at the upper half; during the next half cycle, the current and polarities are reversed. This implies a means of storing the charge, as indicated by the plates at the ends of the current element in Fig. 14.1(c). Therefore, instead of the oscillating electric current we might imagine an electric dipole with oscillating *magnitude* at the end plate locations; this is an alternative view of the Hertzian dipole. The end-plate charge is related to the current by $q = \int I\, dt$. The oscillating-current element generates an alternating magnetic field and the oscillating electric charges generate an alternating electric field. Because of the relationship between charge and current, it is only necessary to consider one of them; we will take the current element as the source of the fields.

Rather than calculating the fields directly, it is convenient to consider potential functions of the radiator and to derive the fields from these functions.

14.2 RETARDED POTENTIAL

In Chapter 7, the relationships between magnetic flux density **B**, the magnetic vector potential **A**, and the current element $I\,d\ell$ were found to be

$$\mathbf{A} = \frac{\mu_0}{4\pi} \int \frac{I\,d\ell}{r} \qquad \mathbf{B} = \nabla \times \mathbf{A} \qquad (14.1a, b)$$

where the integral is over all current source elements, and r is the distance from each element to the field point. Consider a small electric dipole at the origin of coordinates, with its orientation long $\mathbf{a}_z$; for an infinitesimal dipole, its total length can be taken to be $d\ell = \ell = \ell \mathbf{a}_z$, so that

$$\mathbf{A} = \frac{\mu_0 I \ell}{4\pi r} = \mathbf{a}_z \frac{\mu_0 I \ell}{4\pi r} \qquad (14.2)$$

If I is constant, then Eq. (14.2) gives the potential at all points in space. If I changes with time, for example, for a sinusoidal variation,

$$I(t) = I_0 e^{j\omega t} \qquad (14.3a)$$

then so does **A**. The change of **A** will be sensed at each field point with a delay due to propagation from the source dipole to the field point. Since electromagnetic propagation is at the speed of light v_0, the dipole current at time t is sensed at a later time, being delayed by $\Delta t = r/v_0$. Instead, consider that, at each instant, the value of **A** at the field point is due to the current value at an earlier time, by Δt. Thus, the current that is effective in generating the potential at each instant at the field point is

$$I\left(t - \frac{r}{v_0}\right) = I_0 e^{j\omega(t - r/v_0)} \tag{14.3b}$$

Substituting this into Eq. (14.2) gives the potential at the field point, including retardation due to propagation:

$$\mathbf{A}(r, t) = \frac{\mu_0 I_0 \ell}{4\pi r} \, \mathbf{a}_z e^{j\omega(t - r/v_0)} \tag{14.4a}$$

A in Eq. (14.4a) is called a *retarded potential*. It contains a factor r in the exponent as well as in the denominator, a fact that complicates the curl operation of Eq. (14.1b) relative to the static cases considered in Chapter 7. However, the operation can still be carried out in spherical coordinates by using the same techniques employed in connection with Eq. (7.7). Thus, changing $\mathbf{a}_z$ to spherical coordinates,

$$\mathbf{A}(r, t) = \frac{\mu_0 I_0 \ell}{4\pi r} e^{j\omega(t - r/v_0)} (\mathbf{a}_r \cos\theta - \mathbf{a}_\theta \sin\theta) \tag{14.4b}$$

$$= \mathbf{a}_r A_r(r, \theta) + \mathbf{a}_\theta A_\theta(r, \theta) \tag{14.4c}$$

With this form, the expression for $\nabla \times \mathbf{A}$ yields neither an $\mathbf{a}_r$ component nor an $\mathbf{a}_\theta$ component and

$$\mathbf{B}(r, t) = \frac{\mu_0 I_0 \ell}{4\pi r} \sin\theta \; e^{j(\omega t - \beta r)} j\beta\left(1 - \frac{j}{\beta r}\right) \mathbf{a}_\phi \tag{14.5a}$$

where, in the exponent, $\beta = \omega/v_0$. The corresponding expression for **E** is found by substituting Eq. (14.5a) into Maxwell's equation, $\nabla \times \mathbf{B} = \mu\varepsilon j\omega \mathbf{E}$, giving

$$j\omega\mu\varepsilon\mathbf{E}(r, t) = \frac{\mathbf{a}_r}{r \sin\theta} \frac{\partial}{\partial\theta}(B_\phi \sin\theta) + \frac{\mathbf{a}_\theta}{r} \frac{\partial}{\partial r}(rB_\phi) \tag{14.5b}$$

$$\mathbf{E}(r, t) = \frac{I_0 \ell}{j4\pi r \omega\varepsilon_0} e^{j(\omega t - \beta r)} \left[\mathbf{a}_r \frac{j\beta}{r}\left(1 - \frac{j}{\beta r}\right) 2\cos\theta \right.$$

$$\left. - \mathbf{a}_\theta \beta^2 \left(1 - \frac{j}{\beta r} - \frac{1}{\beta^2 r^2}\right) \sin\theta \right] \tag{14.5c}$$

14.3 FAR FIELDS

Several features of Eqs. (14.5) are notable. **B** is entirely along $\mathbf{a}_\phi$, that is, it curls about the current. In contrast, **E** has an $\mathbf{a}_r$ component and an $\mathbf{a}_\theta$ component. Various parts of the fields are in phase quadrature (90° different, denoted by the factor j) to others, and different parts vary proportional to $1/r$, or $1/r^2$, or even $1/r^3$. This last feature is particularly significant because radiating systems, such as radio antennas or atomic emitters, are viewed from large distances. As a result, it is generally true that $1/r^3 \ll 1/r^2 \ll 1/r$. More specifically, examination of Eqs. (14.5a) and (14.5c) indicates that the smaller terms are negligible as long as

$$\beta r = 2\pi \frac{r}{\lambda} \gg 1 \qquad (14.6)$$

In the case that this applies we are said to be in the *far field* region and

$$\mathbf{H} = \frac{jI_0\beta\ell}{4\pi r} \mathbf{a}_\phi \sin\theta \; e^{j(\omega t - \beta r)} \qquad (14.7a)$$

$$\mathbf{E} = \frac{jI_0\beta\ell}{4\pi r} \sqrt{\frac{\mu_0}{\varepsilon_0}} \mathbf{a}_\theta \sin\theta \; e^{j(\omega t - \beta r)} \qquad (14.7b)$$

Note that these radiated far fields are orthogonal in space and are in phase. The relation of their magnitudes can be expressed as the impedance

$$\eta = \frac{E_\theta}{H_\phi} = \sqrt{\frac{\mu_0}{\varepsilon_0}} = \eta_0 \qquad (14.8)$$

which is the same as for a plane wave. It thus appears that while the radiation from a small dipole is quite complicated near the dipole, it becomes like a plane wave at large distances.

As indicated by the exponential factor $\omega t - \beta r$, the fields travel radially outward from the antenna and, at a very large radius, where the surface of the sphere of radiation appears locally to be flat, assumes the form of a forward-moving plane wave. Equations (14.7) differ from those of a plane wave in that they contain a factor r in their denominators. The radiated wave, therefore, carries a reminder that it is not exactly a plane wave, although if we examine such waves over local distances, we find, for example, that $|dE/E| = |-dr/r| \ll 1$, so that the change of $|E|$ can be ignored.

We have seen that the near-field of the dipole contains terms that have a higher power dependence on $1/\beta r$ and therefore vanish in the far field region. Thus, the E_r component vanishes, leaving only E_θ. In addition, both $\mathbf{E}$ and $\mathbf{H}$ have contributions which do not carry a factor j, indicating that they differ in phase by 90° from the terms which have the j. These out-of-phase components also vanish in the far field region. Example 14.1 demonstrates the rapidity with which the transition occurs from near to far fields.

Example 14.1

For a small antenna operating at 30 MHz, compare the magnitudes of E_r and E_θ at $r = 5$, 10, 50, 100, and 1000 m. At these distances, compare the components of **B** that are 90° out of phase.

Both E_θ and E_r have contributions that contain the factor j. From Eq. (14.5c) the magnitudes of these complex components are

$$\frac{|E_r|}{|E_\theta|} = \frac{2\dfrac{\beta}{r}\left[1 + \left(\dfrac{1}{\beta r}\right)^2\right]^{1/2}}{\beta^2\left[\left(1 - \dfrac{1}{\beta^2 r^2}\right)^2 + \left(\dfrac{1}{\beta r}\right)^2\right]^{1/2}}$$

In this equation, $\theta = 45°$ has been arbitrarily assumed so that $\sin\theta = \cos\theta$, so these factors cancel. Multiplying numerator and denominator by $\beta^4 r^4$ gives

$$\frac{|E_r|}{|E_\theta|} = \frac{2[(\beta r)^2 + 1]^{1/2}}{[(\beta r)^4 - (\beta r)^2 + 1]^{1/2}}$$

To compare the H-field component which is 90° out of phase with the magnetic far field, Eqs. (14.5a) and (14.7a) give

$$\frac{H_{90°}}{H_{\text{far component}}} = \frac{1}{\beta r}$$

At $f = 30$ MHz, $\lambda = 10$ m, and $\beta r = \pi r/5$. The table shows the results of these comparisons:

r (m)	$H_{90°}/H_{\text{far}}$	E_r/E_θ
5	0.32	0.70
10	0.16	0.33
50	0.032	0.064
100	0.016	0.032
1000	0.0016	0.0032

It is seen from this that the radial component of E vanishes and the total H field approaches its far-field value fairly rapidly, differing by only a few percent at 5 to 10λ (50 to 100 m in this case).

14.4 RADIATION PATTERN

Power flows outward from the antenna and the average radiated intensity is found from Eq. (9.38b) for Poynting's vector:

$$\mathcal{P} = \frac{1}{2} \mathbf{E} \times \mathbf{H}^* = \frac{1}{2}\frac{I_0\beta\ell}{4\pi r^2}\sqrt{\frac{\mu_0}{\varepsilon_0}}\sin^2\theta\, \mathbf{a}_r \qquad (14.9)$$

This shows that the radiated intensity is a function of angle of radiation. The total power radiated by the antenna is obtained by integrating this time-averaged $\mathcal{P}$ over the surface of a sphere surrounding the dipole. Since the dipole is at the origin, the normal to the spherical surface is $\mathbf{a}_r$ and

$$W = \oint \mathcal{P}\cdot\mathbf{a}_r\, dS = \int_{\phi=0}^{2\pi}\int_{\theta=0}^{\pi}\mathcal{P}r^2\sin\theta\, d\theta\, d\phi = 40\pi^2 I_0^2\left(\frac{\ell}{\lambda}\right)^2 \qquad (14.10)$$

where $\eta_0 = \sqrt{\mu_0/\varepsilon_0} = 120\pi$, and $\beta = 2\pi/\lambda$. Note that this result is independent of r, as it must if energy is conserved, so that the same power crosses every enclosing sphere. This conservation of energy is a consequence of cancellation of the geometric r^2 in Eq. (14.10) with the $1/r^2$ in the intensity expression of Eq. (14.9), a term due to the fact that each of the radiated fields varies as $1/r$. The cancellation constitutes a sensitive verification of the interrelatedness of the conservation of energy, the geometry of the universe, and the $1/r$ dependence of radiation fields.

From the viewpoint of the energy source exciting the antenna, for example, through a transmission line, the antenna appears to be a load that absorbs power (the generator doesn't distinguish the fact that it is then radiated), that is, it looks like a resistor. The effective radiation resistance of the antenna is defined as

$$R_{\text{radiation}} = \frac{W}{I_{\text{rms}}^2} = 80\pi^2\left(\frac{\ell}{\lambda}\right)^2 \tag{14.11}$$

where, for a sinusoidal current, $I_{\text{rms}} = I_0/\sqrt{2}$.

Example 14.2

Find the radiation resistance of a small antenna operating at 30 MHz if $\ell = 10$ cm, 1 m, and 10 m.

At this frequency, $\lambda = 10$ m and

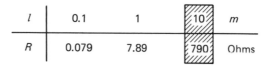

l	0.1	1	10	m
R	0.079	7.89	790	Ohms

Note from Eq. (14.11) that R, and therefore the radiation efficiency, increases as the square of the antenna length. The value of R is small in the region where Eq. (14.11) applies, since ℓ is small. In fact, shading the table value for $\ell = 10$ m indicates that this value is beyond the range of the approximation $\ell/\lambda \ll 1$, that is, for a small antenna.

In addition to radiation losses from the antenna, currents on the surface of the antenna (be it a dipole of the sort considered here, a horn antenna, etc.) experience ohmic electrical resistance that must be added to R_{rad}. For an efficient antenna, we would hope to have this $R_{\text{ohmic}} \ll R_{\text{rad}}$. In addition, for larger antennas, there are phase differences between the currents at different points of the antenna. These appear as inductive and capacitive effects, so that the antenna's input resistance is actually an impedance, that is, it is a complex number. These features will not be discussed here.

The spacial average power per unit area (the average intensity of the wave, $\langle \mathcal{P} \rangle$) is found from Eq. (14.10) to be

$$\langle \mathcal{P} \rangle = \frac{W}{4\pi r^2} = \frac{10 I_0^2 (\beta \ell)^2}{4\pi r^2} \tag{14.12}$$

This is the intensity we would expect at every point from an antenna with the same I_0, β, and ℓ, but which radiates uniformly in all directions. The gain G of any antenna is defined relative to such an omnidirectional antenna. For the electric dipole antenna,

$$G(\theta, \phi) = \frac{\mathcal{P}}{\langle \mathcal{P} \rangle} = 1.5\,\sin^2\theta \tag{14.13}$$

The maximum gain is 1.5 at $\theta = 90°$, implying that 50 percent more power per unit area is radiated in this direction than if an omnidirectional standard (but hypothetical) antenna were used, radiating the same total power. A graph of $G(\theta, \phi)$, called the *radiation pattern,* is shown in Fig. 14.2(a). Here the radial length is proportional to G at each θ. More commonly, the scale of G is normalized to unity, as shown in Fig. 14.2(b).

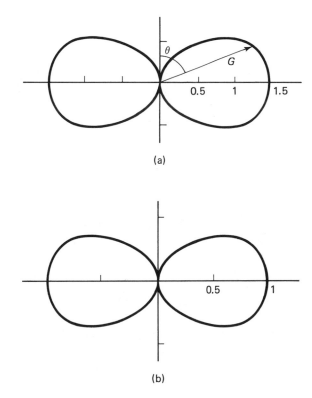

(a)

(b)

Figure 14.2 Dipole-antenna radiation pattern.

14.5 RECEIVING ANTENNAS

Consider two antennas distantly separated in a region of space. Without knowing about the antenna types and properties, we observe that if antenna 1 is excited with current I_1 and antenna 2 with current I_2, certain voltages V_1 and V_2 exist at the antenna inputs. The antennas and intervening space can be regarded as an unknown circuit having two ports, as shown in Fig. 14.3. Such a circuit can be described by the general network relations

$$V_1 = Z_{11}I_1 + Z_{12}I_2$$
$$V_2 = Z_{21}I_1 + Z_{22}I_2 \tag{14.14a}$$

where Z_{11} and Z_{22} are the antennas' self-impedances, each of which would be measured even if there is no excitation current in the other antenna.

The theorem of reciprocity is a very powerful symmetry statement that applies under a broad range of conditions, almost always met in circuit or antenna

Figure 14.3 Two-port network.

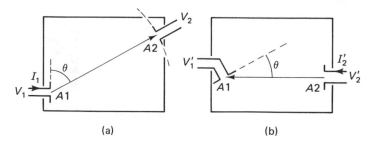

Figure 14.4 Reciprocity determination.

operations, that is, that the circuit defined by the box of Fig. 14.3 does not contain active or nonreversible elements, for example, sources, diodes, magnetic rotators. This theorem specifies that the cross-diagonal impedances of Eq. (14.14a) are equal, that is,

$$Z_{12} = Z_{21} \tag{14.14b}$$

Note, in Eq. (14.14a), that $V_2 = Z_{21}I_1$ gives the signal in antenna 2 (A2) if I_2, the current in A2, is zero (or at least *very* small). This would be the case if only antenna 1 (A1) is directly excited. Assume for the moment, as shown in Fig. 14.4(a), that A1 is stationary and A2 moves tangentially to the circle centered on A1. As the angle θ varies, the value of V_2, and therefore of $Z_{21} = V_2/I_1$, varies according to the radiation pattern of A1. Z_{21} must, therefore, be proportional to the radiation pattern of A1.

Now consider the case that A2 is stationary and excited with current I_2' while I_1 is very small (negligible). As A1 rotates about its axis, as shown in Fig. 14.4(b), the signal in A1 is given by $V_1' = Z_{12}I_2'$. As θ varies, V_1', and therefore $Z_{12} = V_1'/I_2'$, gives the *receiving* pattern of A1, that is, the variation of receiving sensitivity as a function of angle.

If I_1 [in Fig. 14.4(a)] is made equal to I_2' [in Fig. 14.4(b)], then, since $Z_{12} = Z_{21}$, it follows that $V_1 = V_1'$. The reciprocity theorem, therefore, leads to the conclusion that the angular dependence of an antenna's *receiving* sensitivity is identical to its *radiation* pattern, or gain. Therefore, discussions of the radiation properties of any antenna (or of an array, as in the next section) are also presenting its properties as a receiving or detection element. Furthermore, this means that the properties of an antenna can be determined by testing it either as a transmitter or as a receiver.

14.6 PHASED ARRAYS

An extremely useful system is an arrangement of several antennas, that is, an antenna array, where the spacing between individual antennas and the relative amplitudes and phases between their currents are controlled to accomplish a specific purpose. To illustrate this aspect of antenna usage, we shall treat a uniform linear array, where identical equally spaced antennas are arranged in a line and are fed with currents that have the same magnitude but whose phases vary uniformly between neighboring antennas. For simplicity, first consider an array of omnidirectional point

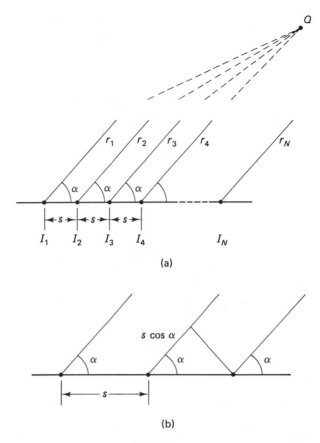

Figure 14.5 Linear dipole array.

(b)

antennas, as shown in Fig. 14.5. This figure might also be regarded as a linear array of electric dipoles that are oriented normal to the plane of the page, so their radiation is uniform in the plane of the page.

Currents feeding the N antennas vary linearly in phase by δ along the array, that is, for the nth antenna,

$$I_n = I_1 e^{-j(n-1)\delta} \qquad n = 1 \text{ to } N \tag{14.15}$$

Consider the radiation to a point Q, which is so distant relative to the array length L, that

$$r_n \gg L \quad \text{where } L = (N-1)s \tag{14.16}$$

Since the rays from all the antennas are, effectively, directed to an infinitely far point, they can be taken to be parallel and assumed to subtend the same angle to the array axis. Then, as shown in Fig. 14.5(b), the path length from each antenna to Q differs from its neighbor by $s \cos \alpha$, so that the distance of each is

$$r_n = r_1 - (n-1)s \cos \alpha \tag{14.17}$$

Because point Q is so distant, the field from each antenna is its far field; from Eqs. (14.7), this field can be written as

$$\mathbf{M}_n = \frac{\mathbf{a}\zeta I_n}{r_n} e^{-j\beta r_n} \tag{14.18}$$

where $\mathbf{M}$ can be either the electric or magnetic field lying along the unit vector $\mathbf{a}$, and ζ includes numerical and geometric factors such as antenna length and frequency. In the case of nonisotropic antennas, for example the electric dipole, it will also include angle-dependent factors, such as the $\sin\theta$ terms of Eqs. (14.7). Substituting Eq. (14.17) in the denominator of Eq. (14.18) and using the binomial expansion yields

$$\frac{1}{r_n} = \frac{1}{r_1}\left[1 - (n-1)\frac{s}{r_1}\cos\alpha\right]^{-1}$$

$$= \frac{1}{r_1}\left[1 + (n-1)\frac{s}{r}\cos\alpha + (n-1)^2\left(\frac{s}{r}\right)^2\cos^2\alpha + \cdots +\right]$$

Since the maximum value of $(n-1)(s/r)\cos\alpha = (L/r)\cos\alpha$, for a very distant point,

$$\frac{1}{r_n} \leq \frac{1}{r_1}\left[1 + \frac{L}{r}\cos\alpha + \left(\frac{L}{r}\right)^2\cos^2\alpha + \cdots +\right]$$

Under the conditions of Eq. (14.16), the terms on the right-hand side are very small and become successively smaller. Retaining only the leading term (unity) yields

$$\frac{1}{r_n} \simeq \frac{1}{r_1} = \frac{1}{r} \tag{14.19}$$

where there is little point in distinguishing the specific element from which r is measured since the difference is negligible for this purpose.

An r_n term also appears in the exponent of Eq. (14.18), but there it cannot be reduced to r as in the denominator. Because it is now multiplied by the factor β, the exponential becomes

$$e^{-j\beta r_n} = e^{-j\beta r_1} e^{j\beta(n-1)s\cos\alpha} \tag{14.20}$$

In this last factor, $\beta = 2\pi/\lambda$, so that

$$e^{j2\pi(n-1)(s/\lambda)\cos\alpha} \leftrightarrow \cos\left[2\pi(n-1)\frac{s}{\lambda}\cos\alpha\right]$$

Whereas we previously had $(n-1)s \leq L \ll r$, the present factor s/λ may not be small; the antenna separation in the array can be comparable to or even larger than the wavelength. As a result, the angle or phase of the second factor of Eq. (14.20) can be significant and cannot be ignored.

The total exponent of Eq. (14.18) is found by combining Eqs. (14.15) and (14.20):

$$I_n e^{-j\beta r_n} = I_1 e^{-j\beta r_1} e^{j(n-1)(\beta s\cos\alpha - \delta)}$$

With Eq. (14.19),

$$\mathbf{M}_n = \mathbf{a}I_1 \frac{\zeta}{r} e^{-j\beta r_1} e^{j(n-1)\psi} \tag{14.21a}$$

where

$$\psi = \beta s \cos \alpha - \delta \tag{14.21b}$$

Note that ψ is the total phase difference between the waves from adjacent radiators as a result of both the current phase difference and the difference of path length.

The field from the entire array is found by adding the individual antenna fields:

$$\mathbf{M} = \sum_{n=1}^{N} \mathbf{M}_n = \mathbf{a}I_1 \frac{\zeta}{r} e^{-j\beta r_1} \sum_{1}^{N} e^{j(n-1)\psi}$$

$$= \mathbf{a}\frac{\zeta}{r} e^{-j\beta r_1} \cdot \frac{1 - e^{jN\psi}}{1 - e^{j\psi}} \tag{14.22}$$

Here the summation is a geometric series that has been evaluated. This result can be put in a more convenient form by factoring.

$$\mathbf{M} = \mathbf{a}\frac{\zeta}{r} e^{-j\beta r_1} \frac{e^{jN\psi/2}}{e^{j\psi/2}} \cdot \frac{e^{-jN\psi/2} - e^{jN\psi/2}}{e^{-j\psi/2} - e^{j\psi/2}}$$

$$= \mathbf{a}\frac{\zeta}{r} e^{-j[\beta r_1 - (N-1)\psi/2]} \frac{\sin \frac{1}{2}N\psi}{\sin \frac{1}{2}\psi} \tag{14.23}$$

The final exponential factor $\beta r_1 - \frac{1}{2}(N - 1)\psi$ is the phase change from the field point Q to the *center* of the array. Finding the radiated power will require taking the absolute value squared of the radiated field, so this term will eventually disappear.

A word is in order here concerning the coefficients in Eq. (14.23). Comparison with Eqs. (14.7) indicates that for the **H** field of an electric dipole,

$$\mathbf{a}\frac{\zeta}{r} = \mathbf{a}_\phi \frac{jI_0\beta\ell}{4\pi r} e^{j\omega t} \sin \theta \tag{14.24}$$

With this observation, note that Eq. (14.23) is the product of three terms:

$\mathbf{a}(\zeta/r)$ is the same term as found for the single isolated antenna.

$e^{-j[\beta r_1 - 2[\beta r_1 - (N-1)\psi/2]}$ is the phase factor previously mentioned.

$\sin(N\psi/2)/\sin(\psi/2)$ is a term that depends only upon the features of the *array* structure, that is, the relation *between* the antennas and their excitations.

14.7 ARRAY FACTOR

Since $\mathbf{M}$ can be either $\mathbf{E}$ or $\mathbf{H}$, the intensity at the distant point Q can be found from Poynting's theorem, and the wave impedance $|\mathbf{E}|/|\mathbf{H}| = \eta$.

$$\mathcal{P} = \tfrac{1}{2}\mathcal{R}\mathbf{E} \times \mathbf{H}^* \propto |\mathbf{M}|^2 = \left(\frac{\zeta}{r}\right)^2 \frac{\sin^2 \frac{1}{2}N\psi}{\sin^2 \frac{1}{2}\psi} \tag{14.25a}$$

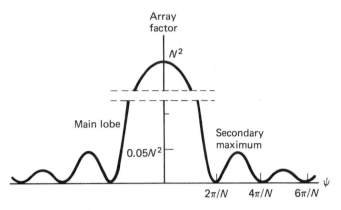

Figure 14.6   Array factor for a large array.

Substituting this in the integral of Eqs. (14.10) and dividing by $4\pi r^2$, as was done for the individual dipole antenna in Eq. (14.12), gives $\langle \mathcal{P} \rangle$ for the array. Then the gain of the array is

$$G = \frac{\mathcal{P}}{\langle \mathcal{P} \rangle} \approx \zeta^2 \frac{\sin^2 \frac{1}{2} N \psi}{\sin^2 \frac{1}{2} \psi} \tag{14.25b}$$

where ζ^2 is the geometric *antenna* factor, which has been discussed with respect to Eq. (14.13) and Fig. 14.2. The second term on the right-hand side of Eq. (14.25b) is called the *array factor*. Some of its features can be analyzed with the help of Fig. 14.6, which shows the array factor in the vicinity of small ψ, for the case of *large N*. Note that:

1. When $\psi = 0$, the array factor has a maximum value that is large compared to the secondary maxima. From Eq. (14.21b), this occurs when

$$\cos \alpha_{max} = \frac{\delta}{\beta s} \qquad \psi = 0 \tag{14.26}$$

At a fixed frequency and for a geometrically fixed antenna, βs is constant and the maximum-gain direction can be varied by controlling the lag angle δ of the successive current inputs. As δ varies from $-\beta s$ through 0 to $+\beta s$, the position of the main peak (the main lobe) changes from 180° through 90° to 0°. This ability to *steer* the radiation beam (or the direction of the antenna's maximum detection sensitivity) electrically, by controlling the phase, has considerable advantages over mechanically rotating a large antenna, for example, greater speed, reduction of mechanical wear, lack of vibration, etc.

2. The maximum gain of the primary lobe is N^2 times that of a single antenna. Even for small N, this gives an array a considerably enhanced gain. This squaring is a valuable consequence of the phase coherence between elements. For N elements with *random* phase relations, the intensity is merely the sum of the individual intensities, that is, it is proportional to N.

3. The first *zero* of the array factor is at $\psi = 2\pi/N$. To see the effect of this, particularly if N is large, consider the following example.

Example 14.3

Determine the full null width (null to null) of an array with $L = 20\lambda$ if there is no interelement phase lag.

From Eq. (14.21b) with $\delta = 0$, the angle of the null is

$$\cos \alpha_0 = \frac{2\pi}{\beta sN} \approx \frac{\lambda}{L} = \frac{1}{20}$$

Therefore, $\alpha_0 = 87.1°$. Since $\alpha_{max} = 90°$ when $\delta = 0$ (a broadside configuration), we have, for this large array, that the first major lobe is quite narrow $(90° - 87.1° = 2.9°)$; the full null width is only $2 \times 2.9° = 5.8°$.

The narrowness of this main lobe of the radiation-gain pattern means that a radiated signal is highly directional and a received signal can be narrowly located.

4. The first minor side-lobe peak occurs when $N\psi/2 = 3\pi/2$. From Eq. (14.25b), the gain of this lobe is easily shown to be $1/\sin^2(3\pi/2N)$. When N is very large, the angle is small and this gain is approximately $(2N/3\pi)^2 = 0.045N^2$, or, from Eq. (11.45b), it is 13.5 dB below the main lobe. This means that a large array is very highly directional, since any radiation going into the side lobe is small, and as a detector, the array is insensitive to signals arriving in the direction of the side lobes, that is, away from the direction in which the array is electrically or mechanically directed.

5. It is very important to note a common misinterpretation of these results. In Fig. 14.5, the angle α is defined with respect to the axis of the linear array. This means that α is the half angle of a cone about that axis. For example, when $\delta = 0$, we have, from Eq. (14.26), that $\alpha_{max} = \pi/2$. The main lobe is then a disc whose plane is normal to the array axis; the radiation is uniform in that equatorial plane. If $\delta \neq 0$, then the plane folds into a cone of peak radiation and peak sensitivity. This is illustrated in Example 14.4, which is to be discussed in the next section. All the secondary peaks and zeros of the array factor of a linear array are also conically distributed about the array axis.

Even with the conical radiation distribution, the phased array can be used to direct a beam with precision (or to locate an emitter with precision) and without ambiguity, making it a most valuable structure. Where the conical pattern is unacceptable, two-dimensional arrays have been built. These can be interpreted as consisting of two orthogonal linear arrays, although frequently the antennas are distributed throughout a plane rather than only on the axes. The two orthogonal conical-radiation patterns intersect in a set of radial direction that correspond to the unique principal radiation directions. This is becoming an important technique in the production of arrays at frequencies above $\sim$30 GHz, where the wavelength is sufficiently small so that arrays with a significant number of elements are not physically large. In some cases, the planar techniques of integrated-circuit technology can be employed in their fabrication.

From Fig. 14.6 and item 2, we see that the maximum intensity radiated from a linear array is N^2, times that of a single antenna. For a two-dimensional $N \times N$ array, this factor is $(N^2)^2 = N^4$, so the radiation is even more greatly enhanced and sharpened, and the secondary peaks are proportionally reduced. Though even greater improvement could be obtained by constructing a three-dimensional array, this has not been practical for microwave or radio-frequency antennas. However, the atoms of a crystal constitute just such a three-dimensional regular array, and the analysis of crystalline radiation, by x-ray, or electron, or neutron diffraction, is a powerful tool in the understanding of material structure, particularly as it relates to physical and chemical properties. In this case, there are $\sim 10^7$ molecular scattering centers per centimeter, so that the peak intensity from a 1-cm^3 sample is enhanced by a factor of $\sim 10^{42}$. Since the total outgoing energy must be conserved, only these sharp coherent peaks can be observed, except in special circumstances. The dynamical theory of diffraction, as developed by P. P. Ewald and M. vonLaue early in this century, is a *tour de force* of mathematical sophistication and physical insight in the application of electromagnetic theory to this case.

14.8 MAJOR SIDE LOBES

In constructing Fig. 14.6, it was not difficult to show that the major maximum occurs when $\psi = 0$. However, it is important to note that the general condition for a maximum is that both the numerator and denominator of the array factor of Eqs. (14.25) are zero, that is, when $\psi/2 = p\pi$, with $p = 0, \pm 1, \pm 2, \ldots$. When $p = 0$, this gives the condition of Eq. (14.26), but, more generally,

$$\cos \alpha_{\max} = \frac{2p\pi + \delta}{\beta s} \qquad p = 0, \pm 1, \pm 2, \pm 3, \ldots \qquad (14.27a)$$

If follows from Eq. (14.27a) that for large array spacings, that is, for large βs, there can be several values of p before $\cos \alpha_{\max}$ exceeds its limits ± 1. As a result, we will find primary, secondary (and tertiary, etc.) values corresponding to major lobes. The maximum value of p for this to happen is, therefore,

$$\frac{-\beta s - \delta}{2\pi} \le p \le \frac{\beta s - \delta}{2\pi} \qquad (14.27b)$$

Example 14.4

Locate the major lobes of a linear array of omnidirectional antennas if $s = 3\lambda/2$, $\delta = \pi/2$, and $N = 6$.

In this case, we have, from Eq. (14.27b), that $-7/4 < p < 5/4$, so that there are three major lobes corresponding to $p = 0, \pm 1$. The locations of these peaks are, from Eq. (14.27a), $\cos \alpha_{m0} = 1/6$, $\cos \alpha_{m1} = 5/6$, and $\cos \alpha_{m(-1)} = -\frac{1}{2}$. From these, we have $\alpha_{m0} = 80.4°$, $\alpha_{m1} = 33.56°$, and $\alpha_{m(-1)} = 120°$. These major lobes are shown in Fig. 14.7.

When more than one major lobe occurs, Fig. 14.6, showing the array factor near the origin of ψ, is not sufficient to visualize the entire radiation-gain pattern. Therefore, the upper portion of Fig. 14.8 presents an expanded view. The lower portion of the figure shows a useful construction to visualize the *window* or effective range of radiation from the antenna array. It is seen that as the angle α changes, the arrow rotates about the circle, and its projection on the horizontal axis traces out

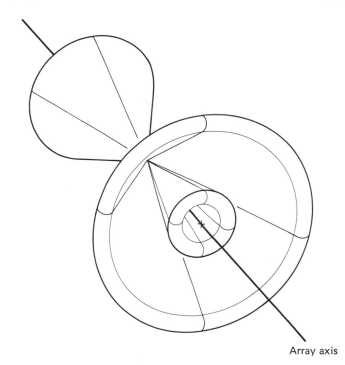

Figure 14.7 Array pattern for
Example 14.4.

Array axis

the full range of ψ. At each α value, a vertical line from the circle intersection to the upper curve gives the corresponding value of gain, so that one can determine the radiation pattern, the diagram of gain as a function of α. Again, it is seen that more than one major lobe can be encountered if βs is sufficiently large.

Example 14.4 (cont.)

> The particular construction of Fig. 14.8 is actually for the parameters of Example 14.4. Examination of this figure shows that the three major peaks do indeed occur at the angles indicated earlier. With this diagram, we can often visualize the array properties. For example, note that if δ is increased by $\pi/2$, a peak appears at $\psi = -4\pi$ and that the end peaks fall along the array axis, $\alpha = 0,\pi$.

Recall that the array factor is to be multiplied by the antenna factor to obtain the total gain. As the array factor generally varies much more rapidly than the antenna factor, it determines the main features of the gain pattern, but the antenna factor cannot be ignored. This is illustrated in the following example.

Example 14.5

> If the individual antennas of Example 14.4 are small vertical electric dipoles (i.e., perpendicular to the plane of Fig. 14.5) instead of isotropic antennas, then the array factor must be multiplied by the antenna factor, $\sin^2\theta$. Note that in this case θ is measured from the antenna axis, which is normal to the array axis, so that θ and α are not simply related. As a result of this additional facto all the major and minor lobes are reduced from the values they have in the horizontal plane, where $\theta = 90°$.

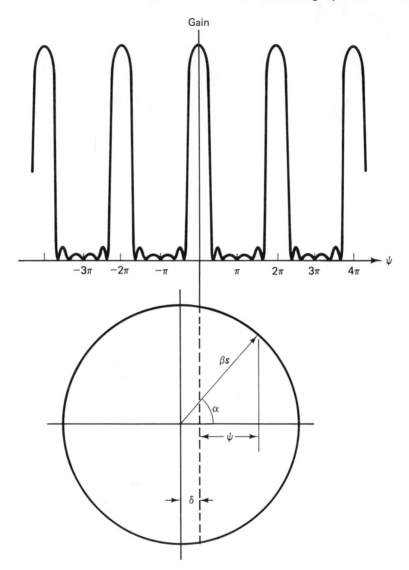

Figure 14.8 Radiation window of a linear array.

For example, in the plane of the antennas (vertical plane) we have $\theta + \alpha = 90°$, the peak lobe, which occurs at $\alpha = 80.4°$, is decreased by a factor $\sin^2 9.6° = 0.028$. The full radiation pattern for this array is shown in Fig. 14.9 in the horizontal and vertical planes. As is customary, these radiation patterns are logarithmic plots. Note that in the vertical plane, the signal at $80.4°$ appears 15.6 dB below the main peaks; it is comparable to the other minor lobes. These small peaks are quite measurable with modern microwave techniques. In Fig. 14.9, the pattern is cut off at -25 dB, which might represent the limit of measurement sensitivity or the noise level in the antenna ambient.

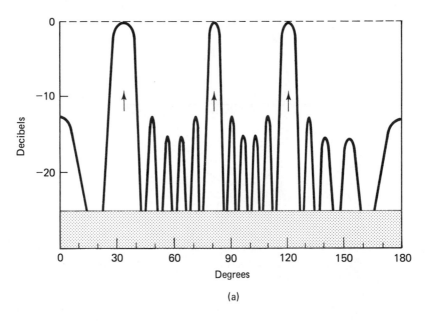

(a)

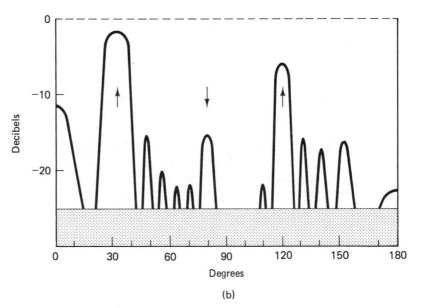

(b)

Figure 14.9 Radiation pattern for Example 14.5. (a) Horizontal plane. (b) Vertical plane.

The preceding discussion concerns arrays that are fairly large and that may contain several elements. Even smaller arrays have useful properties and can be analyzed using the tools developed here.

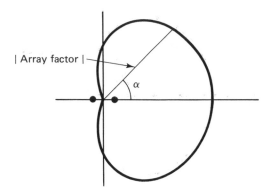

Figure 14.10 Radiation pattern for Example 14.6.

Example 14.6

Determine the radiation pattern in the horizontal plane of two vertical dipole antennas that are separated by $\lambda/4$ and that are excited with currents having a phase difference of $\pi/2$.

For vertical dipoles, the antenna factor in the horizontal plane is independent of angle and can be set equal to unity. For the array factor, we have

$$\psi = \beta s \cos \alpha - \delta = \frac{2\pi}{\lambda} \frac{\lambda}{4} \cos \alpha - \frac{\pi}{2}$$

$$= \frac{\pi}{2}(\cos \alpha - 1)$$

$$\text{Array factor} = \frac{\sin^2\psi}{\sin^2\frac{1}{2}\psi} = 4 \cos^2\tfrac{1}{2}\psi$$

Figure 14.10 shows the array factor, with the antenna orientation shown. Radiation is primarily in the right half plane; the left side is effectively blacked out.

The analysis presented has considered primarily the far fields of antenna arrays. However, the individual antennas reside in the near fields of the other array components and these complicated fields induce currents on the antennas. This alters the input impedances of the elements, affecting the current-phase and -magnitude relationships and, to some extent, their radiation patterns. The problem is particularly severe if one considers two-dimensional arrays. All this indicates that a more careful analysis is necessary, although for the small thin antennas and forcing current sources postulated here, the analysis presented is correct.

14.9 LONG ANTENNAS

Some properties of small antennas have been studied, individually and in arrays. A large antenna is made up of infinitesimal elements of this type, and its radiation field is the sum of the fields of the elements. When considering large antennas, that is, whose lengths are comparable to or greater than λ, additional factors enter the analysis. For example, if a long thin antenna is considered, it is apparent that the current cannot be uniform along its length since, as discussed in the introduction to this chapter, current must flow to and from a reservoir of charge. For a long antenna, the

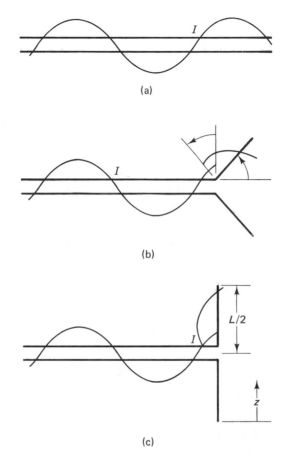

Figure 14.11 Current distribution on a
long antenna.

actual current distribution might be estimated by considering an open-circuited transmission line, as shown in Fig. 14.11(a), which is excited by a source at the left end
of the line. A current pattern (standing wave) is established that must have a null at
the open-circuit end. Now imagine that the line is bent out, as shown in Fig. 14.11(b),
and finally, in part (c), producing a linear antenna. Its current distribution can be obtained from the standing-wave expressions in Chapter 11. Using Eqs. (11.24b) and
(11.26b), it can be shown that

$$
\begin{aligned}
I &= I_0 \sin \beta z & 0 < z < L/2 \\
&= I_0 \sin \beta(L - z) & L/2 < z < L
\end{aligned}
\tag{14.28}
$$

where the transmission-line feed-current is $2I_0$ and z is measured from the bottom of
the antenna, as shown in Fig. 14.11c. β on the transmission line is the same as in
free space.

    In using the field or potential expressions for the small electric dipole antenna,
to find the field of the long antenna, this current variation with position must be entered. For the far field of the antenna shown in Figs. 14.11(c) and 14.12,

$$
\mathbf{E} = \mathbf{a}_\theta \frac{j\beta}{4\pi} \sqrt{\frac{\mu}{\varepsilon}} \sin \theta \, e^{j\omega t} \int_0^L \frac{I(z)}{r} e^{-j\beta r(z)} \, dz
\tag{14.29}
$$

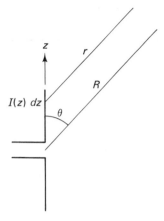

Figure 14.12 Geometric relations for a long antenna.

where the substitution $I\ell = I(z)\,dz$ has been made. In Eq. (14.29), $r(z) = R - (z - L/2)\cos\theta$ must be used in the exponent and $r = R$ in the denominator. It is not exceptionally difficult to do this integration, but we shall not complete the exercise as it would lead the discussion into specialty areas that are not the domain of this text.

It is worth noting that this long antenna can also be regarded as a collinear array of infinitesimal elements, but with the difference from the previous phased array that while the elemental currents are in phase, they differ in magnitude according to Eq. (14.28).

14.10 APERTURE ANTENNAS

Radiation occurs from various kinds of open systems, that is, systems presenting an aperture through which microwave energy passes. Examples, some of which are shown in Fig. 14.13, include a waveguide section with periodic openings in its side wall [Fig. 14.13(d)], a truncated waveguide [either plane, as in Fig. 14.13(c), or flared, as in Fig. 14.13(b)], and an area of a spherical or parabolic reflecting antenna [Fig. 14.13(a)]. For these systems, the radiation is difficult to determine from the generating currents since their distributions over the inner edge and outer metallic surfaces are unknown.

As an aside, note that geometric optics traces the rays that are normal to the wave front, as in Fig. 14.14, and predicts that the beam of radiation will project a sharp image of the aperture. In fact, however, this is not found to be so. The departures are due to interactions with the slot edges (the unknown currents previously mentioned) and the beam is said to be diffracted by the slot. Solution of the aperture antenna problem is equivalent to finding the diffraction pattern of the aperture.

While advanced potential theory can be used to solve such problems, the method is often long and, even so, imprecise. Therefore, a simpler approach will be taken to elucidate the radiation properties, that of Huygens' principle or *physical optics*. Christian Huygens (1629–1695), a Dutch astronomer studying the properties of light, established the basis of modern wave analysis by proposing that each point of

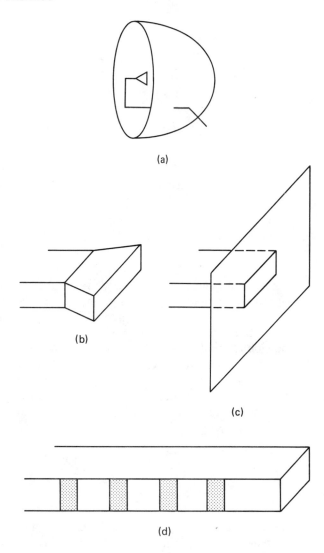

(a)

(b)

(c)

(d)

Figure 14.13 Aperture antennas.

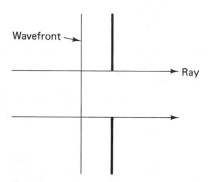

Figure 14.14 Ray construction.

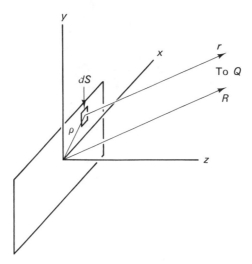

Figure 14.15 Geometric relations for aperture radiation.

a wave is a source of spherical wavelets that radiate in all directions. The radiated pattern can be found by summing the successive individual wavelet contributions.

To illustrate Huygens' method, consider a rectangular hole in a large opaque conducting sheet, as shown in Figs. 14.13(c) and 14.15, with a wave incident from one side (from the left). We wish to find the radiation pattern on the other side of the aperture. A representative radiation area, shown in Fig. 14.15, is $dS = dx'\,dy'$ located at $\boldsymbol{\rho}' = \mathbf{a}_x x' + \mathbf{a}_y y'$. The primed variables distinguish the source location from the unprimed field point. The field point Q is located at $\mathbf{R} = \mathbf{a}_x x + \mathbf{a}_y y + \mathbf{a}_z z$, and the distance from dS to Q is

$$
\begin{aligned}
r &= [(x - x')^2 + (y - y')^2 + z^2]^{1/2} \\
&= \{(x^2 + y^2 + z^2) - 2(xx' + yy') + [(x')^2 + (y')^2]\}^{1/2} \\
&\approx R - \frac{xx' + yy'}{R}
\end{aligned}
\tag{14.30}
$$

where the last equality holds if Q is very distant, so the primed coordinates are small compared to the unprimed, allowing only first-order small terms to be retained in the expansion of the curly bracket in the second line of Eq. (14.30). The field at Q is found, according to Huygens, by adding the wavelets from all the areas at their respective distances r. This corresponds to integrating the contributions from all the areas dS. In doing so, two conclusions of previous discussions are employed.

1. The phase factor must reflect the retardation due to transit time between source and field point. As in Eq. (14.3b), this is included by using the expression for a radial traveling wave

2. The radiated wave amplitude is proportional to $1/r$ at large distances from their source. This is also established, rigorously, in Section 14.12. With these, the radiated field is

$$
\mathbf{E}(x, y, z) = \int \frac{\mathbf{E}(x', y')}{r} e^{j(\omega t - \beta r)}\, dS
\tag{14.31a}
$$

Equation (14.31a) includes the fact that the source field is a function of position in the aperture plane. Equation (14.30) must be substituted in the exponent of Eq. (12.31a), but, from the discussion of Eq. (14.19), recall that it is unnecessary to do so in the denominator:

$$\mathbf{E}(x, y, z) = \int \mathbf{E}_0 \frac{X(x')Y(y')}{R} e^{j(\omega t - \beta R)} e^{j\beta(xx' + yy')/R} \, dx' \, dy' \qquad (14.31b)$$

It has been assumed that the field has the form $\mathbf{E} = \mathbf{E}_0 X(x')Y(y')$, where X is a function of only x', and Y is a function of only y'. (For example, if the aperture were at the end of a TE$_{10}$ waveguide, we expect that $X(x') = \sin \pi x'/a$, $Y(y') = 1$, and $\mathbf{E}_0 = E_0 \mathbf{a}_y$.) If the vector-field direction is constant over the aperture area, then $\mathbf{E}_0$ can be removed from the integral. Equation (14.31b) then conveniently separates into individual x'-dependent and y'-dependent integrals:

$$\mathbf{E}(x, y, z) = \frac{\mathbf{E}_0}{R} e^{j(\omega t - \beta R)} \int X(x') e^{j\beta x x'/R} \, dx' \int Y(y') e^{j\beta y y'/R} \, dy' \qquad (14.31c)$$

It is useful to observe in Eq. (14.31c) that $x/R = \cos \theta_x = \alpha_x$, and $y/R = \cos \theta_y = \alpha_y$ are direction cosines, that is, the θ are the angles between R and the coordinate axes. For those familiar with the Fourier transform, it can be noted that each integral represents the far field as the Fourier transform of the aperture-field component, that is,

$$I_x(\beta \alpha_x) = 2\pi \mathscr{F}(X) = \int X(x') e^{j\beta \alpha_x x'} \, dx' \qquad (14.32)$$

where I_x is a function of α_x. A similar result occurs for $I_y(\beta \alpha_y)$. At a large distance, the field depends on only the α's, that is, the radiation *directions* [except for the $1/R$ dependence of Eq. (10.31c)].

Example 14.7

Determine the radiation pattern for a uniform plane wave incident on the aperture of Fig. 14.15. The aperture dimensions are $a \times b$.

In this case, $\mathbf{E}(x'y') = \mathbf{E}_0$, a constant, so that $X = 1 = Y$, and Eq. (14.32) becomes

$$I_x(\beta \alpha_x) = \int_{-a/2}^{a/2} e^{j\beta \alpha_x x'} \, dx' = a \frac{\sin(\frac{1}{2}\alpha_x \beta a)}{\frac{1}{2}\alpha_x \beta a}$$

There is a similar result for $I_y(\beta x_y)$, giving

$$E(\alpha_x, \alpha_y) = \frac{E_0}{R} ab e^{j(\omega t - \beta R)} \frac{\sin \frac{1}{2}\alpha_x \beta a}{\frac{1}{2}\alpha_x \beta a} \frac{\sin \frac{1}{2}\alpha_y \beta b}{\frac{1}{2}\alpha_y \beta b}$$

Figure 14.16 shows the function $\sin(\frac{1}{2}\alpha\beta c)/(\frac{1}{2}\alpha\beta c)$ for the cases that $c = \lambda/2$ ($\frac{1}{2}\beta c = \frac{1}{2}\pi$) and $c = 10\lambda$ ($\frac{1}{2}\beta c = 10\pi$). From this, it is seen that the narrow aperture ($c = \lambda/2$) has a broad, almost uniform, radiation pattern over the entire range of realizable angles from 90° to 0° (α from 0 to 1). The wide aperture ($c = 10\lambda$) has a narrow main beam, with a first null at 84.3° ($\alpha = 0.1$), that is, a beam width of ±5.7°. In addition, the secondary peaks are small; the maximum *power* density at the second lobe is 0.2^2 of the main lobe, or 14 dB lower. These features are similar to those noted previously for arrays, and that also hold for individual antennas: large radiators (and detectors) have narrower more precise patterns than do small systems.

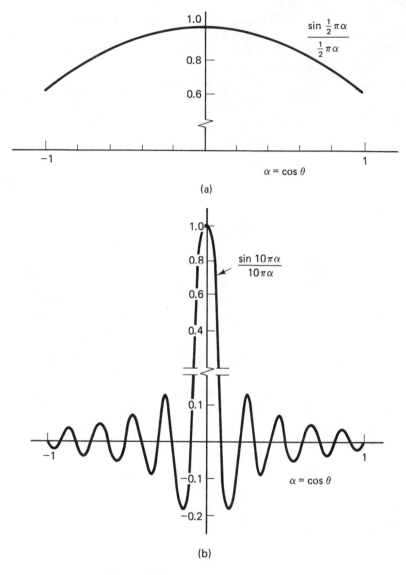

Figure 14.16   Aperture radiation patterns.

Note that the methods of wave analysis presented in this chapter are applicable to all types of electromagnetic waves, including light, microwaves, radio waves, etc. An illustration of this is the transmission and reflection of light by a diffraction grating. There one has an array with values as great as $N \sim 10^5$ and values of $s/\lambda \sim 10$ can be obtained. These give multiple, and very narrow, major lobes, corresponding to the higher-order diffraction peaks. Analogies can also be made with acoustic waves that show interference and diffraction, but there some caution is necessary as these waves are not electromagnetic and can be longitudinal as well as transverse.

14.11 POTENTIAL THEORY

We have considered several types of antennas and derived their radiation patterns from the physical features of each particular antenna. All these patterns have certain features in common and it is worthwhile, before leaving the subject, to see just what behaviors are so general that they can be expected of other antenna types not yet considered. For this, we shall review and apply formal potential theory.

Recall that since the divergence of any curl is zero, Maxwell's equation $\nabla \cdot \mathbf{B} = 0$ is automatically satisfied by the vector potential $\mathbf{B} = \nabla \times \mathbf{A}$. Using this in Maxwell's equation for $\nabla \times \mathbf{E}$ yields

$$\nabla \times \mathbf{E} = -\frac{\partial \mathbf{B}}{\partial t} = -\nabla \times \frac{\partial \mathbf{A}}{\partial t} \qquad (14.33\text{a})$$

or

$$\nabla \times \left(\mathbf{E} + \frac{\partial \mathbf{A}}{\partial t} \right) = 0 \qquad (14.33\text{b})$$

In the last part of Eq. (14.33a), the usual assumption has been made, that the time and space coordinates are independent, so their order of differentiation can be interchanged. Since the curl of any gradient is identically zero, Eq. (14.33b) can be written

$$\mathbf{E} + \frac{\partial \mathbf{A}}{\partial t} = -\nabla V \qquad (14.34)$$

where V is the scalar potential. When the magnetic field is unchanging with time, $\partial \mathbf{A}/\partial t = 0$, so that $\mathbf{E} = -\nabla V$, the static result. In the dynamic case, the electric field induced by a changing magnetic flux (Faraday's law) is directly equal to the (negative) rate of change of $\mathbf{A}$ and this dynamic field is added to any static field from $-\nabla V$.

A relation exists between $\mathbf{A}$ and V that can be found by recalling the static expressions for $\mathbf{A}$ and V in terms of their sources:

$$\mathbf{A} = \frac{\mu}{4\pi} \int \frac{\mathbf{J}}{r} dv \qquad V = \frac{1}{4\pi\varepsilon} \int \frac{\rho}{r} dv \qquad (14.35\text{a, b})$$

The similarity of form between Eqs. (14.35a) and (14.35b) is apparent, and they can be related as a result of the law of conservation of charge:

$$\nabla \cdot \mathbf{J} = -\frac{\partial \rho}{\partial t} \qquad (14.36)$$

While an exact proof is rather complicated, it is at least plausible, by taking the divergence of Eq. (14.35a) and substitution (Eq. 14.36), that

$$\nabla \cdot \mathbf{A} = -\mu\varepsilon \frac{\partial V}{\partial t} \qquad (14.37)$$

This is known as the Lorentz condition on the potentials. Note that in the static case, it reduces to $\nabla \cdot \mathbf{A} = 0$, which was assumed in Chapter 7.

Furthermore, from the curl **H** equation

$$\nabla \times \mathbf{B} = \mu \mathbf{J} + \mu \varepsilon \frac{\partial \mathbf{E}}{\partial t}$$

$$\nabla \times (\nabla \times \mathbf{A}) = \mu \mathbf{J} - \mu \varepsilon \frac{\partial^2 \mathbf{A}}{\partial t^2} - \mu \varepsilon \nabla \frac{\partial V}{\partial t}$$

$$-\nabla^2 \mathbf{A} + \mu \varepsilon \frac{\partial^2 \mathbf{A}}{\partial t^2} = \mu \mathbf{J} \tag{14.38a}$$

where a vector identity for the double curl has been used, along with Eqs. (14.34) and (14.37). Thus, **A** satisfies the inhomogeneous wave equation with the current density as source.

In a similar way, we find, from Maxwell's $\nabla \cdot \mathbf{D}$ equation, that

$$-\nabla^2 V + \mu \varepsilon \frac{\partial^2 V}{\partial t^2} = \frac{\rho}{\varepsilon} \tag{14.38b}$$

Equation (14.38a) is equivalent to three equations, for (A_x, J_x), (A_y, J_y), and (A_z, J_z), each of which has the same form as Eq. (14.38b). Therefore, to understand the behavior of both **A** and V, we will study Eq. (14.38b). [Note, in passing, that in the static case, this reduces to Poisson's equation, $-\nabla^2 V = \rho / \varepsilon$, and with a similar simplification for **A**. The solutions of these equations have already been given.]

14.12 SPHERICAL WAVES

For a sinusoidal time variation, $\mathbf{A} \sim e^{j\omega t}$ and $V \sim e^{j\omega t}$, and Eqs. (14.38) are

$$\nabla^2 \mathbf{A} + \beta^2 \mathbf{A} = -\mu \mathbf{J} \qquad \nabla^2 V + \beta^2 V = -\frac{\rho}{\varepsilon} \tag{14.39a, b}$$

where

$$\beta^2 = \omega^2 \mu \varepsilon \tag{14.40}$$

To illustrate the solution to these equations, consider a sinusoidally varying source at the origin. Moving away from the origin to the region where there are no sources, we have

$$\nabla^2 V + \beta^2 V = 0 \tag{14.41}$$

Expanding $\nabla^2 V$ in spherical coordinates gives

$$\frac{1}{r^2} \frac{\partial}{\partial r}\left(r^2 \frac{\partial V}{\partial r}\right) + \frac{1}{r^2 \sin \theta} \frac{\partial}{\partial \theta}\left(\sin \theta \frac{\partial V}{\partial \theta}\right) + \frac{1}{r^2 \sin^2 \theta} \frac{\partial^2 V}{\partial \phi^2} + \beta^2 V = 0 \tag{14.42}$$

Relative to the distances at which antenna fields are detected, the source can be regarded as being a point, here taken to be at the origin of coordinates. Symmetry then demands that V shall not be a function of the angle variables, so that $V = V(r)$ only, and Eq. (14.42) can be put in the equivalent form:

$$\frac{d^2(rV)}{dr^2} + \beta^2 rV = 0 \tag{14.43}$$

which has the solution

$$V = \frac{C_1}{r} e^{-j\beta r} + \frac{C_2}{r} e^{j\beta r} \qquad (14.44a)$$

The full expression for the potential includes the time factor, so that

$$V = \frac{C_1}{r} e^{j(\omega t - \beta r)} + \frac{C_2}{r} e^{j(\omega t + \beta r)} \qquad (14.44b)$$

which describes waves that are traveling radially outward and radially inward. As the physical problem usually has a source at the origin alone, there are no inward waves, leaving

$$V = \frac{C}{r} e^{j(\omega t - \beta r)} \qquad (14.45)$$

The constant C can be evaluated by referring back to the quasistationary case considered in Chapter 5. This can be approached either by letting the field variation time be long compared to the propagation time over the field region, or by approaching the source so closely that the field changes propagate instantaneously over the field region. In either case the exponential in Eq. 14.45: $e^{j(\omega t - \beta r)} \to 1$ and $V = C/r$. The solutions are then given by Eqs. (14.35), so that the constant C in Eq. (14.45) is obtained by an integral involving the distribution of all the sources. As an example, for the infinitesimal sources $\mathbf{J}\,dv = I\ell$ and $\rho\,dv = Q$, the integrals can be dropped, yielding

$$\mathbf{A} = \frac{\mu_0 I\ell}{4\pi r} \qquad V = \frac{Q}{4\pi\varepsilon_0 r} \qquad (14.46a, b)$$

If Eq. (14.45) is to reduce to Eq. (14.46b) when $\omega t - \beta r \to 0$, then

$$\mathbf{A} = \frac{\mu I\ell}{4\pi r} e^{j(\omega t - \beta r)} \qquad V = \frac{Q}{4\pi\varepsilon r} e^{j(\omega t - \beta r)} \qquad (14.47a, b)$$

are the radiated potentials from small sinusoidally varying sources.

It is seen from this derivation that the amplitude dependence, $\sim 1/r$, is a consequence of the geometry of space, as reflected in the form of the expression for the Laplacian in spherical coordinates; it will appear whenever one is sufficiently far from the source that details of the source distribution are not apparent.

EXERCISES

Extensions of the Theory

14.1. Example 7.4 gives an expression for $\mathbf{A}$ of a static *magnetic* dipole. When converting this to a retarded potential, for the case that $I = I_0 e^{j\omega t}$, we must make a similar correction in the exponential to that made in Eqs. (14.3b). In addition, because it constitutes a geometric array, it can be shown that the static $\mathbf{A}$ must be multiplied by the factor $1 + j\beta r$. With these corrections, calculate the far fields $\mathbf{E}$ and $\mathbf{H}$. Compare them with the far fields from the electric dipole.

14.2. Electric *quadrupoles* can be generated by combining two dipoles in various geometries, and features of their fields can be determined from array theory. Consider two such small dipoles, $I\ell\mathbf{a}_z$ and $-I\ell\mathbf{a}_z$, and determine their radiation pattern if they are located, respectively, at

(a) $\pm\ell/2$ on the z axis

(b) $\pm\ell/2$ on the x axis

In both cases, $I = I_0 e^{j\omega t}$ and $\ell \ll \lambda$.

*14.3. A common measure of the radiation precision of an antenna is the full width of its first lobe, at the points where the power drops to half. Determine this half-power width of the small electric dipole antenna.

14.4. The power P drawn from incident radiation by an aperture antenna equals the intensity of that radiation $|\mathscr{P}|$ multiplied by the aperture area A. From this the effective area of any receiving antenna, including a thin electric dipole, can be defined as $A = P/|\mathscr{P}|$.

For a small electric dipole antenna, the incident electric field generates a voltage in the antenna given by $V_a = \int \mathbf{E} \cdot d\boldsymbol{\ell}$. If, as according to the reciprocity law, this voltage appears across the radiation resistance of the antenna, show that the maximum effective area of the antenna is equal to $(0.691\lambda)^2$.

Problems

14.5. Two-dimensional antenna arrays can be analyzed using the ideas presented in this chapter, either by considering the wave interferences directly or by considering the two-dimensional case as combinations of linear arrays. Using either of these methods, determine the radiation pattern, in the horizontal plane, of four vertical electric dipole antennas that are placed at the corners of a square of side $\lambda/2$ and that are fed with equal in-phase currents.

14.6. (a) Calculate the wave impedance $\eta = |\mathbf{E}|/|\mathbf{H}|$ in the near field of an electric dipole and show that $\eta > \sqrt{\mu/\varepsilon}$. This is called a high impedance field.

(b) Calculate the wave impedance $\eta = |\mathbf{E}|/|\mathbf{H}|$ in the near field of a magnetic dipole (see Exercise 14.1) and show that $\eta < \sqrt{\mu/\varepsilon}$. This is a low impedance field.

14.7. Calculate the average radiated power from the near fields of an electric dipole antenna. Compare your result with the expression derived in the chapter text from the far fields.

14.8. (a) For an omnidirectional antenna, show that at any point, $E(\text{rms}) = \sqrt{30P}/r$.

(b) An electric dipole radiates a total power P. In the equatorial plane of the antenna, show that $E_{(\text{rms})} = \sqrt{45P}/r$. *Hint:* Use the result of (a) and the antenna gain.

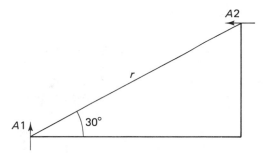

Figure 14.17 For Exercise 14.10.

*14.9. Radio frequency interference (RFI) is a phenomenon whereby radiation from a source is unintentionally detected by another circuit and processed as though it were part of the intended signal, thereby leading to errors, misinterpretations, and failures. This problem presents a crude analysis of this behavior.

  (a) If a small omnidirectional antenna radiates 50 W, determine the electric and magnetic field intensities at a distance of 200 m.

  (b) The magnetic field from the radiation of part (a) is normal to a circuit board on which are printed two parallel lines 1 cm apart and 10 cm long. Determine the emf induced in this circuit if $f = 10^8$ Hz.

14.10. Two identical 1-m long electric dipole antennas are 50 km apart, as shown in Fig. 14.17. Antenna $A1$ is radiating 5000 W of power at a frequency of 6 MHz.

  (a) Show that the intensity at $A2$ is 0.179 μW/m^2.

  (b) Taking the voltage induced in a linear antenna to be $V = \int \mathbf{E} \cdot d\boldsymbol{\ell}$, show that V_{rms} in $A2$ is 4.11 mV.

14.11. An antenna is constructed from a copper rod that is 1 m long and 1 cm in diameter. At what frequency is its small antenna-radiation resistance equal to its electrical resistance? *Note:* You must correct for skin depth at high frequencies.

*14.12. Imagine a small antenna on a lunar experiment that broadcasts its results to the surface of the earth (3.8×10^5 km away). The microwave source has an output power of 10 W and the transmitting and receiving antennas each have a gain of 10 dB along their common axis. What must be the sensitivity of the detector on the receiving antenna if it is to detect the transmitted signal?

*14.13. In a linear phased array of omnidirectional antennas, the current lags by $\pi/2$ between array elements that are 30 cm apart. The array has four antennas and is operated at $f = 1$ GHz. Find the directions of the array-pattern maxima. Determine the null-to-null widths of the main lobes.

14.14. The antennas of the array in Problem 14.13 are moved so that they are 60 cm apart. Find the directions of the array-pattern maxima.

14.15. Verify that the quantitative features shown in Fig. 14.8 do follow from Eqs. (14.27).

14.16. Using the concepts of Fig. 14.8, design an array that has only a single narrow major lobe. Give N, s, and δ. Through what angle can this lobe be steered without introducing significant spurious behavior? *Note:* A second major lobe may not be a significant feature if it is sufficiently separated from the first lobe so that its signal can be suppressed externally or it can be directed to a region where it is not significant.

14.17. The conical radiation pattern of a linear array can be changed to more directed beams by having, for example, a square array of elements. Use the array of Exercise 14.5 and determine how you would control the currents to the elements so that the radiation beam traces out a circle. How would you have it trace out an expanding spiral?

14.18. Draw the array factor, corresponding to Fig. 14.9a, for a phased array with $\delta = \pi/4$ and $N = 3$, for the two cases $\lambda/s = 3/8$ and $\lambda/s = 8/3$.

*14.19. It can be seen that as α changes uniformly in Fig. 14.8, its horizontal projection ψ does not change uniformly. For the array of Example 14.4, show that it requires a larger change of α to trace out the lobe corresponding to $\alpha_{m1} = 33.56°$ than for the lobe at $\alpha_{m0} = 80.4°$, that is, find the full-null widths of these two lobes. *Hint:* The nulls occur at $\psi_{max} \pm 2\pi/N$. Also show that the lobes are not symmetrical.

14.20. Four identical small linear dipole antennas are arranged along a line with their axes parallel to the line. Their currents lag by 180° between antennas. They are separated by 2λ at the frequency of operation.
  (a) Find the antenna factor and the array factor for the radiated field. Show the geometric relations on a diagram.
  (b) Locate all the main peaks.
  (c) Determine the location of the peaks of the first minor side lobe of any of the main lobes.
  (d) What is the approximate amplitude of this side-lobe peak relative to its main peak? *Hint:* For this array, all the side lobes have nearly the same ratio.

14.21. As already mentioned, the array factor must be multiplied by the antenna factor to obtain the total gain. Since the array factor generally varies much more rapidly than the antenna factor, it determines the main features of the gain pattern, but the antenna factor cannot be ignored. In Example 14.5 we showed its effect when the dipoles are normal to the array axis of Example 14.4. Show that the radiation pattern of the array is relatively unchanged if the individual dipoles are aligned along the array axis.

14.22. In analyzing long antennas using Eq. (14.29), it is interesting to try separating the effect of having a finite length from the effect of the current variation along L. To do this, consider the antenna of length L that has a hypothetical uniform current $I(z, t) = I_0 e^{j\omega t}$, where I_0 is constant.
  (a) Calculate E and H in the far-field region for $L = \lambda/2$ and for $L = \lambda$.
  (b) Plot the radiation patterns.

14.23. Repeat Exercise 14.22 using the current distribution given by Eq. (14.28). *Hint:* The sinusoidal current can be represented as a sum of exponentials.

*14.24. For the aperture illumination in Example 14.7, what is the change in angular width of the main lobe (half null width) if the aperture is made three times as large in all dimensions?

14.25. A standard X-band waveguide is used at ~10 GHz. It has internal dimensions $a \times b = 0.9 \times 0.4$ inch and has **E** polarized along the short dimension:

$$\mathbf{E} = \mathbf{a}_y E_0 \sin \frac{\pi x}{a} e^{j(\omega t - \beta z)} \qquad 0 \le x \le a \qquad 0 \le y \le b$$

Determine the distant radiation pattern from a truncated waveguide.

14.26. With reference to the definition in Exercise 14.3, find the radiation width of the following radiating systems.
(a) The array of Example 14.6
(b) The aperture of Exercise 14.25.

14.27. The waveguide in Exercise 14.25 is flared out until its dimensions are $2a \times 3b$. If the field pattern expands proportionally, determine the change in the full width of the main radiation lobe (null to null).

14.28. In Example 14.7, the radiation pattern was found from a uniformly illuminated aperture, corresponding to a spacial step-function field input, that is, E is zero outside of the rectangular area and is a constant value within that area. In Exercise 14.25, we did the same for a half sine input, that is, where the field varies sinusoidally and is greater at the center of the radiating region. Now consider the two cases where the field is sinusoidally greatest at the edges:
(a) $\mathbf{E} = \mathbf{a}_y E_0 \cos(\pi x/a) e^{j(\omega t - \beta z)} \qquad 0 \le x \le a$
(b) $\mathbf{E} = \mathbf{a}_y E_0 [1 + \cos(2\pi x/a)] e^{j(\omega t - \beta z)} \qquad 0 \le x \le a$
Note that the relative phases of the edge fields differ in the two cases. Using your knowledge of the Fourier transform, determine the radiation pattern of these fields. *Hint:* Use superposition.

*14.29 The sun radiates at the rate of 3.8×10^{26} W. Its radius is 6.95×10^8 m and it is 1.49×10^{11} m from the earth, which has a radius of 6.38×10^6 m. Calculate the electric-field intensity at the surface of the sun and at the surface of the earth if all the radiation were at a single frequency. Determine the total radiant power falling on the earth.

14.30. The angular dependence of a radiated electric field is given by

$$E \sim \cos \theta \sin \frac{\phi}{2}$$

where θ and ϕ are the usual spherical–polar coordinate angles.
(a) Find the maximum gain of the antenna producing this field.
(b) At a distance of 10 km from the source, along the z axis, the rms field value is measured as 8 mV/m. Assuming the proper distance dependence, find the rms wave amplitude 1 m from the antenna.
(c) If the antenna is excited with 4/3 A (rms), determine its radiation resistance.

14.31. It was pointed out at the very beginning of the discussion of antennas that a current flow implies an accumulation of charge and, therefore, the presence of an electric potential V. In this event, the electric-field intensity could be obtained from $\mathbf{E} = -\nabla V - \partial \mathbf{A}/\partial t$. In the derivation of the electric dipole antenna, we avoided the necessity of calculating the charge and V by finding $\mathbf{A}$ from the current distribution, calculating $\mathbf{B} = \nabla \times \mathbf{A}$, and then obtaining $\mathbf{E}$ from Maxwell's equation $\nabla \times \mathbf{B} = \mu \, \partial \mathbf{D}/\partial t = j\omega\mu\varepsilon\mathbf{E}$.

(a) It is not difficult now to go back and determine that potential V, for example, from the Lorentz condition that $\nabla \cdot \mathbf{A} = -\mu\varepsilon \, \partial V/\partial t$. Under far-field conditions ($\beta r \gg 1$), derive the expression for V.

(b) The above procedure gives V in terms of the current element $I_0\ell$. It can also be expressed in terms of the charge that results from the sinusoidal current $I = I_0 e^{j\omega t}$. Making this substitution, show that

$$V = \frac{j\beta p}{4\pi\varepsilon r} e^{j(\omega t - \beta z)}$$

where p is the dipole moment of the Hertzian dipole.

14.32. Draw the radiation patterns in the horizontal plane for two vertical electric dipole antennas that are separated by s and excited by currents that differ in phase by δ if

(a) $s = \lambda/2$, $\delta = \pi/2$
(b) $s = \lambda/4$, $\delta = \pi/2$
(c) $s = \lambda/2$, $\delta = \pi$
(d) $s = \lambda$, $\delta = \pi$

14.33. Two identical linear electric dipole antennas are arranged with one at the origin, aligned along the z axis, and one at $x = \lambda/4$, $y = 0$, $z = 0$, and parallel to the y axis. They are fed with equal in-phase currents.

(a) Show that their radiation along the x axis is circularly polarized.

(b) Describe the radiation along the x axis if the second antenna is moved farther from the origin, so that it is at $x = \lambda/2$.

APPENDIX

PROGRAMS
FOR NUMERICAL
EXAMPLES

```
program integral (source, field, fieldplot);

            { written by ted kubasak for e205, prof. hoole, harvey
              mudd college.  this program was written using VAX
              pascal v.3.4.  this program uses the integral method
              to find the equipotential values for a grid which
              surrounds a single loop of wire.  this program sums
              the contributions of each section of the wire, assum-
              ing uniform current density.  all of the values are
              normalized by the value of (mu/2pi).
              this program should give us much better results in
              the region far away from the wire because we don't
              make the value go to zero at the boundaries. however,
              because we treat the current as a point source, our
              results will not be very good near the wire.  }

const  Nnodes = 47;    { the number of nodes we will solve for }
       Ncurr = 32;     { the number of areas the current carrying
                         wires have been broken into. }

type  result = array [1..Nnodes] of real;    { used to store the
                         results for each of the nodes }
```

```
var  i,j         : integer;    { used as counters in the loops }
     soln        : result;     { the array which holds the solutions }
     r,                        { the distance between two points }
     x1,y1,                    { the location of the grid point }
     x2,y2,                    { the location of the point current we
                                 are using in the current calculation }
     current     : real;       { the current in the given region }
     source,                   { the file which contains data on the
                                 two wires we use in the calculations }
     field,                    { the file which contains the location
                                 of the points we arte calculating for }
     fieldplot : text;         { the output file, which contains the
                                 location and result for each point }

begin    { main program }
   open (source, history:=old);   { open input data file. }
   open (field, history:=old);
   open (fieldplot);
   reset (field);
   rewrite (fieldplot);           { prepare file to receive data }

   for i := 1 to Nnodes do begin
      soln [i] := 0;              { initialize each value to zero }
      reset (source);            { go to start of source.dat }
      readln (field, x1, y1);    { read the point's location }

      for j := 1 to Ncurr do begin
         readln (source, x2, y2, current);
                                  { read data for the area in the wire}
      r := sqrt ((x2-x1)**2 + (y2-y1)**2);
                                  { find distance between the points }
      soln [i] := soln [i] - (current * ln(r))
                                  { sum the solution for all the areas}
         end;    { for j }

   writeln (fieldplot, x1:4:1,' ',y1:4:1,' ',soln[i]:6:3)
                                  { send the coordinates and values for
                                    each point to the plot file }
   end      { for i }
end.    { main program }
```

```
program tri (phiplot,triall,genall,allnod);

        { written by ted kubasak for e205, system simulation for
          prof. hoole, harvey mudd college.  this program is written
          in VAX pascal v.3.4.  this program uses the finite element
          method to solve the Poisson equation.  this is the gener-
          alized program which can be used for any finite element
          analysis with only slight modifications.  this program
          uses the procedures Triangle, FirstOrderLocalMats, and
          GlobalPlace given to us by prof. hoole in the "numerical
          methods" handout.  leqt1f is the external procedure which
          is called to solve the the matrix equation which this
          program builds.   }

        { this program can be used to solve for magnetic potentials
          by modifying the format of the input.  in the case of mag-
          netic potentials, the value of eps which the program reads
          should actually be the value of 1/mu for the given tri-
          angle.  also, where the program reads rho, the data should
          input J, the current density.  the program otherwise works
          the same, and the solution A is stored in the vector phi. }

const  max = 300;             { the number of unknown nodes solved for }
       epsnot = 8.8542e-12;
       WkMax = sqr(max);      { the size of the workarea required by
                               leqt1f, at least = max**2 }
       v2 = 10;               { v2 is the voltage of any node which is
                               defined, but not grounded }

type   tridat = array [1..3] of real;             { used to store info
                                                  for triangles }
       triintdat = array [1..3] of integer;       { used to store node
                                                  nos. for triangles }
       locmat = array [1..3,1..3] of real;        { local array for
                                                  each triangle }
       glomat = array [1..max,1..max] of real;    { global matrix for
                                                  all nodes together }
       nodes = array [1..max] of real;            { array for info for
                                                  all the nodes }
       worktype = array [1..WkMax] of real;       { workspace for
                                                  leqt1f }

var  Nunk,                       { number of unknown nodes }
     trianum,                    { number of triangles in region }
     points,                     { total number of nodes in region }
     NodesNotV,                  { number of nodes equal to v2 +
                                 number of unknown nodes }
     i,j,                        { counter variables used in loops }
     M,IER,IDGT   : integer;     { variables used by matrix solver }
     q,                          { vector which holds local of qG }
     x,y,                        { x & y values for tri. vertices }
```

```
   b,c              : tridat;          { values for b and c used to find
                                         phi.  (phi = a + bx + cy)  }
   v                : triintdat;       { holds three nodes of triangle }
   eps,rho,                            { material properties in triangle }
   A                : real;            { area of triangle }
   XG,YG,                              { global vectors for node locs. }
   phi,                                { vector for phi at all nodes }
   qG               : nodes;           { global vector for charge }
   PG               : glomat;          { global matrix to be solved }
   P                : locmat;          { local matrix used to build PG }
   wkarea           : worktype;        { area for solving matrix eqn. }
   phiplot,                            { output data; x, y, potentials }
   triall,                             { nodenumbers for each triangles }
   genall,                             { nodes, Nunk, NodesNotV, trinum }
   allnod           : text;            { holds position of all nodes }

procedure triangle (var b,c: tridat; var A: real; x,y: tridat);
           { calculates the values of b and c used to find phi.
             finds the area of the triangle.  the inputs are x and
             y which contain locations of the triangle's vertices. }
var  delta      : real;
     i,i1,i2    : integer;
begin
   delta := x[2]*y[3] - x[3]*y[2] + x[3]*y[1] - x[1]*y[3] +
            x[1]*y[2] - x[2]*y[1];
   A := abs(delta)/2;
   for i := 1 to 3 do begin
      i1 := (i mod 3) + 1;
      i2 := (i1 mod 3) + 1;
      b[i] := (y[i1] - y[i2])/delta;
      c[i] := (x[i2] - x[i1])/delta
      end
   end;


procedure FirstOrderLocalMats (var P:locmat; var q: tridat;
                               x,y:tridat; eps,rho: real);
           { this procedure computes the submatrices which are used
             to build the global matrices qG and PG.  p is a 3x3
             local element matrix and q is a 3x1 vector.  the inputs
             are x and y (the vertix locations), and eps and rho
             (the material properties of the triangle). }
var  i,j : integer;
begin
   triangle (b,c,A,x,y);
   for i := 1 to 3 do begin
      for j := 1 to 3 do
         P[i,j] := eps*A * (b[i]*b[j] + c[i]*c[j]);
      q[i] := A*rho/3
      end
   end;
```

```
procedure GlobalPlace (var PG: glomat; var qG: nodes; P:locmat;
                   q:tridat; v:triintdat; phi:nodes);
            { this procedure puts the local matrices computed above
              into the global matrices PG and qG.  the inputs are: p
              and q, the local matrices; Nunk, the number of unknown
              nodes; v, a vector containing the nodes of the tri-
              angle; and phi, the solution vector for this problem. }
var  row,col : integer;
begin
   for row := 1 to 3 do
      if v[row] <= Nunk then begin
             { the else corresponds to a known node so ignore it }
         qG[v[row]] := qG[v[row]] + q[row];
         for col := 1 to 3 do
            if v[col] <= Nunk
              then PG[v[row],v[col]] := PG[v[row],v[col]]+P[row,col]
              else qG[v[row]] := qG[v[row]] - P[row,col]*phi[v[col]]
                  { this corresponds to a known node so subtract the
                    value here from the solution vector qG. }
      end
   end;

[unbound] procedure leqt1f (
            { this procedure calculates the solution to the matrix
              eqn. [PG][phi] = [qG].  the inputs are PG which corres-
              ponds to A, the width of the vector qG which is M, the
              number of unknown nodes, which corresponds to N, the
              maximum size of the array A, the vector qG which cor-
              responds to B, and several other variables used by this
              mathlib procedure.
                the output is the vector phi which is place into the
              vector B.  this means that B is modified and becomes
              our solution for the problem.
                this is a fortran procedure in the imsl library of
              math routines at harvey mudd college. }
         %ref A : [unsafe] array [l1..u1 :integer] of
                          array [l2..u2 :integer] of real;
         M : integer;
         N : integer;
         MAX : integer;
         %ref B : [unsafe] array [l3..u3 :integer] of
                          array [l4..u4 :integer] of real;
         IDGT : integer;
         %ref WKAREA : [unsafe] array [l5..u5 :integer] of real;
   var   IER : integer) ; extern;

begin                                   { main program }
  M := 1;                               { sets width of qG (and phi)
                                          to one for leqt1f }
   IDGT := 0;                           { sets output precision }
   open (triall,history :=old);         { open input data file }
```

```
reset (triall);                           { prepare it for access }
open (genall,history :=old);
reset (genall);
open (allnod,history :=old);
reset (allnod);
read (genall, Nunk, NodesNotV, points, trianum);
close (genall);

for j := 1 to points do
   readln (allnod, XG[j], YG[j]); { read positions of all nodes }
close (allnod);

for i := 1 to Nunk do begin
   QG[i] := 0;                      { initialize global vector }
   for j := 1 to Nunk do
      PG[i,j] := 0                  { initialize global matrix }
   end;
for i := (Nunk + 1) to NodesNotV do
   phi[i] := v2;                    { all nodes that have phi=v2 }
for i := (NodesNotV + 1) to points do
   phi[i] := 0;                     { nodes which are grounded }

for j := 1 to trianum do begin
   readln (triall,v[1],v[2],v[3],eps,rho);
                                    { read info for each triangle }
   eps := eps * epsnot;
   for i := 1 to 3 do begin
      x[i] := XG[v[i]];             { get position of each node }
      y[i] := YG[v[i]]             { for use in FirstOrder... }
      end;     { for i }
   FirstOrderLocalMats (P,q,x,y,eps,rho);
   GlobalPlace (PG,qG,P,q,v,phi);
   end;      { for j }

close (triall);
open (phiplot);
rewrite (phiplot);

leqt1f (PG, M, Nunk, max, qG, IDGT, wkarea, IER);
for j := 1 to Nunk do
   phi[j] := qG[j];
for i := 1 to points do
   writeln (phiplot,' ',XG[i]:6:4,' ',YG[i]:6:4,' ',phi[i]);
                        { phiplot is a data file which contains
                          the position and equipotential value
                          for each node.  phiplot is used by a
                          plotting routine to make the equi-
                          potential plot for the data. }
close (phiplot)
end.
```

ANSWERS TO SELECTED PROBLEMS

Chapter 1

1.7

	x	y	z	ρ	ϕ	z	r	θ	ϕ
(a)	3	4	5	5	53.1°	5	7.07	45°	53.1°
(b)	0	4.33	2.5	4.33	90°	2.5	5	60°	90°
(c)	6.06	3.5	2	7	$\pi/2$	2	7.28	1.29	$\pi/6$

1.8

	$\mathbf{A} + \mathbf{B}$	$\mathbf{A} - \mathbf{B}$	$\mathbf{A} \cdot \mathbf{B}$	$\mathbf{A} \times \mathbf{B}$
(b)	$2\mathbf{a}_x$	$-6\mathbf{a}_y + 4\mathbf{a}_z$	-12	$\mathbf{a}_x + 4\mathbf{a}_y + 6\mathbf{a}_z$
(c)	$3\mathbf{a}_\rho + 4\mathbf{a}_z + (5\pi/6)\mathbf{a}_\phi$	$\mathbf{a}_\rho - (\pi/6)\mathbf{a}_\phi$	7.64	$-(\pi/3)\mathbf{a}_\rho - 2\mathbf{a}_\phi + (2\pi/3)\mathbf{a}_z$

1.9 (a) $\mathbf{A} \cdot \mathbf{B} = 0$,   (c) 90°   **1.10** (b) $\mathbf{B} \times \mathbf{C} = 0$ **1.11** Zero

1.14 $\nabla f(0,0,0) = 0$, $\nabla f(2,1,5) = 20\mathbf{a}_x + 10\mathbf{a}_y - 2\mathbf{a}_z$

1.15 $\nabla f = z\mathbf{a}_x + (2y - z)\mathbf{a}_y + (x - y)\mathbf{a}_z$, $\mathbf{n} = \nabla f/|\nabla f|$ **1.36** $\nabla^2 \mathbf{A} = 0$

1.40 (a) $\nabla \times \mathbf{A} = -0.433\mathbf{a}_x + 1.732\mathbf{a}_y - 1\mathbf{a}_z$, $\nabla \cdot \mathbf{A} = 1.732$

      (c) $\nabla \times \mathbf{A} = -\mathbf{a}_\theta 0.866 + \mathbf{a}_\phi 0.732$, $\nabla \cdot \mathbf{A} = 1.5$

Chapter 3

3.1 $D = \mathbf{n}\rho_s$ **3.3** $D = \rho_s$ between, $D = 0$ outside

3.5 (a) $D = 0$, (b) $\mathbf{D}$ is parallel to $\mathbf{a}_x$ **3.8** $D = 0$, $\rho < a$; $\mathbf{D} = \mathbf{a}_\rho \rho_s a/\rho$, $\rho > a$

3.9 $\rho_v = 3\varepsilon_0\rho_0$, $Q = 3\varepsilon_0\rho_0\pi/2$ **3.15** (a) zero, (b) no

3.16 (a) parallel to $\mathbf{a}_x$, (b) parallel to $\mathbf{a}_z$ **3.17** $q = 9.8 \times 10^{-6}$ C

3.18 8.1×10^{-27} **3.21** $\partial E/\partial x = \rho/\varepsilon = $ constant, $\partial E/\partial y = 0 = \partial E/\partial z$

3.23 $Q = -\varepsilon_0/2$

Chapter 4

4.6 single dipole: $\tau_i = m_i B$; $M = \Sigma m_i/v$; $\tau = \Sigma \tau_i = vMB = A\ell MB$　　　　**4.7** $H = I/2R$

4.8 (a) $I = \pi J_0 a^2/2$　(b) $\rho < a$: $\mathbf{H} = \mathbf{a}_\phi (J_0 \rho/2)(1 - \rho^2/2a^2)$; $\rho > a$: $\mathbf{H} = \mathbf{a}_\phi J_0 a^2/4\rho$

4.11 (a) $\mathbf{J} = -\mathbf{a}_x k H_0 e^{-ky}$　(b) $I = H_0 a(1 - e^{-ka})$　　　　**4.13** $\Phi = (\mu_0 I\ell/2\pi)\ln(\rho_2/\rho_1)$

4.14 $\mathbf{G}_1$: $\rho_v = \varepsilon(y^2 + x^2 + z^2)$, $\mathbf{G}_2$: $\mathbf{J} = \nabla \times \mathbf{G}_2/\mu$　　　　**4.19** $B = 1.41(10)^6 \mu_0$ T

4.21 $K_{ind}/K_{free} = 317$

Chapter 5

5.3 $M = n_1 n_2 \times v_2$, $v_2 = $ vol. of coil 2　　　　**5.6** $F = -1.95 \times 10^4$ N

5.8 (a) $U = Q^2/32\pi^2 \varepsilon_0 r^4$ J/m^3, $r > r_0$; $U = 0$, $r < r_0$; $W = Q^2/8\pi\varepsilon_0 r_0$

　　(b) $W = Q^2/4\pi\varepsilon_0 r_0$

5.10 (a) $E = 5000$ V/m, $D = 5000\varepsilon_0$ C; (b) $D = 5000\varepsilon_0$ C, $E = 1250$ V/m

　　(c) $C_{with}/C_{without} = 4$

5.14 For 3.27: $C = 12\varepsilon_0$ F, For 3.31: $C = 8\varepsilon_0 w\ell/t$

5.19 (a) $\Phi = 7.9\mu_0$ Wb (b) $V = 9.9$ volts (c) Terminal (a) is positive　　　　**5.21** CCW

5.22 If axis is along $\mathbf{a}_\rho$, $V = 0$. If axis is along $\mathbf{a}_\phi$, then $V = 0.5$ volt.

5.24 $V = (\mu_0 \omega I_0/2\pi)\ln(9/5)\cos\omega t$　　　　**5.25** (a) $V = 197$ volts. (b) $V(1) > V(2)$

5.28 (a) $V_i = 2NA\mu_0 H_0 \omega \sin\omega t$, $V_{ii} = 0$

　　(b) $V_i = 11NA\mu_0 H_0 \omega \sin\omega t$, $V_{ii} = 9NA\mu_0 H_0 \omega \sin\omega t$

5.31 (d) $M = 0$.　(e) $M = (\mu_0 s/\pi)\ln(5/3)$

5.32 $M_1 = 0 = M_2 = M_3$, $M_4 = \mu_0 (nA)_{toroid}\omega$

5.33 $M = (1/2\pi)\ln(9/8)$ h/m

Chapter 6

6.10 $V = r_0 \rho_s/2\varepsilon_0$

6.12 (a) $\mathbf{E} = -\mathbf{a}_y \dfrac{\rho_s a^2}{\pi\varepsilon_0(a^2 + z^2)^{3/2}}$　(b) $\mathbf{E} = \mathbf{a}_z \dfrac{\rho_s a}{\varepsilon_0(a^2 + z^2)^{3/2}}$

6.14 In both cases $V = 0$.

6.16 $V = (\rho_0/3\pi\varepsilon_0 a^2)[(2a^2 + z^2)^{3/2} - 2(a^2 + z^2)^{3/2} + z^3]$

　　　$\mathbf{E} = \mathbf{a}_z(\rho_0 z/\pi\varepsilon_0 a^2)[2(a^2 + z^2)^{1/2} - (2a^2 + z^2)^{3/2} - z]$

6.18 $W = CV^2/2 + \varepsilon_0 AV^2/2x$　　　　**6.25** $V = -(\rho_0 z/6\varepsilon_0)[5z_0 - 6z + (z^2/z_0)] + V_0 z/z_0$

6.30 $V = V_0 e^{-k_2 y}\sin k_2 x$, $k_2 = 2\pi/b$

6.31 V is given by solution to Example 6.6 without the exponential in y.

6.34 $V = V_0 \ln(\rho/b)/\ln(b/a)$.

6.35 Taking vertical direction as angle reference, $V = E_0 a \cos\phi \, [(a/\rho) - (\rho/a)]$

6.37 $V = V_0[1 - (6\phi/\pi)]$　　　　**6.38** (b) $V = V_0 \cos\phi \, [(\rho/a) - (a/\rho)]/[(b/a) - (a/b)]$

Chapter 7

7.3 $\mathbf{A} = \mathbf{a}_z(\mu_0 I/4\pi)\ln\{[z_1 + (z_1^2 + \rho^2)^{1/2}]/[z_2 + (z_2^2 + \rho^2)^{1/2}]$, where z_1 is the z axis distance from the field point to the distant end of the wire segment, z_2 is to the near end.

7.6 $\mathbf{H} = \mathbf{a}_z Ia/2r^2$　　　　**7.8** (a) $\mathbf{K} = \mathbf{a}_\phi \omega \rho_s \rho$

　　(b) $H_z = \frac{1}{2}K[\sinh^{-1}(a/z) - a/(a^2 + z^2)^{1/2}]$　　　　**7.12** Zero

7.17 There is negligible attenuation due to the hull. From drive shell $H_{sensitivity} < 160$ Oe.

7.18 $\psi = -(K/2)y + C_2$, C_2 is indeterminate.

7.20 $F = 50\mu_0(nI)^2$ directed to move slug into coil.

Chapter 9

9.2 (a) $\mathbf{E} = 1\mathbf{a}_z \exp\{j(2\pi \times 10^9 - 50x - 40y)\}$ V/m, $v = 9.8(10)^7$ m/sec.

9.8 (a) $v = 1.886 \times 10^8$ m/sec. (b) $\lambda = 0.0377$ m (c) $|\mathbf{H}| = 0.042$ A/m
 (d) $\mathbf{H} = \mathbf{a}_y 0.042 \cos(10\pi \times 10^9 t - 166.6z)$

9.9 (a) $v = 3 \times 10^8/\sqrt{\varepsilon_r} = \omega/\beta$ (b) $\mathbf{H} = -\mathbf{a}_x 2(10)^{-4} \cos2\pi(10^8 - 2z/3)$ A/m
 (c) $\mathcal{P} = \mathbf{a}_z[377 + 377 \cos4\pi(10^8 - 2z/3)]10^{-8}$ W/m^2

9.11 $f = 8 \times 10^8$ Hz, $\varepsilon_r = 2.25$, $\mathbf{E} = \mathbf{a}_z 10 \cos8\pi[2(10)^8 t - x]$ V/m,
 $\mathbf{H} = -\mathbf{a}_y 0.040 \cos8\pi[2(10)^8 t - x]$ A/m

9.14 (b) $\mathbf{E} = \mathbf{a}_y 4 \cos(2\pi \times 10^7 t + \pi/6)\mathbf{a}_y$
 (c) $\langle \mathcal{P} \rangle = \frac{1}{2}\mathfrak{R}\{\mathbf{E} \times \mathbf{H}^*\} = -\mathbf{a}_z 1.5$

9.17 (a) $E = 346$ V/m, $H = 0.918$ A/m (b) $E = 4.9 \times 10^8$ V/m, $H = 1.3 \times 10^6$ A/m
 (c) $t = 5.55$ hrs.

9.18 Negatively rotating elliptical polarization: direction of rotation is about $-\mathbf{a}_z$ as wave moves along $+\mathbf{a}_z$. Elliptic axes are $x:y = 5:4$.

9.19 (a) $\sigma/\omega\varepsilon \ll 1 \rightarrow$ insulator. (b) $\lambda = 0.548$ m, $\beta = 11.47/$m
 (c) $|\mathcal{P}| = |E_0^2|/2\eta = 1.45(10)^{-10}$W/m^2
 (d) $\mathbf{H} = (\mathbf{a}_y - \mathbf{a}_x)1.45(10)^{-7} \cos(2\pi \times 10^7 - 11.47z)$ A/m

9.23 (a) $\sigma/\omega\varepsilon < 0.03 \rightarrow f > 9.4 \times 10^8$ Hz. (b) $\sigma/\omega\varepsilon > 30 \rightarrow f < 9.4 \times 10^5$ Hz
 (c) $10^7 \le f \le 10^8$ (using factor of 10)   **9.24** $\mathcal{P} = \mathcal{P}_0 e^{-2z/\delta}$

Chapter 10

10.3 $-df/f = hf/Mc^2$ **10.8** For ρ': $\phi_i = 85.1°$ **10.9** $SWR = 1.25$
10.18 $n = 5.83$, $\eta_2 = 2197$ Ω **10.20** $P_{refl}(30°)/P_{refl}(45°) = 3$

Chapter 11

11.5 $Z_{in} = 50$ Ω. $V = 33.3e^{j30°}$ volts   **11.6** $\eta_L = 2.16 - j4.64 \rightarrow Z_L = 216 - 464j$ Ω
11.8 $Z_{in} = 225$ Ω **11.9** (a) $VSWR = 3.70$ (b) $VSWR = \infty$ (c) $\ell_{short} = 0.3$ m
 (d) $VSWR = 1$
11.10 $\ell = 0.089\lambda$ plus $n\lambda/4$
11.13 $\eta_L = 0.322 - 0.486j \rightarrow Z_L = 322 - 486j$ Ω

Chapter 12

12.3 (a) $6.667 \le f \le 13.33$ GHz (b) $6.122 \le f \le 13.33$ GHz
12.11 $\eta(10) = 500$ Ω, $\eta(01) = -j347$ Ω (capacitive)
12.13 (a) $a = 0.75$ cm, at $2.25:1 \rightarrow b = 0.33$ cm (b) $e^{-2.56} = 0.077$
 (c) α of propagating mode is also high.
12.15 $\Delta\phi(20) = \beta\ell = 334°$; For (11): $e^{-5\alpha} = 0.0052$
12.17 $a > 0.72$ cm. If W. G. dimensions are $>2:1$ then next mode is 20, and $a < 0.96$ cm. Therefore, for example, $a = 0.84$ cm, $b = 0.4$ cm
12.18 (a) $f_c(10) = 15$ GHz, $f_c(20) = 30$ GHz (b) $\beta(10) = 7.766$, $\beta(20) = 5.541$ cm^{-1}
 (c) $\lambda(10) = 0.809$ cm, $\lambda(20) = 1.134$ cm
 (d) $v(10) = 3.236 \times 10^{10}$, $v(20) = 4.536 \times 10^{10}$ cm/sec

Chapter 14

14.3 Full half-power width is 90°.

14.9 (a) $E = 0.194$ V/m, $H = 5.14 \times 10^{-4}$ A/m   (b) $V = 4.06 \times 10^{-4}$ volts

14.12 $\mathscr{P}_{\text{sensitivity}} = 5.51 \times 10^{-16}$ W/m$^2 \rightarrow E_{\text{sensitivity}} = 4.56 \times 10^{-7}$ V/m

14.13 $\alpha(p = 0) = 75.52°$, $\alpha(p = -1) = 138.59°$

$\alpha_0(p = 0) = 60°$ or $2\Delta\alpha \approx 31°$, $\alpha_0(p = -1) = 180°$ or $2\Delta\alpha \approx 83°$

14.19 For $p = 0$, $\cos\alpha_0 = 0.0555$ and $\cos\alpha_0 = 0.2777$; half widths are 6.4° and 6.5°. For $p = 1$, half widths are 14.4° and 10.2°

14.24 Beam width is now $\pm 1.9°$.

14.29 (a) With the (incorrect) assumption of a single frequency signal, $E_{\text{sun}} = 1.88 \times 10^5$ V/m, $E_{\text{earth}} = 877.6$ V/m. $P_{\text{earth}} = 1.74 \times 10^{17}$ W corresponds to ~ 0.14 W/cm^2

INDEX